Analog Circuits

World Class Designs

Newnes World Class Designs *Series*

Analog Circuits: World Class Designs
Robert A. Pease
ISBN: 978-0-7506-8627-3

Embedded Systems: World Class Designs
Jack Ganssle
ISBN: 978-0-7506-8625-9

Power Sources and Supplies: World Class Designs
Marty Brown
ISBN:978-0-7506-8626-6

For more information on these and other Newnes titles, visit: **www.newnespress.com**

Analog Circuits

World Class Designs

Robert A. Pease, Editor

with

Bonnie Baker
Richard S. Burwen
Sergio Franco
Phil Perkins
Marc Thompson
Jim Williams
Steve Winder

ELSEVIER

AMSTERDAM • BOSTON • HEIDELBERG • LONDON
NEW YORK • OXFORD • PARIS • SAN DIEGO
SAN FRANCISCO • SINGAPORE • SYDNEY • TOKYO

Newnes is an imprint of Elsevier

Newnes

Newnes is an imprint of Elsevier
30 Corporate Drive, Suite 400, Burlington, MA 01803, USA
Linacre House, Jordan Hill, Oxford OX2 8DP, UK

 Recognizing the importance of preserving what has been written,
Elsevier prints its books on acid-free paper whenever possible.

Library of Congress Cataloging-in-Publication Data
Application Submitted

British Library Cataloguing-in-Publication Data
A Catalogue record for this book is available from the British Library.

ISBN: 978-0-7506-8627-3

For information on all Newnes publications
visit our Web site at www.books.elsevier.com

Transferred to Digital Printing, 2010

Typeset by Charon Tec Ltd (A Macmillan Company), Chennai, India
www. charontec. com

Printed and bound in the United Kingdom

Table of Contents

Preface

Comments on "World-Class" Analog Design

Achieving excellence in analog circuit design has always been challenging. These days it is still not always easy, so we want to help with some general advice. All the authors of these chapters have presented their best ideas as the kinds of things a good analog circuit designer must know to consistently accomplish very good circuits.

These days so much of analog circuit design can be done using operational amplifiers (op-amps) with a small number of discrete resistors and capacitors. It is often very easy to slap in resistors and the circuit works well. However, this is still not trivial. You might have to pick sets of matched resistors or add a trimpot. Even these days some young engineers have to ask, "So, should I make a 1-ohm/1-ohm unity-gain inverter?" Some kids really don't know how to pick appropriate resistor values; they have never done any practical work or lab work. So we have to teach them about practical circuits. We have to teach them about error budgets. Sometimes 1% resistors are quite appropriate; other times 5%, 10%, 0.1%, or 0.01% might be right. Richard Burwen has good comments on resistors. More on error budgets later.

Recently a guy showed me his design with eight precision op-amps and sixteen precision resistors. After I did some whittling out, we got it down to two precision resistors and one precision op-amp and a greatly improved error budget. More on error budgets later.

Once upon a time, in the 1950s, there were no operational amplifiers that you could buy. The engineers at Philbrick Researches wrote a twenty-eight-page *Applications Manual for Octal Plug-In Computing Amplifiers* (such as the K2-W, see **Figure P-1**). With a little advice from this pamphlet, you could design analog computing circuits and some simple instrumentation, too. I came to work at Philbrick about that time (1960). I studied operational amplifiers based on vacuum tubes and then high-performance solid-state amplifiers.

Applications Notes

Then about 1965, the new arts and applications using transistorized op-amps showed the need for a comprehensive *Applications Manual for Computing Amplifiers for Modeling,*

Figure P-1: Philbrick K2-W, 1952 to 1992.

Measuring, Manipulating, and Much Else. Dan Sheingold, George Philbrick, Bruce Seddon, and several others wrote a lot. I contributed a small bit. This book was *very* useful. My theory is that when Bob Widlar brought out the μA709, he couldn't have *given* it away, but Philbrick had sold and given away many thousands of these books, which made it reasonably easy to apply those IC op-amps. This book was sold for several years for $3. Recently, a good copy sold on e-bay for $300+. It's darn near worth it. Can you get the basic info off the Web? I'll have to look it up on Analog Devices' Website.

Other companies such as Burr Brown, Analog Devices, and TI wrote lots of App Notes and books on op-amp applications. I was never very impressed with them; they were not good explainers. NSC published lots of App Notes. Not all were well documented, but they were pretty good circuits.

Which Op-Amp?

Even for experienced engineers, this can be a bewildering question. There are many low-voltage and high-voltage op-amps; low-voltage noise and high-voltage noise; low-power and high-speed amplifiers; and cheap and expensive ones. Let's see what insights Bonnie Baker can offer.

Precision Capacitors?

How many kid engineers know the price of 1% capacitors? Precision capacitors are rarely justifiable. Yet not all 1% capacitors are really high priced. Sometimes a dime will get you that; other times, it could take a dollar or two. And sometimes a circuit really does *need* 1% capacitors. I just got a *thick* Digikey Catalog the other day, and it has 2% and 1% tolerance polypropylene capacitors at surprisingly reasonable prices, even in small quantities!

Inductors?

Inductors are specialized animals that may be required for filters and for switch-mode regulators. Usually the designer of the switcher provides detailed advice on what to buy. If not, then designing with inductors, or redesigning to adjust the inductor type or values, is a special advanced area of expertise. Most schools don't teach much of this. The design of switchers can be either a high-tech specialty or a monkey-see, monkey-do exercise. The latter might not be as cheap, but it usually does work well.

Diodes

Diodes can be a truly bewildering field. Some can carry small milliamperes; some can leak less than a picoampere; some rectifiers can carry amperes without overheating. But the big ones (such as 1N4005) often cannot be used at high frequencies. The 2N5819 Schottky rectifier can carry a couple amperes, but it is somewhat leaky. Still, it can rectify up to 1 MHz without misbehaving. Who's going to teach everybody about diodes?

Especially tricky is the fact that some good, fast small-signal diodes (1N4148/1N914) do turn on and off quickly—faster than 1 nanosecond sometimes—but at low rep rates, some of them sort of forget how to turn on and have a bad overshoot. That's annoying.

Transistors and Designing With Them

Now, when you get to transistors, this becomes complicated. Designing with transistors is a whole 'nother game. Even experienced analog designers try to minimize that when they can. But sometimes you have to use transistors. Sometimes the transistor's inherent log characteristics are very important. Can you buy a logger? Yes, several companies make and sell loggers. But loggers can be designed for special cases, which a store-bought logger cannot handle, such as low voltage. I've done a couple of these in the last year. I still design low-noise and high-speed amplifiers occasionally using selected transistors, such as 2N3904 and LM394. I often use the curves from "What's All This V_{BE} Stuff, Anyhow?" Or you might merely need to use a transistor as a switch—a crude one or a precision switch.

Filters

When you need a filter, it might not be hard to figure out what is needed; other times more research is needed. Can you avoid inductors? Can you avoid expensive op-amps?

Can you avoid high impedances *or* large capacitors? As with all of analog design, this covers a *huge* dynamic range, and there is usually nothing simple about it. Yet it gets done.

SPICE

I usually try to avoid using SPICE. I use pen and paper; I call it "back-of-envelope SPICE." I do mostly hand computations, and good approximations, using my slide rule or by doing the math in my head. You might say I am in agreement with Dick Burwen's chapter, "How to Design Analog Circuits Without a Computer." Other people think that SPICE is acceptable over a wide range of applications. That makes me nervous. I find that you can use SPICE to save an hour of computation every day for a month and then discover that SPICE has made a costly mistake that wastes all the time you thought you saved. Some people agree with me on that.

Also, when people use monolithic transistor models (such as the ones in the monolithic array, LM3046), that is different from designing with discrete transistors. I mean, who will give you a free model of a 2N3904 that is worth what you paid for it? And in what regimes do you trust it? I would trust it for *only* the crudest noncritical applications.

Some people say they like to trust SPICE. If they get good models and they know what they are doing, good luck to them.

I will mention a few particular places where SPICE models do not usually work well:

- At low values of *Vce* (or *Vds*), where the transistors are starting to saturate.

- At high frequencies at low values of *Vce* (or *Vds*), where the frequency response of the transistor does not ring true.

- Monolithic transistors are *often* badly modeled where they saturate (or start to saturate) since the substrate currents get large.

- Sometimes when an op-amp's inputs get reversed, it will still appear to work like an op-amp without saturating. *Some* kinds of SPICE do work right in this situation, but not all.

- If somebody gives you a bad model, you might have problems. Even when you make your own model, it could have problems.

- Sometimes SPICE fails to converge and wastes a lot of your time.

- Sometimes SPICE gives an absurd answer, such as saying that a 10 exp-25 ampere current step has a real risetime. How can a "current" that consists of 1 electron per day show a "risetime"?

- Usually in a band-gap reference, the fine details of a temperature characteristic do not go in the right direction. SPICE cannot lead you to a better answer. My old LM131 from 1977 had (and still has) a good tempco because it was based on good breadboards. When I tried to run it in SPICE many years later, SPICE said it did not work and could not be made to work. It's a good thing I didn't try it in SPICE in 1977. SPICE was wrong.

- In any circuit where transistors are heated or self-heated, the temp rise of the transistors is *very* hard to model, especially in a distributed layout.

- And sometimes SPICE just *lies*. Sometimes it just gives incorrect answers.

I've had debates with many "SPICE experts" and they try to tell me I am wrong. But I have seen too many cases where I was right and SPICE was wrong. I say this because people bring me their problems when their circuit does not work. I can see through the errors of SPICE; I use special test techniques (mostly in the time domain or in thought experiments) to show why a circuit is misbehaving. SPICE is not only *no help*, it leads to "computer-hindered design."

How Many?

How many are you going to build? If you are going to build large numbers or small, it makes a difference how you engineer it, for minimum overall cost and maximum output.

Low Noise?

Many general-purpose op-amps are pretty quiet, but some that are quiet at low impedance are noisy at high impedances. Others that are quiet at high impedance are noisy at low impedance. Let's see what comments Bonnie Baker has on this topic.

Troubleshooting?

Once you get your circuit built, you apply power and then it does (or does not) work correctly. How do you do the troubleshooting? Better yet, how do you plan in advance a way that you can easily do the needed troubleshooting?

Check out the Bob Pease book *Troubleshooting Analog Circuits*. With 39,000 copies in print in six languages, it has *legs*—and that's because analog circuit troubles do not go away by wishing and sometimes not even by engineering. Sometimes they are solved only by real troubleshooting. But planning ahead can help. See www.national.com/rap/Book/0,1565,0,00.html.

The Future?

People often ask, "Would you encourage your son or daughter to go into engineering?" I reply, "Yes, if it is analog circuit design." They say, "Explain!"

I respond, "My friends and I know many analog design techniques, tricks, and secrets. They cannot be learned from SPICE. Every year there are 200,000 Chinese engineering graduates, and they don't know what we know. We can solve problems they cannot."

I rest my case. /rap

—Robert A. Pease
Staff Scientist, NSC
Santa Clara, CA
August 2007
rap@galaxy.nsc.com

P.S. One of the authors of a chapter in this book said that he took a "well-designed" system and put a good model of it into SPICE. When he ran it, he was surprised to find a sneaky sampling error. So we should not say that SPICE cannot be helpful. We just have to be cautious about trusting SPICE—in any positive or negative way.

About the Editor

Robert A. Pease (Chapters 16, 17, and Appendix B)

Robert A. Pease attended Mt. Hermon School, and graduated from MIT in 1961 with a BSEE. He worked at Philbrick Researches up to 1975 and designed many OpAmps and Analog Computing Modules.

He joined National Semiconductor in 1976. He has designed about 24 analog ICs including power regulators, voltage references, and temp sensors. He has written 65+ magazine articles and holds about 21 US patents. Pease is the self-declared Czar of Bandgaps since 1986. He enjoys hiking and trekking in Nepal, and ferroequinology. His position at NSC is Staff Scientist. He is a Senior Member of the IEEE.

Pease wrote the definitive book, "Troubleshooting Analog Circuits", now in its 18th printing. It has been translated into French, German, Dutch, Russian, and Polish. Pease is a columnist in *Electronic Design* magazine, with over 240 columns published. The column, "Pease Porridge", covers a wide range of technical topics.

He also has posted many technical and semi-technical items on his main web-site: http://www.national.com/rap. Many of Pease's recent columns are accessible there.

Pease was inducted into the E.E. Hall Of Fame in 2002. Refer to: http://www.elecdesign.com/Articles/Index.cfm?ArticleID=17269&Extension=pdf. See Pease's other web site at http://www.transtronix.com. He can be contacted at rap@galaxy.nsc.com.

P.S. Pease is also the self-declared Czar of Proofreading, for ~20 years. He has proof-read several books and many technical articles. Without his sharp eye, this book would have been hard to bring out with fully accurate information, as there are so many opportunities for errors in a technical document of this magnitude.

About the Authors

Bonnie Baker (Chapters 4, 8, 13, 14, 15, and Appendix A) writes the monthly "Baker's Best" for *EDN* magazine. She has been involved with analog and digital designs and systems for over 20 years. Bonnie started as a Manufacturing Product Engineer supporting analog products at Burr-Brown. From there, Bonnie moved up to IC Design, Analog Division Strategic Marketer, and then Corporate Applications Engineering Manager. In 1998, she joined Microchip Technology and served as their analog division Analog/mixed signal Applications Engineering manager and Staff Architect Engineer for one of their PICmicro divisions. This expanded her background to not only include analog applications, but also the microcontroller. She is now, back in the Burr-Brown fold, working for Texas Instruments in their Precision Analog Division.

Along with her expertise in analog design, Bonnie has a drive to share her knowledge and experience and has written over 250 articles, design notes, and application notes. In addition to being an *EDN* columnist, she is also a frequent presenter at technical conferences and shows.

Richard S. Burwen (Chapter 9) received a S.B. (cum laude) in physics in 1949 and an A.M. in engineering sciences and applied physics in 1950 from Harvard. He was one of three founders of Analog Devices and worked as a consultant to the company, designing several of the circuits for its initial product lines. Other companies with which he was associated in their beginning phases included Mark Levinson Audio Systems, Cello Ltd., Novametrix Medical Systems, and KLH Burwen Research. He became a founder of Copley Controls in 1984 and designed many of the company's products. In the case of all the companies he helped start, Richard maintained his independence by working as a consultant in his own laboratory. He designed his home and laboratory in 1965, in Lexington, Massachusetts, around his 20,000 watt, 169-speaker, 5-channel recording and reproducing studio. Since retiring from circuit design consulting in 2002, he has been even more active consolidating his 63 years of audio development into audio digital signal processing software described at www.burwenaudio.com and www.burwenbobcat.com.

Sergio Franco (Chapter 12) is a professor of electrical engineering at San Francisco State University, where he teaches microelectronics courses and acts as an industry consultant. Prior to assuming his current professorship, Sergio was employed at Zeltron, Zanussi's Electronics Institute (Udine, Italy). He received a B.S. in physics from the University of Rome, a M.S. in physics from Clark University, and a Ph.D. in computer science from the University of Illinois. Sergio is a member of the IEEE, and in his spare time enjoys classical music, gardening, and mountain hiking.

Phil Perkins (Chapter 2) is a Fellow of LTX Corporation, Norwood, Massachusetts. He was a cofounder of LTX in 1976. Before LTX he was an engineer at Teradyne, Inc., Boston, Massachusetts. His work includes designing analog instrumentation for the LTX semiconductor test systems. His designs include V/I Sources, Test Heads, and DSP measuring instruments. He holds a patent for "Mixed signal device under test board interface". He received Bachelor's, Master, and Engineer degrees in Electrical Engineering from Massachusetts Institute of Technology.

Phil's interests include walking in the woods looking for wildflowers, church activities, home computer hobbying plus consulting for friends. He lives in Needham, Massachusetts with his lovely wife, Laurie. Phil can be contacted at phil_perkins@ltx.com.

Dr. Marc Thompson (Chapters 1, 3, and 5) was born on Vinalhaven Island, Maine. He specializes in custom R/D, analysis, and failure investigations into multi-disciplinary electrical, magnetic, and electronic systems at his engineering consulting company Thompson Consulting, Inc. in Harvard, Massachusetts. He is also an Adjunct Professor in the Electrical and Computer Engineering Department of Worcester Polytechnic Institute where he teaches graduate-level courses in advanced analog circuit design, power electronics, electric motors, and power distribution.

Dr. Thompson is author of a textbook entitled "Intuitive Analog Circuit Design", published in 2006 by Elsevier Science/Newnes. Another text entitled "Power Quality in Electronic Systems", was co-authored with Dr. Alexander Kusko, and was published by McGraw-Hill in 2007.

Dr. Thompson has seven U.S. patents and is a Firefighter with the Harvard, Massachusetts Fire Department, and has the B.S., M.S., and Ph.D. degrees in electrical engineering from the Massachusetts Institute of Technology. In his spare time he enjoys biking, travel, and repairing his c. 1899 vintage house in Maine.

Jim Williams (Chapter 18) was at the Massachusetts Institute of Technology from 1968 to 1979, concentrating exclusively on analog circuit design. His teaching and research interests involved applications of analog circuit techniques to biochemical and biomedical problems.

Concurrently, he consulted for U.S. and foreign concerns and governments, specializing in analog circuits. In 1979, he moved to National Semiconductor Corporation, continuing his work in the analog area with the Linear Integrated Circuits Group. In 1982, he joined Linear Technology Corporation as staff scientist, where he is presently employed. Interests include product definition, development, and support. Jim has authored over 350 publications relating to analog circuit design. Awards include the 1992 Innovator of the Year Award from *EDN* magazine and election to the Electronic Design Hall of Fame in 2002.

His spare-time interests include sports cars, collecting antique scientific instruments, art, and restoring and using old Tektronix oscilloscopes. He lives in Palo Alto, CA with his wife, son, and 84 Tektronix oscilloscopes.

Steve Winder (Chapters 6, 7, 10, and 11) is now a European Field Applications Engineer for Supertex Inc. Steve works alongside design engineers throughout Europe to design circuits using components made by Supertex, a US-based manufacturer of high voltage MOSFETs and CMOS ICs.

Prior to joining Supertex in 2002, Steve was, for many years, a team leader at British Telecom research laboratories. There he designed analog circuits for wideband transmission systems, mostly high frequency, and designed many active and passive filters.

CHAPTER 1

Review of Feedback Systems

Marc Thompson

Dr. Marc Thompson leads us to an appreciation of how the world has learned about FEEDBACK (negative) over the years, so we can understand how to do better feedback in our systems. /rap

Introduction and Some Early History of Feedback Control

A feedback system is one that compares its output to a desired input and takes corrective action to force the output to follow the input. Arguably, the beginnings of automatic feedback control[1] can be traced back to the work of James Watt in the 1700s. Watt did lots of work on steam engines, and he adapted[2] a centrifugal *governor* to automatically control the speed of a steam engine. The governor comprised two rotating metal balls that would fly out due to centrifugal force. The amount of "fly-out" was then used to regulate the speed of the steam engine by adjusting a throttle. This was an example of proportional control.

The steam engines of Watt's day worked well with the governor, but as steam engines became larger and better engineered, it was found that there could be stability problems in the engine speed. One of the problems was *hunting*, or an engine speed that would surge and decrease, apparently hunting for a stable operating point. This phenomenon was not well understood until the latter part of the 19th century, when James Maxwell[3] (yes, the same Maxwell famous for all those equations) developed the mathematics of the stability of the Watt governor using differential equations.

---

[1] Others may argue that the origins of feedback control trace back to the water clocks and float regulators of the ancients. See, e.g., Otto Mayr's *The Origins of Feedback Control*, The MIT Press, 1970.

[2] The centrifugal governor was invented by Thomas Mead c. 1787, for which he received British Patent #1628.

[3] James C. Maxwell, "On Governors," *Proceedings of the Royal Society*, 1867, pp. 270–283.

Invention of the Negative Feedback Amplifier

We now jump forward to the 20th century. In the early days of the telephone, practical difficulties were encountered in building a transcontinental telephone line. The first transcontinental telephone system, built in 1914, used #8 copper wire weighing about 1000 pounds per mile. Loss due to the resistance of the wire was approximately 60dB.[4] Several vacuum tube amplifiers were used to boost the amplitude. These amplifiers have limited bandwidth and significant nonlinear distortion. The effects of cascading amplifiers (**Figure 1-1**) resulted in intolerable amounts of signal distortion.

Harold Black graduated from Worcester Polytechnic Institute in 1921 and joined Bell Laboratory. At this time, a major task facing AT&T was the improvement of the telephone system and the problem of distortion in cascaded amplifiers. In 1927, Black[5] was considering the problem of distortion in amplifiers and came up with the idea of the negative feedback amplifier (**Figure 1-2**).

> "Then came the morning of Tuesday, August 2, 1927, when the concept of the negative feedback amplifier came to me in a flash while I was crossing the Hudson River on the Lackawanna Ferry, on the way to work. For more than 50 years I have pondered how and why the idea came, and I can't say any more today than I could that morning. All I know is that after several years of hard work on the problem, I suddenly realized that if I fed the amplifier output back to the input, in reverse phase, and kept the device from oscillating (singing, as we called it then), I would have exactly what I wanted: a means of canceling out the distortion in the output. I opened my morning newspaper and on a page of *The New York Times* I sketched a simple diagram of a negative feedback amplifier plus the equations for the amplification with feedback. I signed the sketch, and 20 minutes later, when I reached the laboratory at 463 West Street, it was witnessed, understood, and signed by the late Earl C. Bleassing.
>
> I envisioned this circuit as leading to extremely linear amplifiers (40 to 50 dB of negative feedback), but an important question is: How did I know I could avoid self-oscillations over very wide frequency bands when many people doubted such circuits would be stable? My confidence stemmed from work that I had done two years earlier on certain novel oscillator circuits and three years earlier in designing the terminal circuits, including the filters, and developing the mathematics for a carrier telephone system for short toll circuits."

---

[4] William McC. Siebert, *Circuits, Signals and Systems*, The MIT Press, 1986.
[5] Harold Black, "Inventing the Negative Feedback Amplifier," *IEEE Spectrum*, December 1977, pp. 55–60. See also Harold Black's U.S. Patent #2,102, 671, "Wave Translation System," filed April 22, 1932, and issued December 21, 1937, and Black's early paper "Stabilized Feed-Back Amplifiers," *Bell System Technical Journal*, 1934.

A typical closed-loop negative feedback system as is commonly implemented is shown in **Figure 1-3**. The "plant" in this diagram might represent, for instance, the power stage in an audio amplifier. A properly designed control system can maintain the output at a desired level in the face of external disturbances and uncertainties in the model of the plant. The goal of the feedback system is to force the output to track the input, perhaps with some gain and frequency-response shaping.

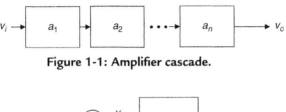

Figure 1-1: Amplifier cascade.

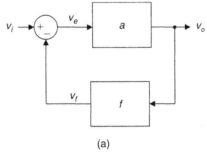

(a)

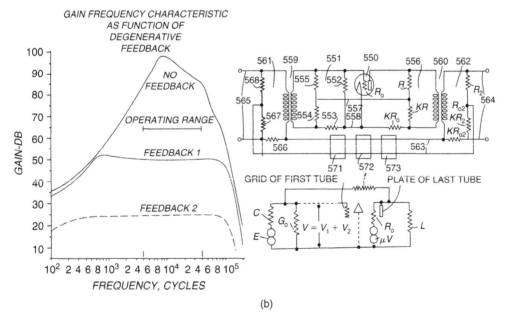

(b)

Figure 1-2: Classical single-input, single-output control loop, as envisioned by Black. (a) Block diagram form. (b) Excerpt from Black's U.S. patent #2,102,671, issued in 1937.

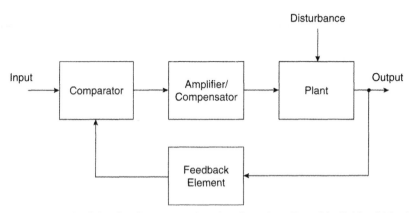

Figure 1-3: Typical feedback system showing functionality of individual blocks.

In this configuration, the output signal is fed back to the input, where it is compared with the desired input. The difference between the two signals is amplified and applied to the plant input.

In order to design a successful feedback system, several issues must be resolved:

- First, how do you generate the model of the plant, given that many systems do not have well-defined transfer functions?

- Once you have the model of the plant, how do you close the loop, resulting in a stable system with a desired gain and bandwidth?

Control System Basics

A classical feedback loop, as envisioned by Black, is shown in **Figure 1-4**. Note that there is an external disturbance in this system, the voltage v_d.

In this system, a is the forward path gain and f is the feedback gain. The forward gain a and feedback factor f may have frequency dependence (and, hence, the plant should be denoted as $a(s)$), but for notational simplicity we'll drop the Laplace variable s.

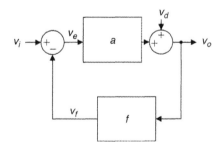

Figure 1-4: Classical single-input, single-output control loop, with input voltage v_i, output voltage v_o, and external disturbance v_d.

Initially, let's set the disturbance v_d to zero. The "error" term v_e is the difference between the input and the fed-back portion of the output. We can solve for the transfer function with the result:

$$v_o = av_e$$
$$v_e = v_i - v_f \qquad [1\text{-}1]$$
$$v_f = fv_o$$

The closed-loop gain is ($A \equiv$ closed-loop gain):

$$A = \frac{v_o}{v_i} = \frac{a}{1+af} \qquad [1\text{-}2]$$

or:

$$A = \frac{FORWARD\ GAIN}{1 - LOOP\ TRANSMISSION} \qquad [1\text{-}3]$$

Note what happens in the limit of $af \gg 1$:

$$A \approx \frac{1}{f} \qquad [1\text{-}4]$$

This is the key to designing a successful feedback system; if you can guarantee that $af \gg 1$ for the frequencies that you are interested in, *then your closed-loop gain will not be dependent on the details of the plant gain $a(s)$*. This is very useful, since in some cases the feedback function f can be implemented with a simple resistive divider . . . which can be cheap and accurate.

Loop Transmission and Disturbance Rejection

The term in the denominator of the gain equation is $1 + af$, where the term $-af$ is called the *loop transmission*, or *L.T.* This term is the gain going around the whole feedback loop; you can find the L.T. by doing a thought experiment: Cut the feedback loop in one place, inject a signal, and find out what returns where you cut. The gain around the loop is the loop transmission.

Now let's find the gain from the disturbance input to the output:

$$\frac{v_o}{v_d} = \frac{1}{1+af} = \frac{1}{1 - L.T.} \qquad [1\text{-}5]$$

Note that if the loop transmission is large at frequencies of interest, then the output due to the disturbance will be small. The term $(1 + af)$ is called the *desensitivity* of the system. Let's figure out the fractional change in closed-loop gain A due to a change in forward-path gain a.

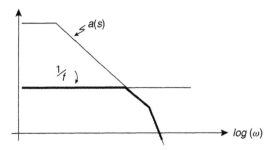

Figure 1-5: Plot for estimating closed-loop transfer function graphically. The curve *a(s)*
depicts the frequency dependence of the forward-path gain. The line 1/*f* is the inverse of the
feedback gain, shown here for resistive feedback. The thick line indicates our estimate for
closed-loop transfer function. For *a(s)f* ≫ 1, the closed-loop gain is approximately 1/*f*. For
a(s)f ≪ 1, the closed-loop gain is approximately *a(s)*.

$$A = \frac{a}{1+af}$$

$$\frac{dA}{da} = \frac{(1+af)-af}{(1+af)^2} = \left(\frac{1}{1+af}\right)\left(\frac{1}{1+af}\right) = \frac{A}{a}\left(\frac{1}{1+af}\right) \qquad [1\text{-}6]$$

$$\therefore \frac{dA}{A} = \frac{da}{a}\left(\frac{1}{1+af}\right)$$

This result means that if $af \gg 1$, then the fractional change in closed-loop gain (dA/A) is
much smaller than the fractional change in forward-path gain (da/a).

We can make a couple of approximations in the limit of large and small loop transmission.
For large loop transmission ($af \gg 1$), as we've shown before, the closed-loop gain
$A \approx 1/f$. For small loop transmission ($af \ll 1$), the closed-loop gain is approximately
$a(s)$. If we plot $a(s)$ and $1/f$ on the same set of axes, we can find an approximation for the
closed-loop gain as the lower of the two curves, as shown in **Figure 1-5**.

Stability

So far, we haven't discussed the issue regarding the stability of closed-loop systems.
There are many definitions of stability in the literature, but we'll consider *BIBO*
stability. In other words, we'll consider the stability problem given that we'll only
excite our system with bounded inputs. The system is BIBO stable if *bounded inputs*
generate *bounded outputs*, a condition that is met if all poles are in the left-half plane
(**Figure 1-6**).

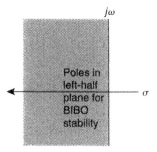

Figure 1-6: Closed-loop pole locations in the left-half plane for bounded input, bounded output (BIBO) stability.

Consider the feedback system with a first-order plant and unity feedback (**Figure 1-7**). The input/output transfer function is:

$$\frac{v_o}{v_i} = \frac{A/s}{A/s + 1} = \frac{A}{s + A} = \frac{1}{\frac{s}{A} + 1}$$

[1-7]

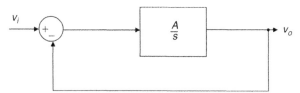

Figure 1-7: First-order system comprising an integrator inside a negative-feedback loop.

Note that as the forward-path gain A increases, the closed-loop bandwidth increases as well, with the closed-loop pole staying on the real axis at $s = -A$. As long as A is positive, this system is BIBO stable for any values of A.

The second-order system (**Figure 1-8a**) is also easy to work out, with transfer function:

$$\frac{v_o}{v_i} = \frac{\dfrac{K}{(\tau_a s + 1)(\tau_b s + 1)}}{1 + \dfrac{K}{(\tau_a s + 1)(\tau_b s + 1)}} = \frac{K}{K + (\tau_a s + 1)(\tau_b s + 1)}$$

$$= \left(\frac{K}{K+1}\right) \frac{1}{\left(\dfrac{\tau_a \tau_b}{1+K}\right) s^2 + \left(\dfrac{\tau_a + \tau_b}{1+K}\right) s + 1}$$

[1-8]

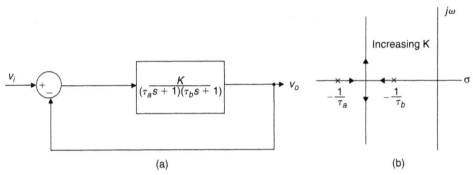

Figure 1-8: Second-order system with negative-feedback loop. (a) Block diagram.
(b) Root locus as *K* increases.

The pole locations are plotted in **Figure 1-8b**, with the locus of closed-loop poles shown for *K* increasing. Note the fundamental trade-off between high DC open-loop gain (which means a small closed-loop DC error) and loop stability. For *K* approaching infinity, the closed-loop poles are very underdamped.

Routh Stability Criterion

The Routh test is a mathematical test that can be used to determine how many roots of the characteristic equation lie in the right-half plane. When we use the Routh test, we don't calculate the location of the roots—rather, we determine whether there are any roots at all in the right-half plane, without explicitly determining where they are.

The procedure for using the Routh test is as follows:

1. Write the characteristic polynomial:

$$1 - L.T. = a_o s^n + a_1 s^{n-1} + \cdots + a_n \qquad [1\text{-}9]$$

    Note that since we've written the characteristic polynomial $(1 - L.T.)$, we now are interested in finding whether there are *zeros* of $(1 - L.T.)$ in the right-half plane. Zeros of $(1 - L.T.)$ in the right-half plane correspond to closed-loop poles in the right-half plane. Furthermore, we assume that $a_n \neq 0$ for the analysis to proceed.

2. Next, we see if any of the coefficients are zero or have a different sign from the others. A necessary (but not sufficient) condition for stability is that there are no nonzero coefficients in the characteristic equation and that all coefficients have the same sign.

3. If all coefficients have the same sign, we next form a matrix with rows and columns in the following pattern, which is shown for n even.[6] The table is filled horizontally and vertically until zeros are obtained in the rows. The third row and following rows are calculated from the previous two rows.

$$
\begin{array}{cccccc}
a_0 & a_2 & a_4 & \dots & \dots & \dots \\
a_1 & a_3 & a_5 & \dots & \dots & \dots \\
b_1 & b_2 & b_3 & \dots & \dots & \dots \\
c_1 & c_2 & c_3 & \dots & \dots & \dots \\
\dots & \dots & \dots & \dots & \dots & \dots \\
0 & 0 & 0 & 0 & 0 & 0
\end{array}
$$

$$
b_1 = \frac{-\begin{vmatrix} a_0 & a_2 \\ a_1 & a_3 \end{vmatrix}}{a_1} = \frac{a_1 a_2 - a_0 a_3}{a_1}
$$

$$
b_2 = \frac{-\begin{vmatrix} a_0 & a_4 \\ a_1 & a_5 \end{vmatrix}}{a_1} = \frac{a_1 a_4 - a_0 a_5}{a_1}
$$

$$
b_3 = \frac{-\begin{vmatrix} a_0 & a_6 \\ a_1 & a_7 \end{vmatrix}}{a_1} = \frac{a_1 a_6 - a_0 a_7}{a_1}
$$

$$
c_1 = \frac{-\begin{vmatrix} a_1 & a_3 \\ b_1 & b_2 \end{vmatrix}}{b_1} = \frac{b_1 a_3 - a_1 b_2}{b_1}
$$

$$
c_2 = \frac{-\begin{vmatrix} a_1 & a_5 \\ b_1 & b_3 \end{vmatrix}}{b_1} = \frac{a_5 b_1 - a_1 b_3}{b_1}
$$

[1-10]

4. The number of poles in the right-half plane is equal to the number of sign changes in the first column of the Routh matrix.

Let's apply the Routh test to the transfer function:

$$
H(s) = \frac{1}{(s+1)(s+2)(s+3)(s-2)} = \frac{1}{s^4 + 4s^3 - s^2 - 16s - 12}
$$

[1-11]

---

[6] For n odd, a_n terminates the second row.

In this case, we already know that there is one right-half-plane pole at $s = +2$ radians/second, but we'll use the Routh test to verify this. The Routh matrix is:

$$\begin{pmatrix} 1 & -1 & -12 & 0 \\ 4 & -16 & 0 & 0 \\ \dfrac{-\begin{vmatrix} 1 & -1 \\ 4 & -16 \end{vmatrix}}{-1} = -12 & \dfrac{-\begin{vmatrix} 1 & -12 \\ 4 & 0 \end{vmatrix}}{-1} = 48 & \dfrac{-\begin{vmatrix} 1 & 0 \\ 4 & 0 \end{vmatrix}}{-1} = 0 & 0 \\ \dfrac{-\begin{vmatrix} 4 & -16 \\ -12 & 48 \end{vmatrix}}{-12} = 0 & \dfrac{-\begin{vmatrix} 4 & 0 \\ -12 & 0 \end{vmatrix}}{-12} = 0 & 0 & 0 \end{pmatrix}$$ [1-12]

We see that there is one sign change in the first column, with the elements of the matrix changing from $+4$ to -12. Hence, there is one right-half-plane pole, as expected.

Let's next apply the Routh test to a system with three poles inside a unity-feedback loop (**Figure 1-9**). We'll use the Routh test to determine the values of K that result in stable operation of this feedback loop. The closed-loop transfer function for this system is:

$$\frac{v_o(s)}{v_i(s)} = \frac{\dfrac{K}{(s+1)^3}}{1 + \dfrac{K}{(s+1)^3}} = \left(\frac{K}{1+K}\right)\left(\frac{1}{\dfrac{s^3}{K+1} + \dfrac{3s^2}{K+1} + \dfrac{3s}{1+K} + 1}\right)$$ [1-13]

The denominator polynomial is:

$$D(s) = a_0 s^3 + a_1 s^2 + a_2 s + a_3 = \left(\frac{1}{K+1}\right)s^3 + \left(\frac{3}{K+1}\right)s^2 + \left(\frac{3}{K+1}\right)s + 1$$ [1-14]

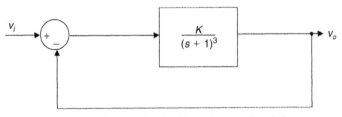

Figure 1-9: Three poles inside unity-feedback loop.

The Routh matrix is:

$$
\begin{array}{cc}
\left(\dfrac{1}{1+K}\right) & \left(\dfrac{3}{1+K}\right) \\[3ex]
\left(\dfrac{3}{1+K}\right) & 1 \\[3ex]
\dfrac{\left(\dfrac{3}{1+K}\right)^2 - \left(\dfrac{1}{1+K}\right)}{1+K} = \dfrac{8-K}{(1+K)^2} & 0 \\[3ex]
1 & 0 \\[1ex]
0 & 0
\end{array}
\qquad\qquad [1\text{-}15]
$$

Note that if $K > 8$, there are two sign changes in the first column. Therefore, for $K = 8$, we expect two poles on the $j\omega$ axis, and for $K > 8$ the system is unstable with two poles in the right-half plane. For $K < 8$, the system is stable with all three poles in the left-half plane.

The Phase Margin and Gain Margin Tests

The previous analyses tell us what the bandwidth and DC gain of a closed-loop system is but don't consider the question of whether a system will oscillate or have large amounts of overshoot. Using a simple Bode plot technique and a method known as the *phase margin method*, we can determine the relative stability of a feedback system. Phase margin is a very useful measure of the stability of a feedback system. The method for finding phase margin for a negative feedback system is as follows (**Figure 1-10**):

1. Plot the magnitude and angle of the negative of the loop transmission, or $-a(s)f(s)$.

2. Find the frequency where the magnitude of $a(s)f(s)$ drops to $+1$. This is the *crossover frequency* ω_c.

3. The difference between the angle at the crossover frequency and $-180°$ is the *phase margin* ϕ_m.

4. The *gain margin* (G.M.) is defined as the change in open-loop gain required to make the system unstable. Systems with greater gain margins can withstand greater changes in system parameters before becoming unstable in closed loop.

5. The *phase margin* is defined as the negative change in open-loop phase shift required to make a closed-loop system unstable.

6. In general, a well-designed feedback loop has a phase margin of at least $45°$ and a G.M. > 3 or so.

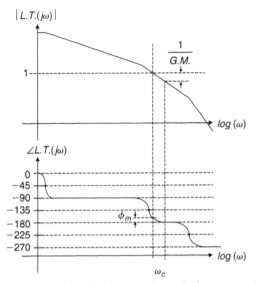

Figure 1-10: Plot of gain margin and phase margin.

Relationship Between Damping Ratio and Phase Margin

The damping ratio and phase margin are directly related. For a second-order system, a low phase margin in general implies a low damping ratio. For a standard second-order system with damping ratio < 0.6, the relationship is approximately:

$$\zeta \approx \frac{\phi_m}{100} \qquad\qquad [1\text{-}16]$$

Therefore, a damping ratio of 0.6 corresponds to a phase margin of 60°. The actual relationship over the range of damping ratios $0 < \zeta < 2$ is shown in **Figure 1-11**.

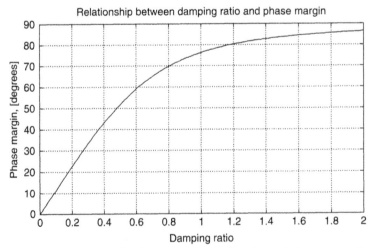

Figure 1-11: Relationship between phase margin and damping ratio.

Loop Compensation Techniques—Lead and Lag Networks

Several networks are available to compensate feedback networks. These networks can be added in series to the plant to modify the closed-loop transfer function or be placed in other locations in the feedback system. Shown next is a quick look at "lead" and "lag" networks.

The lag network (**Figure 1-12a**) is often used to reduce the gain of the loop transmission so that crossover occurs at a benign frequency. The transfer function of the lag network is:

$$H(s) = \frac{v_o(s)}{v_i(s)} = \frac{R_2Cs + 1}{(R_1 + R_2)Cs + 1} = \frac{\tau s + 1}{\alpha \tau s + 1}$$

$$\alpha = \frac{R_1 + R_2}{R_2}$$

$$\tau = R_2C$$

[1-17]

The Bode plot of the lag network (**Figure 1-12b**) shows that the network produces magnitude reduction at frequencies between the pole and the zero. When using a lag network, you'll typically place the lag zero well below the crossover frequency of the loop. This ensures that the lag network doesn't provide too much negative phase shift at crossover.

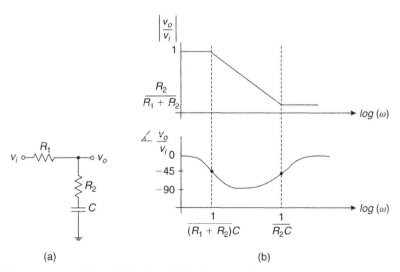

(a)   (b)

Figure 1-12: Lag network. (a) Circuit. (b) Bode plot of magnitude and phase angle of the frequency response of the lag network.

$$H(s) = \frac{v_o(s)}{v_i(s)} = \frac{R_2 Cs + 1}{(R_1 + R_2)Cs + 1} = \frac{\tau s + 1}{\alpha \tau s + 1}$$

$$\alpha = \frac{R_1 + R_2}{R_2}$$ [1-18]

$$\tau = R_2 C$$

The lead network (**Figure 1-13a**) is used to provide positive phase shift in the vicinity of the crossover frequency. The transfer function of the lead network is:

$$H(s) = \frac{v_o(s)}{v_i(s)} = \left(\frac{R_2}{R_1 + R_2}\right)\left(\frac{R_1 Cs + 1}{\dfrac{R_1 R_2}{R_1 + R_2}Cs + 1}\right) = \frac{1}{\alpha}\frac{\alpha \tau s + 1}{\tau s + 1}$$

$$\alpha = \frac{R_1 + R_2}{R_2}$$ [1-19]

$$\tau = (R_1 \| R_2)C$$

The Bode plot of the lead (**Figure 1-13b**) shows that the lead provided +45 degrees of positive phase shift at the zero frequency, while at the zero there is only +3 dB of gain increase. When using a lead network, one generally places the lead zero near the crossover frequency of the loop to take advantage of the positive phase shift provided by the lead. The lead pole is then above crossover.

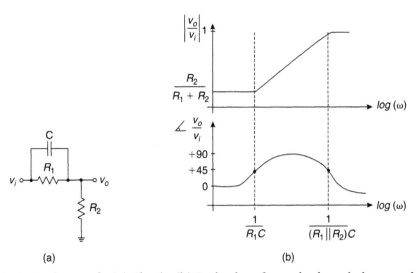

(a)   (b)

Figure 1-13: Lead network. (a) Circuit. (b) Bode plot of magnitude and phase angle of the frequency response of the lead network.

Parenthetical Comment on Some Interesting Feedback Loops

The inquisitive student may wonder whether a system that has a loop transmission magnitude greater than unity where the loop transmission angle is $-180°$ can be stable or not. In using the gain margin/phase margin test, we look at the frequency at which the magnitude drops to unity, and we don't concern ourselves with other frequencies. By example, we'll show next that a system that has a loop transmission magnitude greater than unity where the loop transmission angle is $-180°$ *can* be stable. It's understood that this is not necessarily an intuitive result, but we'll run with it anyway. Consider the system of **Figure 1-14a**, which is a unity-feedback system with two zeros and three poles in the forward path.

The negative of the loop transmission for this system is:

$$-L.T. = \frac{100(s+1)^2}{s^3} \qquad \text{[1-20]}$$

A plot of the negative of the loop transmission is shown in **Figure 1-14b**. Note that the loop transmission magnitude is greater than unity at the frequency when the angle of the negative of the loop transmission is $-180°$. In this case, the angle is less than $-180°$ up to 1 radian/second.

We'll next use the Routh test to determine the stability of this system. The closed-loop transfer function for this system is:

$$\frac{v_o(s)}{v_i(s)} = \frac{\dfrac{100(s+1)^2}{s^3}}{1+\dfrac{100(s+1)^2}{s^3}} = \frac{100(s^2+2s+1)}{s^3+100s^2+200s+100} \qquad \text{[1-21]}$$

We can use the Routh criteria or we can factor the denominator of this transfer function to determine stability. The denominator polynomial is:

$$D(s) = s^3 + 100s^2 + 200s + 100 \qquad \text{[1-22]}$$

The Routh matrix is as follows:

$$
\begin{array}{cc}
1 & 200 \\
100 & 100 \\
\dfrac{(100)(200)-(1)(100)}{100} = 199 & 0 \\
\dfrac{(199)(100)-(100)(0)}{199} = 100 & 0 \\
0 & 0
\end{array}
\qquad \text{[1-23]}
$$

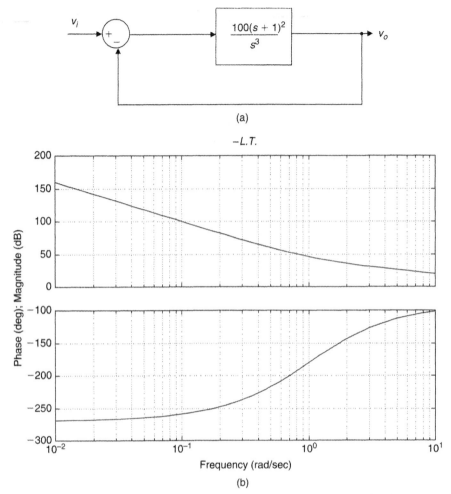

(a)

Figure 1-14: Unity-feedback system that has a loop transmission magnitude greater than unity where the loop transmission angle is −180°. (a) System. (b) Bode plot of −L.T. showing the magnitude is greater than 1 when the angle is more negative than −180°.

The Routh test shows that this system is BIBO stable since there are no sign changes in the first column of the Routh matrix. Numerical analysis shows that the closed-loop poles and zeros are at frequencies:

Zeros:	2 Zeros at −1 r/s
Poles:	Poles at −97.97 r/s, −1.12 r/s and −0.92 r/s

Therefore, all poles are in the left-half plane and the system is BIBO stable. The closed-loop step response (**Figure 1-15**) confirms that the system is stable. Note the

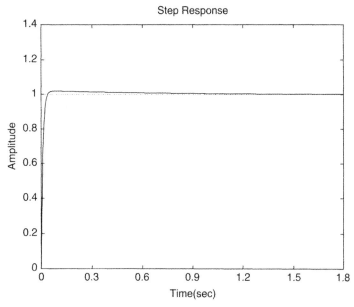

Figure 1-15: Step response of feedback system that has a loop transmission magnitude greater than unity where the loop transmission angle is −180°.

long decaying "tail" of the step response while it settles to unity gain. This long tail is characteristic of systems with closely spaced poles and zeros.[7]

Example 1-1: Gain of +1 amplifier

Consider an operational amplifier with a DC gain of 10^5, a low-frequency pole at 10 radians/second, and a high-frequency pole at 10^6 radians/second. This transfer function $a(s)$ is representative of many commercially available operational amplifiers and is expressed as:

$$a(s) = \frac{10^5}{(0.1s + 1)(10^{-6}s + 1)} \qquad [1\text{-}24]$$

What is the bandwidth and risetime when this op-amp is configured as a gain of +1 amplifier?

Shown in **Figure 1-16** is the Bode plot of the forward path gain $a(s)$. Note that the DC gain is 10^5 (100 dB) and the phase starts at 0° and falls asymptotically to −180° at frequencies much higher than 10^6 radians per second (above the second pole).

---

[7] Also known as a pole/zero "doublet."

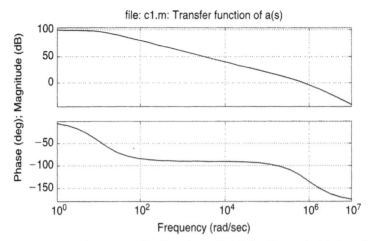

Figure 1-16: Open-loop transfer function of *a(s)* for Examples 1-1 and 1-2.

The gain of +1 op-amp circuit is shown in **Figure 1-17a** where the op-amp has unity feedback. The block diagram of this circuit is shown in **Figure 1-17b**.

For this circuit, the negative of the loop transmission is:

$$-L.T. = a(s)f = \frac{10^5}{(0.1s+1)(10^{-6}s+1)} \qquad [1\text{-}25]$$

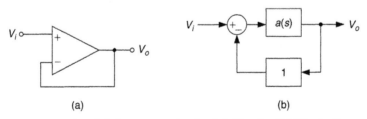

(a)   (b)

Figure 1-17: Gain of +1 op-amp circuit. (a) Circuit. (b) Block diagram.

which is the same as the open-loop transfer function of the op-amp *a(s)*. Using MATLAB, the bandwidth and phase margin are calculated, as shown in **Figure 1-18**. The results show a phase margin of 52° and a crossover frequency of 786,150 radians/second (125 kHz). From this analysis, we expect some overshoot in the step response (since the phase margin results in a damping ratio of ~0.5), some overshoot in the frequency response, and a 10–90% risetime[8] of approximately 0.35/125000 = 2.8 µs (**Figure 1-19**).

---

[8] In general, an estimate of 10–90% risetime is 0.35/f_c, where f_c is the crossover frequency in Hz.

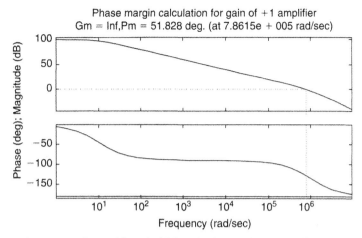

Figure 1-18: Phase margin and bandwidth calculation for gain of +1 op-amp circuit.

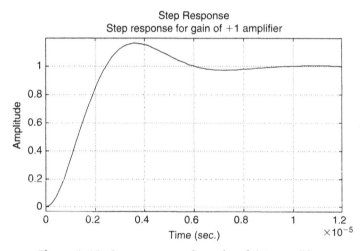

Figure 1-19: Step response for gain of +1 amplifier.

Example 1-2: Gain of +10 amplifier

What is the bandwidth and risetime when this same op-amp is configured as a gain of +10 amplifier? The gain of +10 op-amp circuit is shown in **Figure 1-20a**, and the block diagram is shown in **Figure 1-20b**. Note that the 9 R/R divider gives a feedback factor $f = 0.1$.

For this circuit, the negative of the loop transmission is:

$$-L.T. = a(s)f = \frac{10^4}{(0.1s + 1)(10^{-6}s + 1)} \qquad [1\text{-}26]$$

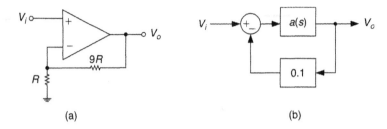

Figure 1-20: Gain of +10 op-amp circuit. (a) Circuit. (b) Block diagram.

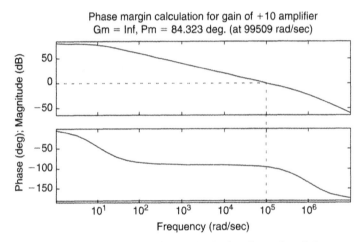

Figure 1-21: Phase margin and bandwidth calculation for gain of +10 op-amp circuit.

When we plot the loop transmission magnitude and phase (**Figure 1-21**), the results show a phase margin of 84° and a crossover frequency of 99,509 radians/second (15.8 kHz). From this analysis, we expect no overshoot in the step response and frequency response (since the phase margin is close to 90°) and a 10–90% risetime of approximately 0.35/15,800 = 22 μs, as shown in **Figure 1-22**. Therefore the system is well damped and there is no overshoot in the step response (**Figure 1-23**).

Example 1-3: Integral control of reactive load

Consider the op-amp driving a reactive load as shown in **Figure 1-24**. Assume that the op-amp is ideal; it has infinite bandwidth and can source and sink infinite current. Given this, the transfer function is:

$$H(s) = \frac{v_o(s)}{v_i(s)} = \frac{1}{LCs^2 + \dfrac{L}{R}s + 1} = \frac{1}{10^{-10}s^2 + 10^{-6}s + 1} \qquad [1\text{-}27]$$

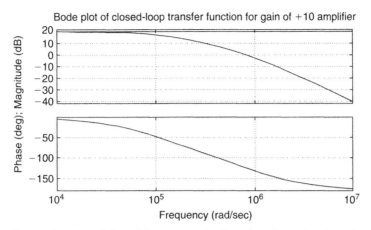

Figure 1-22: Bode plot of closed-loop transfer function for gain of +10 amplifier.

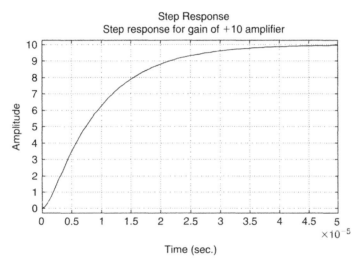

Figure 1-23: Step response for gain of +10 amplifier.

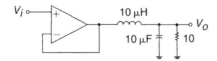

Figure 1-24: Ideal op-amp driving a reactive load.

This second-order reactive load $H(s)$ has the following:

- Poles: $-5 \times 10^3 \pm 9.99 \times 10^4 j$

- Damping ratio: 0.05

- $Z_o = 1\Omega$

- Q: 10

When plotting this transfer function $H(s)$, we see the underdamped response of **Figure 1-25.**

In this example, we'll design a closed-loop controller to regulate the output voltage (**Figure 1-26**). We'll assume that the forward path compensator $G_c(s)$ includes an integrator so that there will be zero DC error[9] in V_o.

As a first attempt, let's try a compensator transfer function of the form:

$$G_c(s) = \frac{4 \times 10^3}{s} \qquad\qquad [1\text{-}28]$$

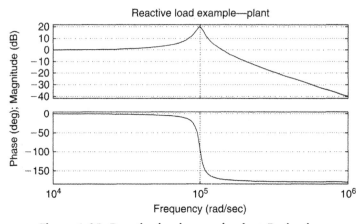

Figure 1-25: Reactive load example plant Bode plot.

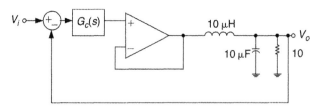

Figure 1-26: Closed-loop controller for Example 1-3.

---

[9] Having an integrator in $G_c(s)$ ensures that the DC error of the loop is zero. If there were an error, the integrator would forever integrate to infinity. Hence, there must be zero DC error.

The resultant loop transmission is:

$$-L.T. = \left(\frac{1}{10^{-10}s^2 + 10^{-6}s + 1} \right)\left(\frac{4 \times 10^3}{s} \right)$$ [1-29]

Plotting this loop transmission magnitude and phase yields some interesting results, as shown in **Figure 1-27**.

Results are:

- Crossover frequency: 4006 radians/sec (637 Hz)

- Phase margin: 90°

- Given this, we expect a well-controlled step response, with risetime ~0.5 ms.

The resultant step response of the closed-loop system is shown in **Figure 1-28**.

Things look OK except for the oscillatory behavior on the rising edge. What's going on here? The Bode plot of the loop transmission phase and magnitude tells the story. Although the phase margin is fine, the gain margin isn't so great, due to the underdamped pole pair. In order to help this oscillation problem, we might attempt to add a pole above crossover to damp out the underdamped pair. Let's try adding a pole at 5×10^4 radians/second to damp the complex poles at 10^5 radians/sec. This results in:

$$G_c(s) = \left(\frac{4 \times 10^3}{s} \right)\left(\frac{1}{2 \times 10^{-5}s + 1} \right)$$ [1-30]

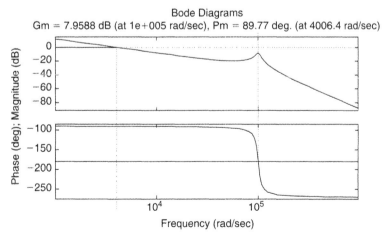

Figure 1-27: Plot of magnitude and phase of the loop transmission, Attempt #1, showing a crossover frequency of 4006 radians/second and a phase margin of 89.8°.

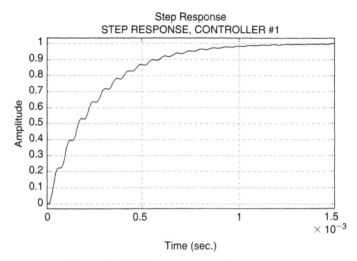

Figure 1-28: Step response, Attempt #1.

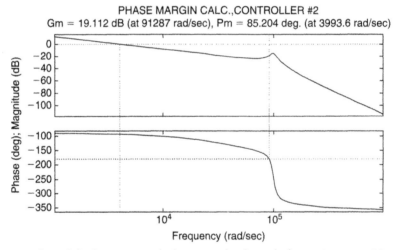

Figure 1-29: Plot of the loop transmission magnitude and phase, Attempt #2, showing crossover frequency of 3993 radians/second and a phase margin of 85°.

The results of a MATLAB crossover frequency and phase margin test are shown in **Figure 1-29**. Note that by adding the low-pass filter, we've significantly improved the gain margin of this circuit. Results show:

- Crossover frequency: 3993 radians/sec (635 Hz)

- Phase margin: 85°

- Better-behaved step response, as shown in **Figure 1-30**.

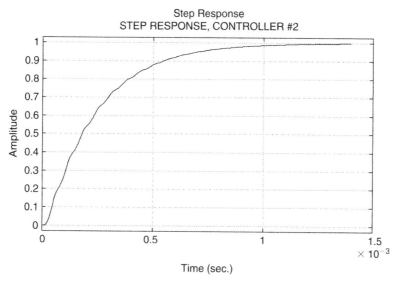

Figure 1-30: Step response, Attempt #2.

Example 1-4: Photodiode amplifier

A photodiode amplifier is shown in **Figure 1-31a**. The photodiode puts out a current proportional to the light hitting it, and the *transimpedance* connection of the op-amp converts this photodiode current into an output voltage. The input-to-output ideal transfer function is:

$$\frac{v_o}{i_p} = -R_f \qquad [1\text{-}31]$$

When this circuit is modeled, the photodiode is modeled as a current source in parallel with a parasitic capacitance C_p, as shown in **Figure 1-31b**.

The block diagram for this system is shown in **Figure 1-32**.

The closed-loop transfer function of this system is:

$$\frac{v_o}{i_p} = -\frac{R_f}{R_f C_p s + 1} \left(\frac{a(s)}{1 + \dfrac{a(s)}{R_f C_p s + 1}} \right) \qquad [1\text{-}32]$$

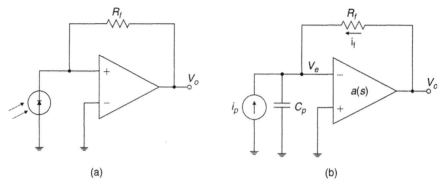

Figure 1-31: Photodiode amplifier. (a) Circuit. (b) Model, showing equivalent model of photodiode that includes a current source i_p and parasitic capacitance C_p.

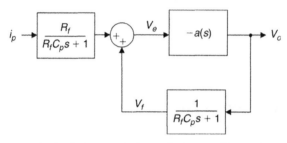

Figure 1-32: Block diagram of photodiode amplifier. The block $a(s)$ is the gain of the operational amplifier.

Note that if the loop transmission is much larger than 1 the approximate transfer function is $-R_f$. The loop transmission of this system, easily found by inspection, is:

$$-L.T. = \frac{a(s)}{R_f C_p s + 1} \qquad [1\text{-}33]$$

In general, an operational amplifier will have a dominant pole and a second pole near the crossover frequency, resulting in:

$$-L.T. = \frac{a_o}{s(\tau s + 1)(R_f C_p s + 1)} \qquad [1\text{-}34]$$

This means that there are three poles (at least) in the loop transmission. If we attempt to close a feedback loop with a bandwidth greater than $1/R_f C_p$ there are potential problems with stability.

A model of an actual system, using the CLC426 op-amp, was created, including dominant pole, second pole, and output resistance (**Figure 1-33**).

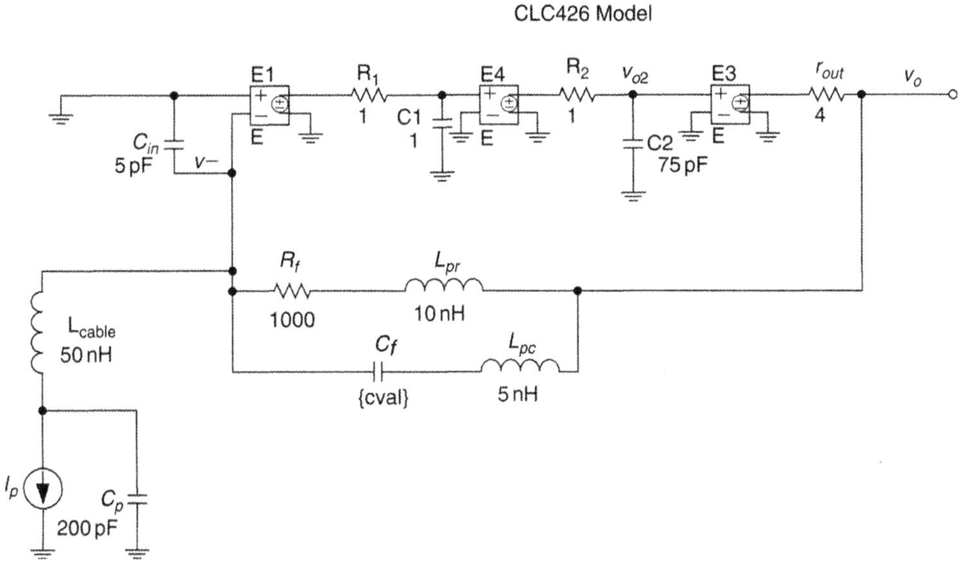

CLC426 Model

Figure 1-33: Photodiode amplifier model.

Parasitics in this circuit are as follows:

- L_{cable}: Inductance of cable connecting photodiode to PC board. Approximate inductance is 10 nanohenries per centimeter of length. A value of 50 nanohenries was used for all simulations.

- L_{pr}: Inductance in series with feedback resistance, approximately 10 nH.

- L_{pc}: Inductance in series with feedback capacitor, approximately 5 nH.

- C_p: Capacitance of photodiode, approximately 60 pF.

- C_{in}: Input capacitance of op-amp, approximately 5 pF.

The $R_f C_p$ combination results in a pole inside the feedback loop that results in potential instability. Plotted in **Figure 1-34** is frequency response of v_o/i_p of the original circuit with $R_f = 15\,\mathrm{k}$, showing that the response is very underdamped. Other unmodeled poles can result in oscillation.

By adding a capacitance across the feedback resistor, a *lead* transfer function is created (**Figure 1-35**). The added lead zero can create positive phase shift near the crossover frequency, hence improving stability. The feedback factor becomes:

$$f(s) = \frac{\dfrac{1}{C_p s}}{\dfrac{1}{C_p s} + \dfrac{R_f}{R_f C_f s + 1}} = \frac{1}{1 + \dfrac{R_f C_p s}{R_f C_f s + 1}} = \frac{R_f C_f s + 1}{R_f (C_f + C_p)s + 1} \qquad [1\text{-}35]$$

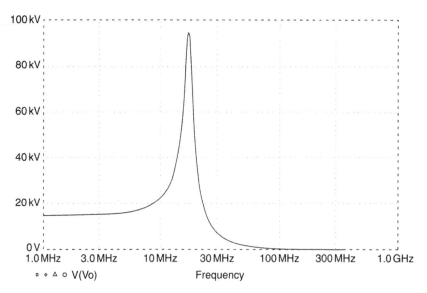

Figure 1-34: PSPICE results of frequency response of the photodiode amplifier, showing potential instability near 15 MHz.

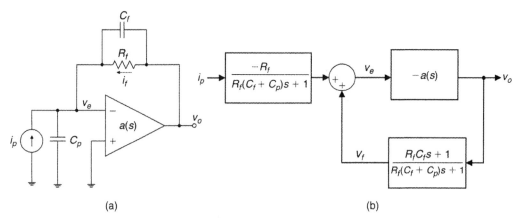

(a)   (b)

Figure 1-35: Photodiode amplifier with lead compensation. (a) Circuit. (b) Block diagram.

Simulations were modified for $R_f = 1$ k and C_f adjustable from 2 pF to 10 pF. Results (**Figure 1-36**) show that it may be possible to achieve ~50 MHz bandwidth by proper adjustment of C_f. This, of course, depends on the accuracy of these simulations. Measurements should be made on the prototype to verify or refute these models.

Example 1-5: MOSFET current source

In **Figure 1-37** is a MOSFET current source with an op-amp used in a negative feedback configuration to maintain control of the MOSFET drain current. If the op-amp is

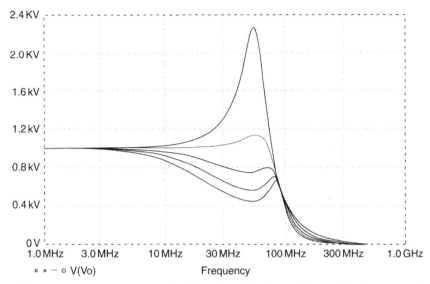

Figure 1-36: Frequency response results for photodiode amplifier with C_f:2, 4, 6, 8, 10 pF, with $R_f = 1k$ and $C_p = 60$ pF.

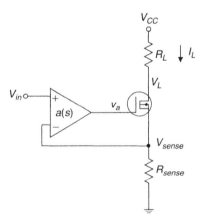

Figure 1-37: MOSFET current source. The voltage V_{sense} senses MOSFET drain current.

ideal and if the MOSFET is operating in the linear region, the input-output transfer function is:

$$\frac{I_L}{V_{in}} \approx \frac{1}{R_{sense}}$$

[1-36]

This result is contingent on the feedback control system being stable (i.e., not oscillating). A small-signal model is shown in **Figure 1-38**. Following are the parameters:

τ_h Time constant of op-amp high-frequency pole

r_{out} Output resistance of op-amp

C_{gs} MOSFET gate-source capacitance

C_{gd} MOSFET gate-drain capacitance

g_m MOSFET transconductance

R_L Load resistance

The transfer function of the MOSFET source follower is estimated using the method of open-circuit time constants. This method assumes that a single pole dominates the transfer function; the resultant transfer function from the output of the op-amp to v_{sense} is:

$$\frac{v_{sense}}{v_a} \approx \frac{A_o}{\tau s + 1}$$

$$A_o = \frac{g_m R_{sense}}{1 + g_m R_{sense}}$$

$$\tau = \left[\frac{r_{out} + R_{sense}}{1 + g_m R_{sense}}\right]C_{gs} + [r_{out} + R_L + G_M r_{out} R_L]C_{gd}$$

$$G_M = \frac{g_m}{1 + g_m R_{sense}}$$

[1-37]

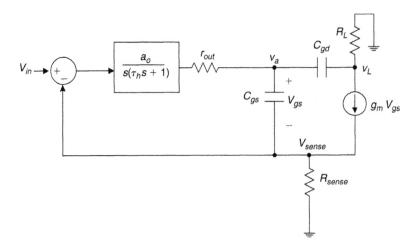

Figure 1-38: MOSFET current source—small-signal model.

The system was modeled assuming a TLO84 op-amp (with gain-bandwidth product of 4 MHz) and an IRF7403 MOSFET with $g_m = 10$ A/V, $C_{gs} = 1040$ pF and $C_{gd} = 160$ pF and a load resistance $R_L = 1\,\Omega$. In **Figure 1-39** is 6.7 Mrad/second and a phase margin of 28°. Due to the low phase margin, significant overshoot in the step response is expected. This is shown in **Figure 1-40**.

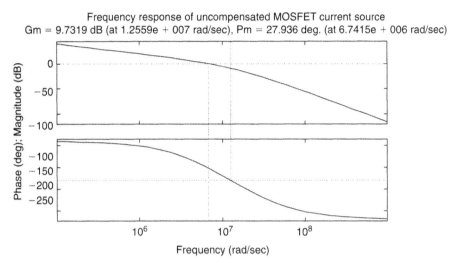

Figure 1-39: MOSFET current source—Bode plot of loop transmission showing crossover frequency of 6.7 Mrad/sec and a phase margin of 28°.

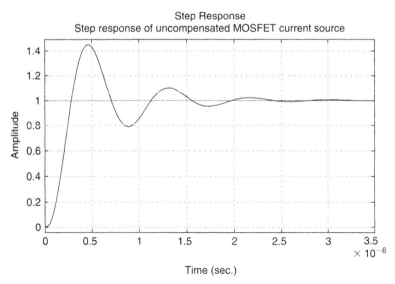

Figure 1-40: MOSFET current source—step response.

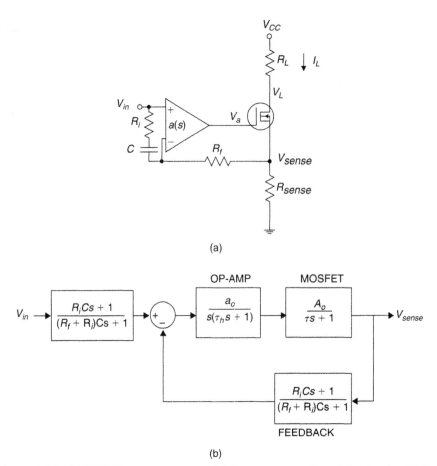

(a)

(b)

**Figure 1-41: MOSFET current source with lag compensation. (a) Circuit with lag
components R_i , C, and R_f added. (b) Block diagram.**

By adding lag compensation (**Figure 1-41a**) the system can be stabilized to provide
more phase margin. Shown in **Figure 1-41b** is the block diagram of the system. The lag
compensation adds a zero in the loop transmission at $-R_iC$ and a lag pole at $-(R_f + R_i)C$.

Step response results for the resultant system with the following parameters are shown in
Figure 1-42.

- $R_i = 47$ k

- $R_f = 470$ k

- $C = 1000$ pF

Note that the system has a better-behaved step response, with no overshoot, but that the
10–90% risetime has significantly increased compared to the uncompensated case.

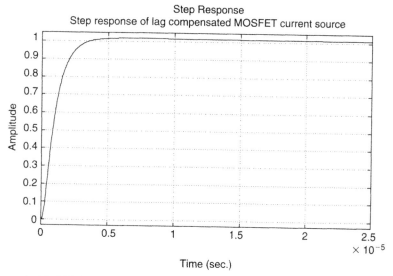

Figure 1-42: MOSFET current source with lag compensation, step response.

Example 1-6: Maglev example

Maglev[10] systems using superconducting magnets in the magnetic suspension have low damping. Furthermore, it has been demonstrated that these electrodynamic suspensions (EDS) may have slightly negative damping under certain operating conditions (with poles in the right-half plane). Therefore, a control system is needed to prevent underdamped or unstable vertical oscillations.

In order to suspend the magnet statically, the downward gravitational pull is canceled by an upward magnetic force. The magnet is levitated stably; that is, a deviation from the equilibrium position results in a restoring force, similar to a mass and a spring. In an electromagnetic suspension (such as a steel ball suspended in a magnetic field) there is no stable equilibrium for DC excitation of the magnet.

The magnetic levitating force acting on the magnet is given by:

$$f_z = -k_m z = -Ci_M^2 z \qquad \text{[1-38]}$$

where k_m is the equivalent spring constant, z is the vertical distance with reference to the magnet *null position*, i_M is magnet current, and C is a constant that accounts for magnet and coil geometry and relative velocity between the magnet and levitating coils.

---

[10] "Maglev" is a term used generically for a number of systems that use magnetic suspensions for ground transportation. As of 2004 there was one revenue-producing Maglev system operating in China, and others are proposed.

Assuming that there are incremental changes in forces, magnet vertical position, and magnet currents, a linearized model can be generated relating incremental changes in magnet vertical position to changes in incremental magnet current. Vertical force, vertical position, and magnet current are given as the sum of a DC component and an incremental component:

$$
\begin{aligned}
f_Z &= F_Z + \tilde{f}_z \\
z &= Z_o + \tilde{z} \\
i_M &= I_M + \tilde{i}_m
\end{aligned}
\qquad [1\text{-}39]
$$

Putting this into the force equation results in:

$$
f_Z \approx -CI_M^2 Z_o - CI_M^2 \tilde{z} - 2CI_M Z_o \tilde{i}_m \qquad [1\text{-}40]
$$

where second-order and higher terms have been neglected. At equilibrium, there is a resultant magnetic force that balances the force of gravity:

$$
F_Z = Mg = -CI_M^2 Z_o \qquad [1\text{-}41]
$$

Newton's law applied to the magnet results in:

$$
M \frac{d^2 \tilde{z}}{dt^2} = f_Z - Mg = -CI_M^2 \tilde{z} - 2CI_M Z_o \tilde{i} \qquad [1\text{-}42]
$$

resulting in:

$$
\frac{M}{CI_M^2} \frac{d^2 \tilde{z}}{dt^2} + \tilde{z} = \frac{2Mg}{k_m I_M} \tilde{i}_m \qquad [1\text{-}43]
$$

Using the spring constant k and converting the equation to the frequency domain results in:

$$
\left(\frac{M}{k_m} s^2 + 1 \right) z(s) = \frac{2Mg}{k_m I_M} i_m(s) \qquad [1\text{-}44]
$$

resulting in the transfer function between magnet position and magnet control current:

$$
\frac{z(s)}{i_m(s)} = \frac{2Mg}{k_m I_m \left(\dfrac{M}{k_m} s^2 + 1 \right)} \qquad [1\text{-}45]
$$

This result shows that this suspension has two $j\omega$-axis poles, as in a simple, lossless mass-spring system:

$$s_{p1,2} = \pm j\sqrt{\frac{k_m}{M}} \qquad [1\text{-}46]$$

Such a system can have the closed-loop poles arbitrarily adjusted by applying position velocity. By adjusting the parameters K_f, K_v, and K_p the poles can be placed in the left-half plane with sufficient damping to achieve good ride quality (**Figure 1-43**).

Representative numbers for a Maglev suspension magnet section are as follows:

$M = 10,000\,\text{kg}$

$k_m = 10^5\,\text{Newtons/cm} = 10^7\,\text{Newtons/meter}$

$I_M = 10^4\,\text{amps}$

This results in a resonant frequency of $\omega_o = 31.6\,\text{r/s}$ (with $f_o = 5\,\text{Hz}$), and a plant transfer function:

$$a(s) = \frac{z(s)}{i_m(s)} = \frac{1.96 \times 10^{-6}}{(10^{-3}s^2 + 1)} = \frac{A}{\dfrac{s^2}{\omega_o^2} + 1} \qquad [1\text{-}47]$$

The poles are on the $j\omega$ axis, corresponding to the underdamped suspension (**Figure 1-44**). In order to improve ride quality for the passengers, the suspension poles must be moved into the left-half plane by selection of suitable compensation.

The system is compensated using velocity feedback, as shown in **Figure 1-45**. Velocity feedback is equivalent to adding damping to the system.

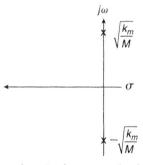

Figure 1-43: Pole plot of plant for Maglev example showing poles on the $j\omega$ axis.

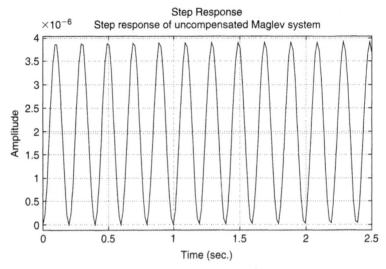

Figure 1-44: Step response of uncompensated Maglev suspension.

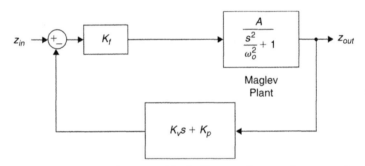

Figure 1-45: Control system block diagram for Maglev example.

A control system with $K_v = 10^5$ and $K_p = 10^4$ results in a system with closed-loop transfer function:

$$H(s) = \frac{1.64 \times 10^{-6}}{8.36 \times 10^{-4} s^2 + 1.64 \times 10^{-2} s + 1}$$

[1-48]

The closed-loop poles have a damping ratio of $\zeta = 0.28$ and at pole locations $-9.8 \pm j(33.2)$ radians/second. Therefore, we expect some oscillation near 33 radians/second (5.2 Hz). The resultant control system results in a much more well-behaved step response (**Figure 1-46**).

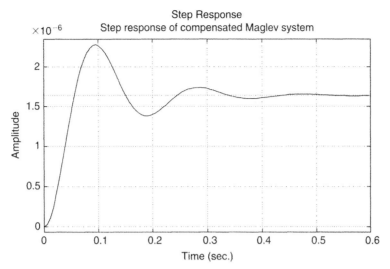

Figure 1-46: Step response of compensated Maglev example.

Appendix: MATLAB Scripts

MATLAB Script for Gain of +1 and +10 Amplifiers

```
function c1
% Control system example #1
% Calculates parameters for gain of +1 and gain of +10 amplifiers
% Marc Thompson, 10/22/99

% Open loop transfer function a(s)
ao=1e5;                    % DC gain
=conv([0.1 1],[1e-6 1]);   % Poles
a=tf(ao,d);                % Create transfer function a(s)
bode(a)                    % Plot Bode plot of a(s)
title('file: c1.m: Transfer function of a(s)')
figure;
pzmap(a);                  % Plot pole/zero map of a(s)
title('file: c1.m: Pole map of a(s)');
damp(a)                    % Find natural frequency and damping
ratio of
a(s)


% gain of +1
f=1;                       % Feedback gain of +1
margin(a*f)                % Find phase and gain margin
title('Phase margin calculation for gain of +1 amplifier');figure;
```

```
f=tf(f,1);                       % Create feedback f(s)
cloop=feedback(a,f)              % Close the loop, find transfer
                                   function H(s)
bode(cloop)
title('Bode plot of closed-loop transfer function for gain of +1
amplifier');figure
Step(cloop)
title('Step response for gain of +1 amplifier');grid;figure
pzmap(cloop)
title('Pole map of closed-loop gain of +1 amplifier')
;grid;figure;
damp(cloop)                           % Find natural frequency and damping
                                        ratio of

H(s)


% gain of +10
f=0.1;
margin(ao*f,d);title('Phase margin calculation for gain of +10
amplifier');figure;
f=tf(f,1);
cloop=feedback(a,f);cloopgainof10=cloop
bode(cloop);title('Bode plot of closed-loop transfer function for
gain of
+10 amplifier');figure
step(cloop);grid;title('Step response for gain of +10 amplifier');
pzmap(cloop);title('Pole map of closed-loop gain of +10
amplifier');grid;
Damp(cloop)
```

MATLAB Script for Integral Control Example

```
function c3
% Control example 3
% driving reactive load

L=10e-6;
C=10e-6;
R=10;
Zo = sqrt(L/C);                  % Characteristic Impedance
Q=R/Zo

% Calculate PLANT
num=1;
denom=[L*C L/R 1];
plant=tf(num,denom)
damp(plant)                      % Find poles and damping ratio
bode(plant); title('file: c3.m; REACTIVE LOAD EXAMPLE')
```

```
% Integral control, attempt #1
Gain=4e3;                       % Integrator gain
denom=[1 0];
Gc=tf(Gain,denom);              % Form Gc(s)
Forw=series(plant,Gc);          % Cascade with plant
margin(Forw);                   % Find gain and phase margin
F=tf(1,1);
Cloop=feedback(Forw,F,-1)
figure; step(Cloop);title('STEP RESPONSE, CONTROLLER #1');grid


%Integral control, attempt #2
figure
d=[1/5e4 1];
LPF=tf(1,d);                    % Add lowpass filter to damp complex
                                pole pair
Gc=series(LPF,Gc);
Forw=series(plant,Gc);
margin(Forw); title('PHASE MARGIN CALC., CONTROLLER #2');
F=tf(1,1);
Cloop=feedback(Forw,F,-1)
figure; step(Cloop);title('STEP RESPONSE, CONTROLLER #2');grid
```

MATLAB Script for MOSFET Current Source Example

```
function moscursource
% Analysis of MOSFET current source
% Marc Thompson, 3/28/00

% LOAD
RL=1;

% MOSFET model
gm = 8.6;                       % transconductance
Cgs = 1040e-12;
Cgd = 160e-12;

% OPAMP model
rout = 100;                     % output resistance of opamp
ao=2*pi*4*1e6;                  % GBP = 4 MHz
denom=[1/ao 0];
highpole=[1/ao 1];
d=conv(denom,highpole);
opamp=tf(1,d)

% MOSFET follower model
Rsense = 0.08;                  % current sense resistor
Ao=gm*Rsense/(1+gm*Rsense);     % gain of follower
```

```
% MOSFET OCTC calculation
Rgs=(rout+Rsense)/(1+gm*Rsense);
Tgs=Rgs*Cgs;
GM=gm/(1+gm*Rsense);
Rgd=rout+RL+(GM*rout*RL);
Tgd=Rgd*Cgd;
T=Tgs+Tgd % sum of OCTCs
mosfetpole=1/T
mosfet=tf(Ao,[T 1])

% Find loop transmission
LT=series(opamp,mosfet)
margin(LT);
title('moscursource. FREQUENCY RESPONSE OF UNCOMPENSATED MOSFET
CURRENT SOURCE')figure;

% close the loop
f=tf(1,1)
uncomp=feedback(LT,f,-1);
step(uncomp); grid;

title('moscursource. STEP RESPONSE OF UNCOMPENSATED MOSFET CURRENT
SOURCE')
figure

% add lag compensation
Rf=470000; C=1e-9;
Ri=47000;
numlag=[Ri*C 1];
denomlag=[(Ri+Rf)*C 1];
f=tf(numlag,denomlag)
comp=feedback(LT,f,-1);
comp=series(f,comp);
step(comp); grid;
title('moscursource. STEP RESPONSE OF LAG COMPENSATED MOSFET
CURRENT
SOURCE')
```

MATLAB Script for Maglev Example

```
function maglev
% Maglev example
% Marc Thompson 4/3/00

% Maglev plant
wn=sqrt(1e3);
num=1.96e-6;
denom=[1/wn^2 0 1];
plant=tf(num,denom);
```

```
step(plant);
title('Step response of uncompensated Maglev system');grid;
figure

% feedback compensation
Kp=1e5;
Kv=1e4;
num=[Kv Kp];
f=tf(num,1);

% Closed-loop
sys=feedback(plant,f,-1)
damp(sys)
step(sys);
grid;
title('Step response of compensated Maglev system')
```

References

Abramovich, D., "Phase-locked loops: a control centric tutorial," *Proceedings of the 2002 American Control Conference*, May 8–10, 2002, pp. 1–15.

———, "The outrigger: a prehistoric feedback mechanism," *Proceedings of the 42nd IEEE Conference on Decision and Control*, December 9–12, 2003, pp. 2000–2009.

Abramovitch, D., and Franklin, G., "A brief history of disk drive control," *IEEE Control Systems Magazine*, vol. 22, no. 3, June 2002, pp. 28–42.

Bennett, S., "Development of the PID controller," *IEEE Control Systems Magazine*, vol. 13, no. 6, December 1993, pp. 58–62, 64–65.

Bernstein, D.S., "Feedback control: An invisible thread in the history of technology," *IEEE Control Systems Magazine*, vol. 22, no. 2, April 2002, pp. 53–68.

Black, Harold S., "Stabilized feed-back amplifiers," *Electrical Engineering*, vol. 53, no. 1, 1934, pp. 114–120, reprinted in *Proceedings of the IEEE*, vol. 87, no. 2, February 1999, pp. 379–385.

———, "Inventing the negative feedback amplifier," *IEEE Spectrum*, December 1977, pp. 55–60.

———, United States Patent #2,102,671, "Wave Translation System," issued December 21, 1937, available from *www.uspto.gov*.

Calleja, H., "An approach to amplifier frequency compensation," *IEEE Transactions on Education*, vol. 46, no. 1, February 2003, pp. 43–49.

Denny, M., "Watt steam governor stability," *European Journal of Physics*, vol. 23, 2002, pp. 339–351.

Desoer, C., "In Memoriam: Harold Stephen Black (1898–1983)," *IEEE Transactions on Automatic Control*, vol. 29, no. 8, August 1984, pp. 673–674.

Fasol, K.H., "A short history of hydropower control," *IEEE Control Systems Magazine*, vol. 22, no. 4, August 2002, pp. 68–76.

Headrick, M.V., "Origin and evolution of the anchor clock escapement," *IEEE Control Systems Magazine*, vol. 22, no. 2, April 2002, pp. 41–52.

Herwald, S., "Recollections of the early development of servomechanisms and control systems," *IEEE Control Systems Magazine*, vol. 4, no. 4, November 1984, pp. 29–32.

Jury, E., "On the history and progress of sampled-data systems," *IEEE Control Systems Magazine*, vol. 7, no. 1, February 1987, pp. 16–21.

Kline, R., "Harold Black and the negative-feedback amplifier," *IEEE Control Systems Magazine*, vol. 13, no. 4, August 1993, pp. 82–85.

Lepschy, A.M., Mian, G.A. and Viaro, U., "Feedback control in ancient water and mechanical clocks," *IEEE Transactions on Education*, vol. 35, no. 1, February 1992, pp. 3–10.

Lewis, F.L., *Applied Optimal Control and Estimation*, Prentice Hall, 1992.

Lundberg, K., "Internal and external op-amp compensation: A control-centric tutorial," *ACC* 2004.

Lundberg, K.H., and Roberge, J.K., "Classical dual-inverted-pendulum control," *Proceeding of the* IEEE CDC 2003, pp. 4399–4404, December 9–12, 2003, Maui, Hawaii.

Mancini, R., "The saga of Harry Black," *EDN Magazine*, March 15, 2001, p. 34.

Maxwell, J.C., "On governors," *Proceedings of the Royal Society*, 1867, pp. 270–283.

Mayr, O., *The Origins of Feedback Control*, The MIT Press, 1970.

Michel, A.N., "Stability: The common thread in the evolution of feedback control," *IEEE Control Systems Magazine*, vol. 16, no. 3, June 1996, pp. 50–60.

Pidhayny, D., "The origins of feedback control," *IEEE Transactions on Automatic Control*, vol. 17, no. 2, April 1972, pp. 283–284.

Roberge, J.K., *Operational Amplifiers: Theory and Practice*, John Wiley, 1975.

Siebert, W. McC., *Circuits, Signals and Systems*, The MIT Press, 1986.

Thompson, Marc T., *Intuitive Analog Circuit Design*, Amsterdam, Elsevier-Newnes, 2006.

Tilbury, D., Luntz, J., and Messner, W., "Controls education on the WWW: Tutorials for MATLAB and Simulink," *Proceedings of the American Control Conference*, Philadelphia PA, June 1998, pp. 1304–1308.

CHAPTER 2

My Approach to Feedback Loop Design

Phil Perkins

Phil works on test circuits such as a V-I card, which can force a V or I into a test load and can also read the I or V on the load at the same time. I have worked on and designed this kind of V-I circuit, and it is a good challenge. Phil leads us to appreciate the problems in making such a tester "universal" and tolerant of every kind of load. Not easy, not impossible, but it does indicate the interaction among the circuit engineer, the customer, and the customer's system. How do you know what your customers need?/rap

I like designing feedback loops. I have been designing and building feedback controlled systems for audio and low-frequency control since high school. My interest in high-fidelity audio started in the late 1950s. Transistors were scarce and not very good, so I worked with vacuum tube circuits. I learned that negative feedback would improve just about every characteristic of an audio amplifier. I built Heathkits and modified them to suit my own preferences. I experienced oscillation when there was too much feedback.

For a freshman project at MIT, I learned how negative feedback could transform an unstable device into a stable one. I built a device to suspend a steel ball a fraction of an inch below an electromagnet driven by a tube amplifier. The position of the ball was sensed by a light shining on a photocell. The ball would partially interrupt the light as it was pulled higher. The photocell output was fed back to the amplifier input to control the magnet. After I got the hookup right, the first thing the circuit did was oscillate. I tried out my newly acquired capacitor substitution box and discovered a network that would tame the oscillation. I later learned that it was called a lead-lag network. I was developing an intuitive feel for what to do to make feedback stable.

During my studies at MIT, I learned about circuit theory, circuit analysis, and feedback circuit analysis. MIT taught methods for analyzing a circuit "by inspection" as well as the usual loop and node equations and mathematical analysis. I learned the theory of analyzing circuits and transforming between the time domain and the frequency domain. Then I could relate my early experiences to the theory. Along with learning the theory, I really appreciated learning methods of analyzing a circuit by inspection to get approximate results.

Much of the feedback loop design work I do is satisfied during the design phase with only rough approximations of performance. Actually, there are so many variables and effects you cannot consider during the design of a circuit, it is often useless to analyze with great precision. (You don't need a micrometer to measure your feet for a new pair of shoes.)

Since graduating from MIT, I have worked in the semiconductor automatic test equipment (ATE) field, designing instrumentation and other parts of ATE systems. First I worked for Teradyne and now I work for LTX. At Teradyne and LTX I have designed several programmable power sources. These programmable sources make heavy use of feedback loops. I have developed a method for design and analysis, which I describe here. I work and communicate better when I use a specific example to illustrate what I am designing or describing. The example I use here is a programmable voltage/current source I designed for LTX. The drawings are based on sketches I made in my notebook during the development of that product.

My Approach to Design

First, I need a specification of the instrument I am going to design. Then I make a block diagram of the circuit. Also, I draw a "front panel" of the instrument to show its functions and how they are controlled. This front panel has knobs and switches and dials, even though the finished product may be software controlled. The front panel helps to evaluate the functions that were specified and to investigate interactions between functions. In other words, does it do what you wanted, the way you want it to?

After I have a block diagram, I like to start the circuit design with a greatly simplified schematic made with a few basic building blocks of ideal characteristics. These blocks are simplified models of the real circuit elements I have to work with. I prefer to design the final circuit with blocks that work similarly to these basic blocks. The basic circuit blocks I like to use include:

- Amplifier with flat frequency response

- Ideal op-amp

- Ideal diode

- Ideal zener (voltage clamp)

- Voltage output DAC

To analyze a specific aspect of a design, I make a new schematic that eliminates anything that won't significantly affect the results. I want to be able to analyze by inspection and sketches. After reaching a conclusion, I might check my assumptions on a more complete schematic, or I might build a prototype and take measurements.

I use a notebook to record, develop, and analyze my designs. I draw block diagrams, schematics (from simple to detailed), and sketches of wave forms and frequency responses. I keep the notebook in chronological order and draw new drawings or sketches when there is a significant change to consider. During the preliminary design phase of a project, I might draw dozens of similar sketches as I develop my ideas and the design. I draw in pencil so I can make small changes or corrections without having to redraw the whole thing. I date most pages in my notebook, and I usually redraw a diagram if I invent a change on a later day. I also record the results of experiments and other measurements.

I have my notebooks going back to the beginning of my work at LTX. They have been a great source of history and ideas. Sometimes a question comes up that can be answered by going back to my notebooks rather than by making new calculations or experiments. There is real value in having diagrams, sketches, notes, and test results all in one place. Recently, though, I have been using various CAD and CAE systems to record some of my design developments. Sometimes the precision that the computer insists on has been helpful, and other times it's a hindrance. With CAE results, I now have two or three places where parts of the design process are documented. I need to develop a new system for keeping all the historical data in one place. Even with CAE, I don't expect to ever give up handwriting a substantial part of my design development notes.

What Is a V/I Source?

Integrated circuits need to be tested at several stages of their manufacture. Electrical testing is done with automatic test equipment (ATE). One of the instruments of an ATE system is the programmable voltage source. It is used to apply power or bias voltage to a pin on the device under test (DUT) or to a point in the DUT's test circuit. Programmable voltage sources usually can measure the current drawn at the output and sometimes include the capability of forcing current instead of voltage. In that case, the instrument is called a *V/I source*.

A V/I source I designed at LTX is called the *device power source*, or DPS. It is part of the Synchromaster line of linear and mixed-signal test systems. The DPS can force voltage or current and measure voltage or current. Its output capability is ±16 V at 1 A to ±64 V at 0.25 A. The current measure ranges are 62 μA full scale to 1 A full scale. There is great opportunity for analog design sophistication and tricks in designing a V/I source. Some typical performance requirements are:

- 0.1% forcing and measuring accuracy (at the end of 20 feet of cable)

- 100 μsec settling time after value programming or load change

- Minimum overshoot and ringing (sometimes 1 V overshoot is acceptable, other times overshoot must be zero)

There are many interesting aspects to the design of a V/I source. Perhaps the most challenging is to design a V/I source that can force voltage with acceptable performance into a wide range of capacitive loads. Why capacitive loads? In many (perhaps most) situations in which a voltage source is connected to the DUT or its test circuit, there needs to be a bypass capacitor to ground. This is the same bypass capacitor you need when actually using the DUT in a real end-use circuit. Occasionally, bypass capacitors need to be gigantic, as in the case of an audio power amplifier IC that requires 1000 μF.

At the same time the V/I source is forcing voltage, the test may require measuring the current drawn by the DUT. The current measure function usually requires a selection of resistors in series with the source output. The current is measured as a voltage across the resistor.

An Ideal V/I Source

It would be ideal if the capacitive loading on the V/I source output had no affect on the force voltage response to programming changes. However, one response effect we should accept is a reduction of the output voltage slew rate as the capacitive load increases, due to the limited amount of current available from the source. Since the V/I source must have some slew rate limit even at no load, the ideal voltage wave form would look like the one in **Figure 2-1.**

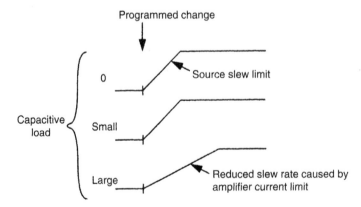

Figure 2-1: Ideal voltage wave form.

In this ideal case, the voltage will settle to the programmed value just as soon as it is finished slewing. In practice, there is probably no way to achieve the ideal case. The design of a V/I source involves lots of compromises.

Designing a V/I Source

To illustrate my approach to feedback loop design, I will describe a circuit that I developed for the LTX V/I source called DPS. A feature of the DPS is compensation for capacitive loading. With capacitive load compensation, the DPS can drive loads of any reasonable amount of capacitance with good stability and without overshoot.

The figures in this section are very close to the kind of drawings I do in my notebook. Here is a first model, shown in **Figure 2-2**, of the circuit to consider. I have simplified it to concentrate on the force voltage mode and the response when driving capacitive loads. This model meets the performance requirements listed above.

I picked $10\,\text{k}\Omega$ for *R1* because that value works well in real circuits, and calculations will be easy. I like to have a real value in mind just to simplify the analysis. *R2* sets the output voltage full-scale ranges of $2\,\text{V}$ to $64\,\text{V}$. *R3* varies from $1\,\Omega$ at $1\,\text{A}$ full scale to $1\,\text{k}\Omega$ at $100\,\mu\text{A}$ full scale. The $\times 1$ amplifier *A2* eliminates the loading effect of *R2* on the current measure function. With CL possibly as large as $1000\,\mu\text{F}$, I expect that the rolloff caused by *R3* and *CL* will be a major source of stability problems.

To investigate the effects of capacitive load on this model, I will simplify it further (see **Figure 2-3**). One of the objectives of a model is to reduce the problem to the basics and make approximate performance estimates easy (or at least possible). Figure 2-3 shows the simplified version.

I have given the simplified model a single voltage and current range. I have left out the load resistor *RL*, retaining only the capacitor load. I have retained the $\times 1$ amplifier to

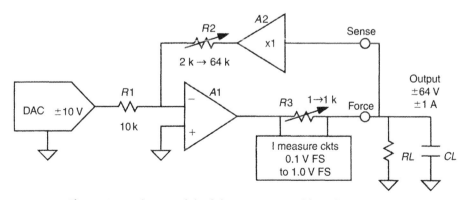

Figure 2-2: First model of the programmable voltage source.

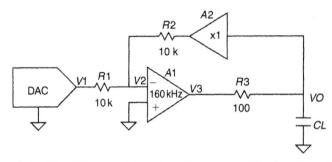

Figure 2-3: Simplified model to investigate capacitive loading effects.

remind me that *R2* does not load the output. If the compensation works in the simple case, then I expect it can be expanded to work on the complete DPS.

I like to design a feedback loop with just one integrator or dominant pole. The other stages are preferably flat frequency response. With this approach in the model, the analysis is easy. I want to design the real circuit the same way. Settling time is related to the lowest frequency pole in the loop. Therefore, extra poles below the loop unity gain frequency (UGF) make the settling time longer. That's bad in ATE applications, where fast settling is very important. Poles below the loop UGF may be needed, but they should be carefully considered.

I have given op-amp *A1* a UGF of 160 kHz. Why 160 kHz? My past experience with programmable sources has shown 100 kHz to be the maximum practical loop bandwidth. More than 100 kHz leads to oscillation due to phase shift from the many amplifiers that will ultimately be in the loop and from the output cable. The output cable? The output cable may be 20 feet long; the cable inductance has a significant effect.

As with other parameters I choose, *A1*'s UGF is a convenient approximation; 160 kHz corresponds to a time constant of 1 μsec. We often need to switch between time domain analysis and frequency domain analysis. I like to remember a simple conversion to eliminate the need to consult a table or calculator while designing. Just remember that 1 μsec corresponds to 160 kHz; you can extrapolate from there—2 μsec → 80 kHz, 100 μsec → 1.6 kHz, etc. Conversely, 1 MHz corresponds to 160 nsec.

Now we need to analyze the model. Bode plots for the individual stages and the complete loop will give us an idea of the loop stability and bandwidth. Let's start with *CL* = 0, as shown in **Figure 2-4**.

The loop gain is unity at 80 kHz and the phase shift is 90°, the "ideal" case for stability.

What happens when *CL* is increased? First, make new Bode plots as shown in **Figure 2-5**.

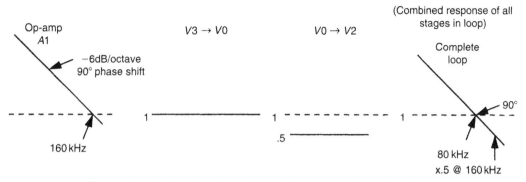

Figure 2-4 : Bode plots for individual stages and complete loop.

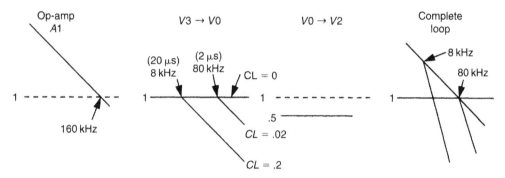

Figure 2-5: Revised Bode plots.

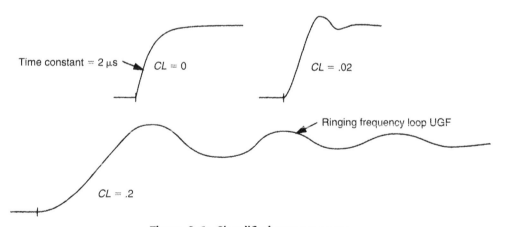

Figure 2-6: Simplified step response.

For $CL = 0.02\,\mu\text{F}$ the R3CL rolloff starts at 80 kHz, adding 45° phase shift at the loop UGF. For $CL = 0.2\,\mu\text{F}$ the added phase shift at 80 kHz is nearly 90°. From experience and by inspection, I would estimate the step response of this circuit to be as shown in **Figure 2-6**.

For $CL = 0$, the only time constant in the loop comes from the op-amp A1. The step response will be a simple decaying exponential of time constant $2\,\mu sec$ (based on loop UGF of 80 kHz). At $CL = 0.02$, there is 45° phase shift added to the loop at the UGF; I estimate a small amount of ringing, perhaps 1 cycle. At $CL = 0.2$, I know the added phase shift is higher (nearly 90°) and expect more ringing.

I don't need to estimate closer than $\pm 30\%$ or so because this circuit is a greatly simplified model. The main thing we need is to get within a factor of 2 or so of the real performance.

Capacitive Load Compensation

In a previous attempt at cap load compensation, I added an *RC* network as shown in **Figure 2-7**. This technique is similar to adding a resistor and capacitor from force to sense on the output of a power supply. That is one way to stop oscillation when there is a long cable to a capacitive load. This V/I source has a programmable choice of four *RC* networks for compensation. The *RC* network can be chosen to eliminate ringing for a range of capacitive loads, at the price of increasing settling time for no load. The overshoot is not reduced, however. Overshoot has become a serious issue in a few cases, so when I designed the DPS I wanted a better compensation technique that would reduce ringing and overshoot.

What Causes the Overshoot?

It will be easier to investigate the cause of the overshoot with a better model. The preceding model is completely linear. It behaves the same for small or large signals. The amplifier in the model is an ideal op-amp. A real amplifier will be limited to some maximum slew rate and will be limited in its current output. Here we can do a thought experiment. Assume instant settling time and perfect stability. Make a feedback loop from

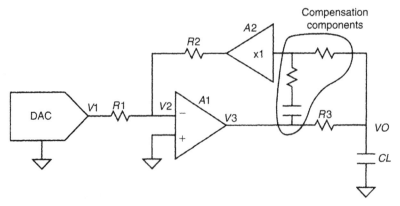

Figure 2-7: Previous attempt at capacitive loading compensation.

an amplifier with a slew rate limit and a current limit. Then the step response would look like the ideal, as shown in **Figure 2-8**.

These wave forms show the ideal effect of capacitive load on a voltage source: no effect until the capacitor current would exceed the amplifier current limit, then the slew rate is decreased according to the current limit. Let's see how close we can get to this ideal.

Improving the Model

To add slew rate limiting to the circuit, we could simply redefine the characteristics of the op-amp *A1* to include a slew rate limit. However, that would add several parameters to remember when analyzing the circuit. I prefer to add a few ideal components to the circuit that would add the slew rate limit and make it easier to visualize the effect.

To model a slew rate limited op-amp, I take an ideal op-amp with very high bandwidth and put components around it. The components show the effect and interaction of slew rate and bandwidth, as shown in **Figure 2-9**.

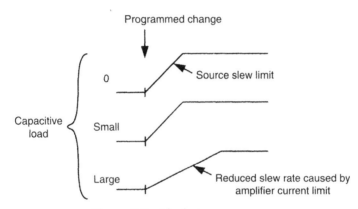

Figure 2-8: Ideal step response.

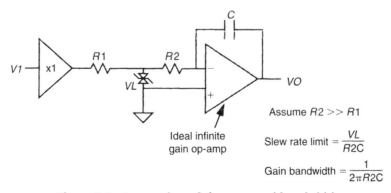

Assume $R2 >> R1$

Slew rate limit $= \dfrac{VL}{R2C}$

Gain bandwidth $= \dfrac{1}{2\pi R2C}$

Figure 2-9: Interaction of slew rate and bandwidth.

The double anode zener represents an ideal bidirectional clamp that limits the voltage to *VL*. I have simplified the model to be an inverting amplifier only, since that is what our circuit needs. A "better" model with differential inputs would just make the analysis by inspection harder.

Note that slew rate and gain-bandwidth (GBW) can be set independently of each other by changing *VL*. Real op-amps show this interaction of parameters. Compare a bipolar op-amp and a FET op-amp of similar GBW. The FET op-amp has higher slew rate but needs a larger input voltage than the bipolar to get to its maximum slew rate. A feedback loop built from this model will be linear when *V1* < *VL* and will be in slew rate limit when *V1* > *VL*.

Model to Investigate Overshoot

I have made a new circuit model to include the slew rate of a real amplifier. I simplified the op-amp model a bit to make the circuit easier to analyze, as shown in **Figure 2-10**. In this case the simplification should have no effect on the accuracy of the analysis. This op-amp model is good here because it is easy to see at which point the circuit becomes nonlinear. When *V2* < *VL*, the circuit is linear. When *V2* is clamped at *VL*, the op-amp is at slew rate limit (see Figure 2-10).

With this model, I can estimate the time response. When *CL* = 0, the output will be at slew rate limit until *V2* becomes smaller than *VL*. From that time on, the circuit is linear and the output is a simple time constant response.

This circuit *must* overshoot with capacitive load. That's because *CL* causes a delay between *V3* and *VO* and there is also a delay from *VO* back to the amplifier output *V3*. When the output is programmed from zero to a positive value, *V2* starts negative and stays negative until *VO* gets all the way to the programmed value. At that point, *V3* is above the programmed value and positive current is still flowing into *CL*. To make *V3* go down, *V2* has to be positive. *VO* has to go above the programmed value in order for *V2* to go positive, and that's overshoot. It's a little like trying to stop a rowboat the instant you get to the dock. There is no brake—all you can do is row backward, which takes time to take effect. Likewise, it's too late to stop *VO* from overshooting if you don't reverse the current in *CL* until after *VO* gets to the value at which you want to stop.

How do you stop the boat at the dock? Start slowing down before you get there. In this model, we need an anticipator to reduce the slew rate of *V3* before *VO* gets to the programmed value. One way to do this, as shown in **Figure 2-11**, is to add a capacitor *C2* across the feedback resistor *R2*. The faster the output is slewing, the more current *C2* adds to the null point (*V2*) and turns down the amplifier input in anticipation of *VO* getting to its programmed value (see Figure 2-11).

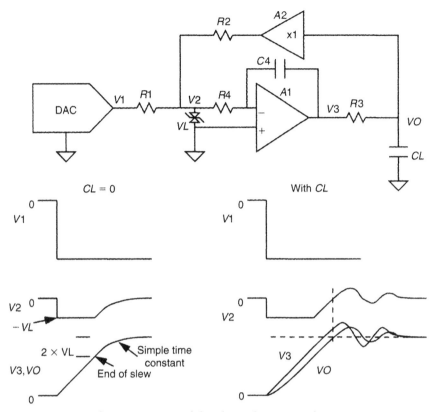

Figure 2-10: Model to investigate overshoot.

During the time that *VO* is slewing at a constant rate, the current in *C2* is constant and provides a fixed offset current into the loop null point at *V2*. Without *C2*, *VO* started to slow down only when it was 2*VL* away from the final value. With *C2*, *VO* starts to slow down when it is an additional (slew rate × *R2C2*) away from the final value. That's the anticipation. What happens with different values for *C2*? If *C2* is too small, *VO* will overshoot. If *C2* is too large, *VO* will be too slow to settle. Since ATE applications need minimum settling time, *C2* needs to be set just right. Therefore, the value of *C2* would have to be programmable according to the load capacitance.

In addition to affecting the transient response, *C2* changes the loop gain. At high frequencies where *C2* is low impedance, the loop gain is increased by a factor of 2. Looking ahead to the real DPS, the closed-loop gain from DAC output to source output will be programmable. With other values for *R1*, *C2* would increase loop gain by a different amount, even though the anticipation was the same. I want to make the frequency response independent of range, so I will put *C2* in an inverting op-amp circuit. This way, the compensation will behave

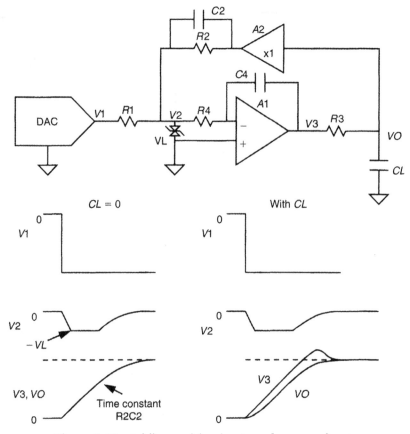

Figure 2-11: Adding anticipation to reduce overshoot.

more ideally. My goal is to have simple, independent circuit blocks that can be implemented in the final circuit (see **Figure 2-12**).

I have added *R6* in series with *C2* because the real *A3* circuit would have a gain limit at high frequency. In the model, *R6* allows us to use an ideal infinite gain op-amp for *A3* and still set a realistic maximum gain for this stage. For frequencies up to several times the loop UGF, I want the gain of each stage to be set by the passive components and not limited by amplifier gain. That way, the frequency response is sufficiently predictable and the amplifier could be changed to another type without serious effect.

Back to the Frequency Domain

At this point, we have an idea that improves the time response of the circuit, but we have ignored loop stability. To get an idea of the small-signal stability, we need to make Bode plots of the new model, as shown in **Figure 2-13**. From a Bode plot, I can estimate phase shifts according to the sketch in **Figure 2-14**.

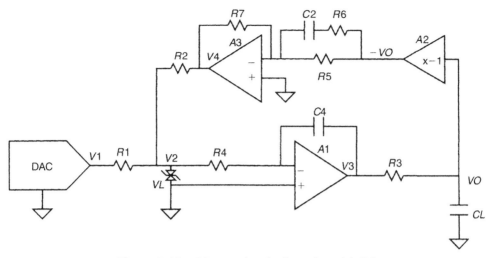

Figure 2-12: *C2* **moved to be in series with** *R6.*

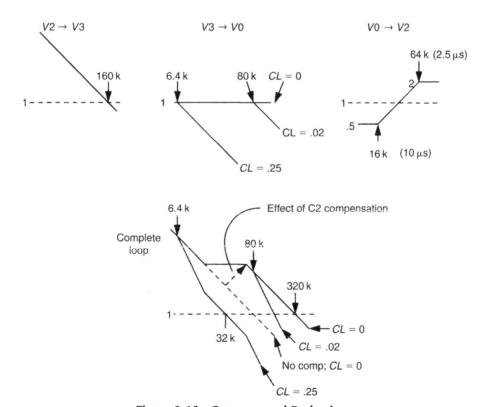

Figure 2-13: Compensated Bode plots.

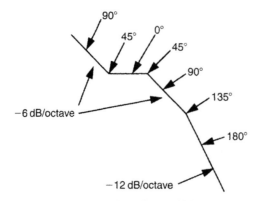

Figure 2-14: Bode plot phase shift estimates.

My criteria for stability of this circuit are:

1. Loop UGF shall be 80 kHz or less.

    Although the Bode plots don't show it, the real circuit will have many poles above 100 kHz. If the loop UGF were higher than 80 kHz, there would be lots of phase shift added by these poles, causing serious instability.

2. Phase shift at UGF is no more than 135° to 150°.

    At 135° the step response has a small amount of ringing. Much more phase shift would increase the ringing unacceptably.

    Applying these stability criteria to the drawing of compensated Bode plots, I conclude:

    When *CL* is zero, the UGF is too high.

    When *CL* is small, both the phase shift and UGF are too high.

    When *CL* is just right, the loop will be stable.

    When *CL* is too large, the phase shift is too high (this case is not shown in the drawing, but it's similar to uncompensated).

Is this a problem? I'm not surprised that making *CL* too large would cause instability. This compensation scheme has a limited range. The big problem is the excessive UGF for small *CL*. Overcompensation is expected to result in slower than optimum time response but should not cause oscillation!

To reduce the UGF when the loop is compensated but *CL* is small or zero, we need to reduce the loop gain. This gain reduction has to be done in the forward part of the loop at *A1*. Changing the gain in the feedback patch would change the closed-loop gain, which

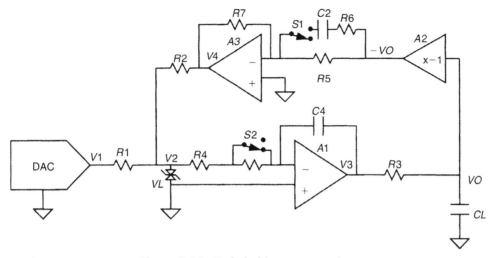

Figure 2-15: Switchable compensation.

sets the full-scale range. Increasing *R4* is the best way to reduce the loop gain. Changing *C4* would also change loop gain, but using solid-state switches to switch capacitors is likely to produce side effects. Increasing *R4* to reduce the loop gain by a factor of 4 shifts the Bode plot down, decreasing the UGF to an acceptable 80 kHz.

By reducing the gain of the compensated loop, we have achieved stability when *CL* = 0. However, the settling time is still seriously affected by the compensation because there is now a pole well below the loop UGF. The compensation should be programmed or switched to an appropriate value according to the load capacitance. In **Figure 2-15** I have drawn a circuit to show switching one value of compensation.

S1 adds the anticipation or frequency compensation and *S2* reduces the loop gain. To achieve both stability and the elimination of overshoot, these functions must be located in the different parts of the loop as shown.

Range of Compensation Required

The basic stability problem of capacitive loading is caused by the time constant of *R3* and *CL*. *R3* is expected to range from 1 Ω to 1 kΩ. *CL* could reasonably range from 1 nF to 1000 μF. That is a time constant range of 1 nsec to 1 sec. Time constants up to about 1 μsec will have little effect since the loop time constant is 2 μsec. Time constants above 100 msec are probably too long to be useful in an ATE system. Therefore, we have a time constant range of 100,000 : 1 to compensate for. The preceding circuit has only a narrow range of good compensation. There would have to be too many programmable choices of compensation networks to cover this wide range. Wouldn't it be neat if there were a way to cover a wider range with one compensation network?

Phase Margin Approach to Loop Compensation

We have been looking at the frequency domain. Now let's consider the capacitive load problem from a phase margin point of view. If the loop frequency response is a single pole, then the loop phase shift is 90° at UGF. A capacitor load adds, depending on frequency, from 0° to 90° additional phase shift to the loop. Phase shift approaching 180° is a problem. What if the no-load loop frequency response were −3 dB/octave instead of −6 dB/octave? That would give the loop a 45° phase shift at no load. Adding a capacitive load would increase the phase shift from 45° to a maximum of 135°. Anywhere in that phase range, stability would be acceptable. This idea sounds like it would work for any value of capacitive load without any switching.

How would we get a loop gain of −3 dB/octave? One way would be to give the *A1* stage −3 dB/octave response and the feedback path a flat response, as shown in **Figure 2-16**.

Replacing *C* in the inverting op-amp model with a series-parallel *RC* network comes close to −3 dB/octave and 45° phase shift over a range of frequencies, as shown in **Figure 2-17**.

This approach to a −3 dB/octave loop does not fix the overshoot problem since it does not allow for the anticipation network we need in the feedback path. A better approach to −3 dB/octave loop response is to leave the forward gain (*A1*) at −6 dB/octave and make the anticipator circuit +3 dB/octave. Adding more series *RC*s to *R6 C2* will give the anticipator the desired +3 dB/octave over a range of frequencies. The more *RC*s, the wider the range. However, each *RC* adds a pole below the loop UGF, and these poles increase settling time. The compensation network does cover a wide range, but it still needs to be switchable to minimize settling time.

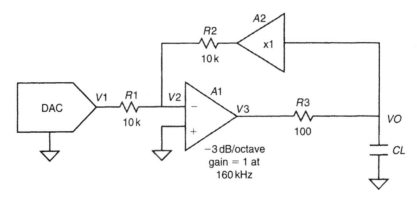

Figure 2-16: −3 dB/octave loop gain.

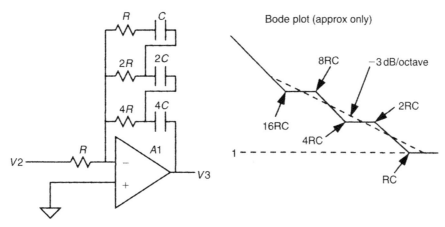

Figure 2-17: −3 dB/octave amplifier.

At this point, analysis by inspection became less reliable for me, so I used SPICE to simulate the multiple *RC* compensation network. I ran a SPICE analysis of a triple *RC* network. Phase analysis is easy with simulation but hard to do by inspection or on the bench. I found it tough to get close to a 45° phase lead out of a small number of components. I decided to shoot for 30° phase lead. SPICE showed me I could get 30° from a double *RC* network over a 50 : 1 frequency range. That covers a sufficiently wide range of capacitive load without making settling time too bad.

LTX Device Power Source (DPS) Performance

The LTX DPS turned out pretty well. The LTX Cadence programming language includes a DPS statement to specify the expected load capacitance and a choice of modes: minimum risetime, minimum settling time, or minimum overshoot. The operating system takes into consideration the voltage range and the current measure range, then selects one of four compensation networks (or no compensation) and the corresponding loop gain settings. The three modes are simply different degrees of compensation. To minimize overshoot takes the greatest compensation and results in the longest settling time. The compensation works well for all practical values of load capacitance. Overcompensation (specifying a larger capacitance than the actual) makes the settling time longer but causes no stability problem.

The DPS contains other functions, more stages, and many more details than I have outlined here. I designed each of the analog functions using the same method. All the stages can be grouped together into blocks that match very closely the simplified blocks on which I based the design. The DPS behaves very much like the block diagram and simplified models would predict.

Summary of My Method

By way of an example, I have shown the method I like to use for analog design, especially for feedback loops. Here's an outline:

1. Draw a "front panel" of the instrument to be designed. "Try out" its functions.

2. Make a simple circuit model for one function or aspect of the instrument. The model should emphasize that one aspect and deemphasize other aspects.

3. Make simplifying assumptions and analyze the circuit by inspection where possible. Go back and forth between time domain and frequency domain analysis. Check your assumptions.

4. Change the model and analyze again until the results are acceptable.

5. Repeat steps 1 through 3 for other aspects of the instrument.

6. Design the full circuit with circuit blocks that behave like the ideal blocks in the models.

7. Test a prototype of the instrument to see if it behaves like the models.

Simple, isn't it?

CHAPTER 3

Basic Operational Amplifier Topologies and a Case Study

Marc Thompson

In this chapter, Marc shows us an example of a real op-amp as well as how to analyze it and use it. This is not a perfect op-amp, but what op-amp is? Answer: None. But this one would be usable. (I've seen worse.) The main point is, how do we apply an op-amp even if it is not perfect? In a real-life application where the capacitive load is not perfect. /rap

In This Chapter

The basic operational amplifier is discussed from a topological point of view. A step-by-step case study illustrates the basic building blocks in a monolithic op-amp. At the end of this chapter we'll consider some of the real-world limitations of operational amplifiers.

Basic Device Operation

The ideal operational amplifier (**Figure 3-1**) has the following characteristics:

- *Differential inputs.* The output is an amplified version of the difference between the + and − terminals.

- *Infinite gain.* The gain is infinite.

- *Infinite bandwidth.* There are no bandwidth limitations.

- *Infinite slew rate.* There is no limit to the rate with which the output can change. In other words, there is no limit to dV_{out}/dt.

- *Zero input current.* The input current to both inputs is zero.

- *Zero output impedance.* The output impedance is zero.

- *Zero power dissipation.* The ideal op-amp doesn't draw or dissipate any power.

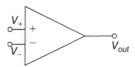

Figure 3-1: Ideal operational amplifier showing differential inputs V+ and V−. The ideal op-amp has zero input current and a gain approaching infinity that amplifies the difference between V+ and V−.

- *Infinite power supply rejection.* The output is not dependent on variations in power supply voltage.

- *Infinite common-mode signal rejection.* The output doesn't depend on the value of the common-mode signal.

The ideal op-amp is, of course, nonexistent, but over the years device manufacturers have done a better and better job of designing devices that approach the ideal. For instance, it is common to find devices with DC gains much higher than 10^6 and/or gain-bandwidth products of greater than 100 MHz.[1]

The usual method for doing first-cut analysis of closed-loop op-amp circuits is to assume a "virtual ground." This term is a bit of a misnomer because the input terminals in general don't need to be at ground potential. However, in an operational amplifier operating with negative feedback, the difference between the two inputs is ideally zero volts. If the op-amp + terminal is at ground, the − terminal will be at approximately ground. If, in a different configuration the + terminal is at +6 V, the − terminal will also be at approximately +6 V.

A basic two-stage op-amp is shown in **Figure 3-2**. This is a two-stage op-amp because it consists of two gain stages: the input differential gain stage followed by a second common-emitter gain stage. The input differential amplifier stage (Q_1 and Q_2) has high differential-mode gain and low common-mode gain. The second gain stage (Q_3) provides additional gain and also provides a DC level shifting function. Compensation capacitor C_c provides a low-frequency pole and by the process of pole splitting causes the next-highest frequency pole to move to a higher frequency. Pole splitting has important ramifications in overall amplifier stability.

Emitter follower Q_4 buffers the high gain node from the output. The output stage (Q_5 and Q_6) is a class AB *push-pull* stage. This output stage can source or sink current. For

---

[1] Compare this to the baseline specs of the 741 op-amp (a device from the 1960s that is still for sale) with nominal DC gain of 200,000 and gain-bandwidth product of 1 MHz. A Linear Technology LT1226 (a 1990s-era device) has a gain-bandwidth product of 1000 MHz.

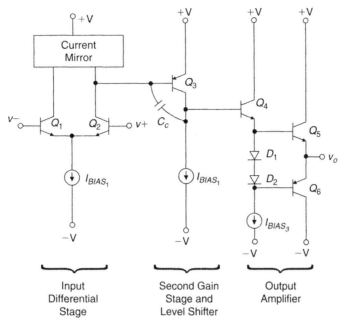

Figure 3-2: Basic two-stage op-amp with an input differential gain stage followed by a second gain and level shifting stage and an output buffer.

instance, when sourcing current Q_5 is On; when sinking current Q_6 is On. The diodes at the bases of Q_5 and Q_6 provide two functions. First, they reduce *crossover distortion* in the output stage. Second, by properly sizing D_1 and D_2[2] relative to the sizes of Q_5 and Q_6 we can set a modest bias current in the output stage transistors, which lowers the incremental output resistance of the output amplifier.

A push-pull output stage without biasing diodes is shown in the PSPICE circuit of **Figure 3-3**. The output waveforms show significant DC offset as well as *crossover distortion* or a dead zone of approximately $+/- 0.6$ V around $v_o = 0$.

This push-pull circuit can be modified to improve voltage offset, as shown in **Figure 3-4.** The diodes between the bases of the transistors set up an approximate 1.2 V bias that begins to turn on the output transistors. Resistors R_1 and R_2 bias the diode string.

The small resistor R_3 is provided to reduce the chances of thermal runaway. Here's how thermal runaway would work in this circuit: Let's say that the output transistors Q_3 and Q_4 are carrying significant current. They will heat up. Since the transistors are now at a higher temperature, they require less base-emitter voltage to sustain the same output

---

[2] In some real-world op-amps, the voltage drop provided by these two diodes is alternately provided by an alternate circuit topology. See, for example, the 741 operational amplifier, which uses a "V_{BE} multiplier" to provide this function.

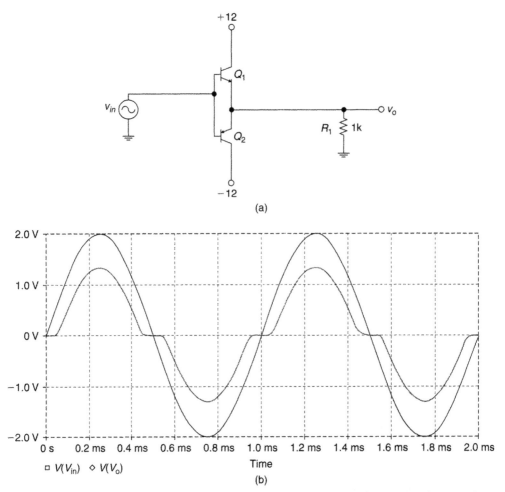

Figure 3-3: Push-pull stage. (a) Basic circuit. (b) Output response for 4 V pp sinewave input exhibiting crossover distortion at the output.

current.[3] Therefore, the current in the output stage increases. The transistors heat up more, and the circuit "runs away" thermally. The inclusion of the small resistor R_3 ensures that if the collector currents begin to increase significantly, the voltage drop across R_3 will become significant, hence stealing away base drive voltage from the transistors.

Let's next study the power dissipation in the basic push-pull amplifier (**Figure 3-5a**), assuming ideal components (i.e., no crossover distortion) and a symmetrical power supply voltage. The NPN transistor is on for positive load current I_L. The power dissipation is:

$$P_D = V_{CE}I_L = (V_S - V_o)I_L = V_SI_L - I_L^2R_L \qquad [3\text{-}1]$$

---

[3] Remember that the temperature coefficient of V_{BE} is approximately -2.2 mV/°C.

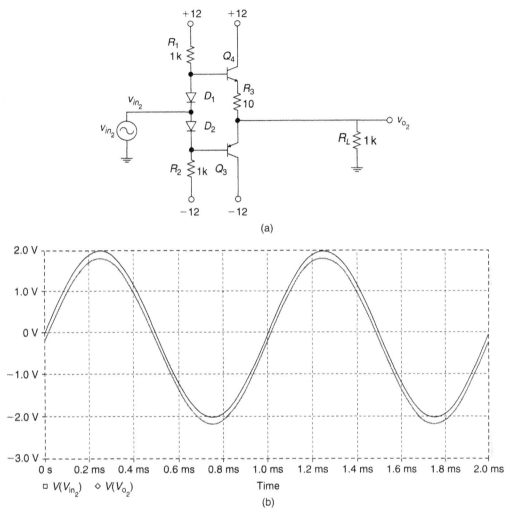

Figure 3-4: Push-pull stage with base biasing diodes. (a) Basic circuit.
(b) Output response for 4 V pp sinewave input.

Therefore, the shape of the power dissipation vs. load current curve has a parabolic shape
(**Figure 3-5b**). The maximum power dissipation is found by taking the derivative of the
power dissipation with respect to load current.

$$\frac{dP_D}{dI_L} = V_s - 2I_L R_L \qquad [3\text{-}2]$$

The maximum power dissipation occurs at load current $I_L = V_S/(2R_L)$, and at an output
voltage which is half the supply voltage.[4]

---

[4] One might ask why the power dissipation is zero at maximum collector current. In this simplified model,
we assume that the saturation voltage of the transistor is zero; hence a finite collector current multiplied by
zero V_{CE} results in zero transistor power dissipation.

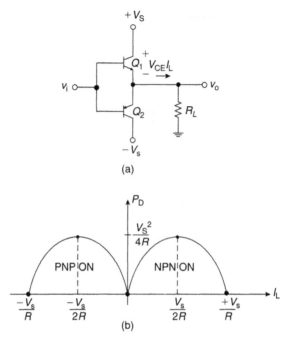

Figure 3-5: Circuit for studying power dissipation in push-pull amplifier. (a) Basic circuit. (b) Transistor power dissipation P_D vs. load current I_L for NPN and PNP transistors.

Example 3-1: Case study: Design, analysis, and simulation of a discrete operational amplifier

Let's start off designing a discrete operational amplifier using discrete components. This resulting design will not be state of the art, but rather the design process illustrates the various building blocks that exist in most monolithic IC op-amps. The design insight afforded by this exercise will help us understand the limitations of real-world op-amps. Op-amp limitations are discussed in more detail later in this chapter.

Differential input stage

In an op-amp, some form of differential input stage is needed. One possible implementation is the current-mirror-loaded differential input stage (**Figure 3-6**). The components of this stage are:

- Differential input transistors Q_1 and Q_2.
- Current mirror transistors Q_3 and Q_4.
- Bias current source Ibias, with details omitted here.

Let's do a DC sweep of v_{in+} while grounding v_{in-} (see **Figure 3-7**). This will give us an idea as to the voltage offset of the input stage. Ideally, we would like the output v_1 to be

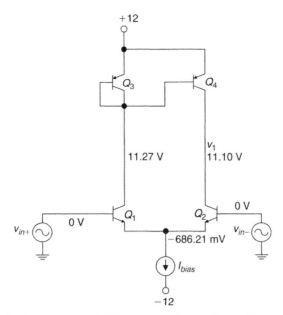

Figure 3-6: Differential input stage with input transistors Q_1 and Q_2 and current mirror Q_3 and Q_4. We take the output voltage to be v_1 at the connection of the collectors of Q_2 and Q_4.

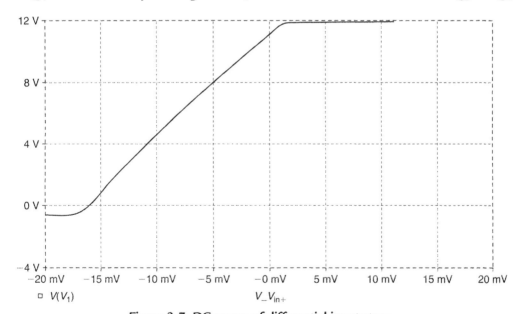

Figure 3-7: DC sweep of differential input stage.

in the middle of the (approximately) linear gain region $v_{in+} = 0\,\text{V}$. In practicality, however, this voltage offset will not be zero due to device mismatches and errors in the Q_3/Q_4 current mirror. A SPICE simulation shows this effect; the output at v_1 transitions from approximately ground to approximately $+12\,\text{V}$, whereas v_{in+} is varied from

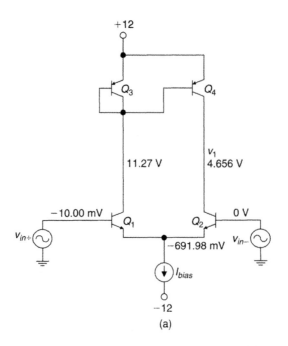

(a)

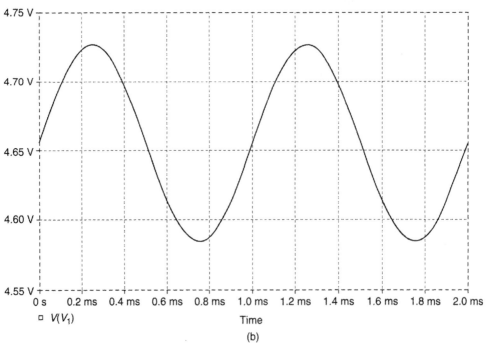

(b)

Figure 3-8: Results of AC sinewave sweep. (a) Circuit.
(b) PSPICE result for a 0.2 mV-pp variation in v_{in+}.

approximately −15 mV to ground. Note that the gain of this stage (gain being v_{o1}/v_{in+}) is approximately 1000.

Let's put a DC bias of −10 mV at v_{in+} (**Figure 3-8a**) and put a 200-μV peak-peak 1-kHz sinewave at the input (**Figure 3-8a**). Note that the DC bias level of v_1 sits at +4.656 V (**Figure 3-8b**), which means that Q_2 and Q_4 are both On and in the forward-active region. Therefore, we expect active gain with the transistors biased under these conditions. The gain from v_{in+} to v_1 is approximately 575.

Continuing with the design, we now recognize that further transistors are needed for buffering and level-shifting functions. The output at v_1 only has positive voltages; in a practical amplifier, we transition both positive and negative voltages at the output. Therefore, we need to level-shift the output voltage from the differential stage. One possible way to do this is with a folded-cascode amplifier, as shown in **Figure 3-9**. This folded cascode (Q_5 and bias current sources I_{bias2} and I_{bias3}) works as follows: First, assume that I_{bias2} and I_{bias3} are the same value. Transistor Q_5 buffers the difference between the collector currents of Q_4 and Q_5, or in this case the small-signal variation in i_{c5} is:

$$i_{c5} \approx I_{C4} - I_{C2} = \left(\frac{I_{BIAS}}{2} + \Delta i \right) - \left(\frac{I_{BIAS}}{2} - \Delta i \right) = 2\Delta i \qquad [3\text{-}3]$$

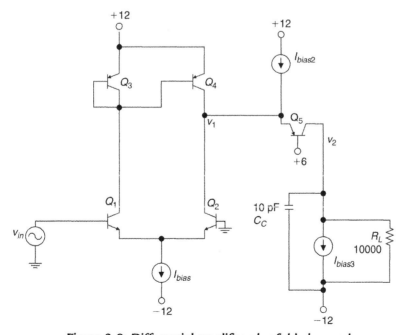

Figure 3-9: Differential amplifier plus folded cascade.

Furthermore, for the differential input stage:

$$\Delta i = g_m v_{in} \qquad [3\text{-}4]$$

It is the $2\Delta i$ current that is converted to voltage v_2 by loading it with current source I_{BIAS3}.

Emitter follower buffering and output push-pull stage

To this input stage, we now need to add an output push-pull stage (**Figure 3-10a**). The output transistors Q_7/Q_8 can sink and source current. Emitter follower Q_6 is added so that the output transistor doesn't significantly load down the high-gain node at the collector of Q_5. We see the input and output of this closed-loop gain-of-1 configuration in **Figure 3-10b**. We note that there is a $-6\,\text{mV}$ offset between input and output.

We next configure the op-amp for a gain-of-20 (**Figure 3-11a**). We note that the output can swing $\pm 4\,\text{V}$ without clipping (**Figure 3-11b**). As we raise the amplitude of the driving signal further, the output clips at $<5\,\text{V}$ (**Figure 3-11c**) but the negative sinewave is OK.

> **Note:** This design is shown for illustrative purposes only. There are many design improvements that could be made, including better biasing, more intelligent topology selection, and the like. So please, no diatribes about how it's a crummy design!

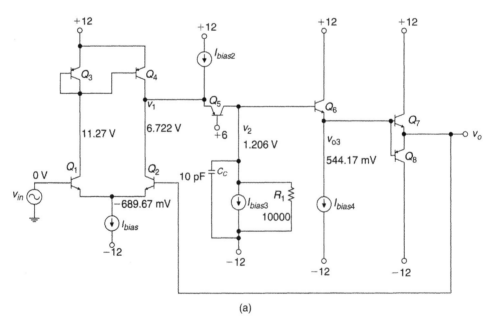

(a)

Figure 3-10: Final amplifier showing unity feedback. (a) Circuit showing some PSPICE-predicted node voltages. (Continued on following page)

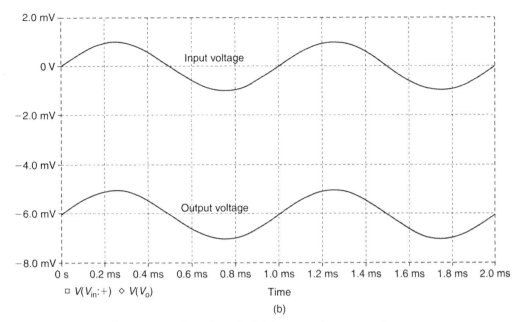

Figure 3-10: (Continued) (b) Input and output voltages.

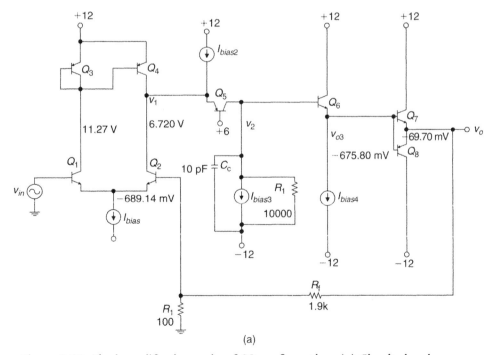

(a)

Figure 3-11: Final amplifier in a gain-of-20 configuration. (a) Circuit showing some
PSPICE-predicted node voltages. (Continued on following page)

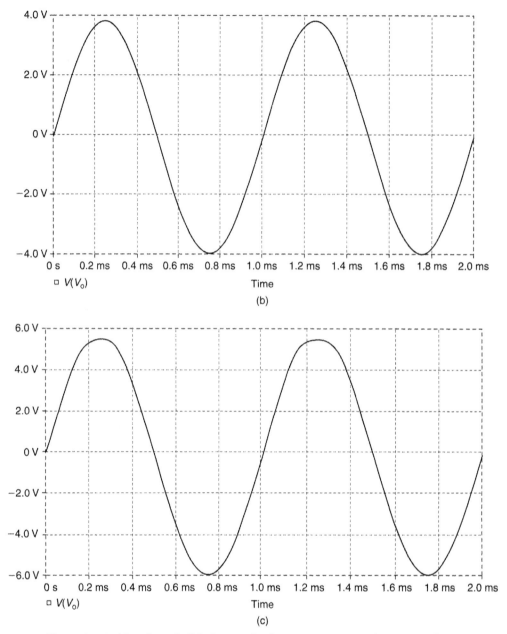

Figure 3-11: (Continued) (b) Output for input 400-mV pp sinewave, 1 kHz.
(c) Output for input 600-mV pp sinewave.

Brief Review of LM741 Op-Amp Schematic

We'll now do a quick review of the topology of the LM741 operational amplifier, made by many companies since the 1960s (**Figure 3-12**). This is a two-stage op-amp with pole splitting.

- *Input differential gain stage*: Q_1, Q_2, Q_3 and Q_4; Q_5 and Q_6 are the current mirror load for the differential stage. Q_7 is a beta helper transistor that improves the gain ratio of the mirror. Q_8 is the bias current source.

- *Bias current mirrors*: Q_8 and Q_9; Q_{10} and Q_{11}; Q_{12} and Q_{13}. These mirrors set the bias current levels throughout the op-amp.

- *Second gain stage*: Comprises common emitter amplifier with 50 Ω emitter degeneration Q_{17}. Transistor Q_{15} provides buffering between the output of the first gain stage and the input of the second gain stage. The 30-picofarad capacitor C_1 provides a pole-splitting function.

- *Output push-pull amplifier*: Transistors Q_{14} and Q_{20} form the heart of the push-pull. A V_{BE} multiplier[5] between the bases of Q_{14} and Q_{20} partially biases the push-pull On. Transistor Q_{16} provides current limiting for positive output currents.

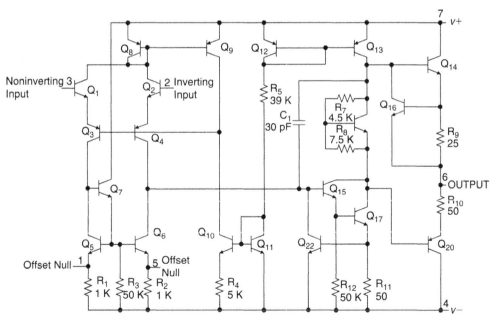

Figure 3-12: Schematic of the LM741 op-amp, from National Semiconductor.
(Reprinted with permission of National Semiconductor Corporation.)

---

[5] This "V_{BE} multiplier" circuit provides approximately 1.6 V_{BE} drop.

Some Real-World Limitations of Operational Amplifiers

Voltage Offset

With a differential amplifier we assumed ideal (i.e., perfectly matched) devices. However, in the real world, there is mismatch between devices, and in the operational amplifier front end, this mismatch manifests itself as a *voltage offset*. The voltage offset is the voltage that must be applied differentially to force the output of the op-amp to zero. In commercially available op-amps, the voltage offset is typically fractions of a millivolt, up to a few millivolts.

We can model voltage offset by adding a voltage generator of value V_{os} to an ideal op-amp, as shown in **Figure 3-13**. The voltage offset can be especially troublesome in high-gain configurations, such as **Figure 3-14**. Note that the offset generator is in series with the input signal v_i and that output voltage of this configuration is:

$$v_o = 1000v_i + 1000V_{os} \qquad [3\text{-}5]$$

For small input signals, the voltage offset can swamp out the input signal.

A datasheet excerpt from National Semiconductor shows the input offset voltage for the LM741 op-amp (**Figure 3-15**). Note that different grades of op-amp have slightly different specifications, but the typical offset is <1 millivolt, whereas the maximum offset is a few millivolts.

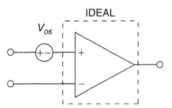

Figure 3-13: Operational amplifier showing voltage source modeling the voltage offset V_{os}.

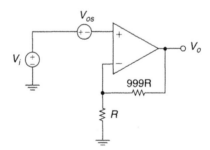

Figure 3-14: Gain of +1000 amplifier showing effects of voltage offset V_{os}.

Parameter	Conditions	LM741A			LM741			LM741C			Units
		Min	Typ	Max	Min	Typ	Max	Min	Typ	Max	
Input Offset Voltage	$T_A = 25°C$										
	$R_S \leq 10\,k\Omega$					1.0	5.0		2.0	6.0	mV
	$R_S \leq 50\,\Omega$		0.8	3.0							mV
	$T_{AMIN} \leq T_A \leq T_{AMAX}$										
	$R_S \leq 50\,\Omega$			4.0							mV
	$R_S \leq 10\,k\Omega$						6.0			7.5	mV

Figure 3-15: Datasheet excerpt from the National Semiconductor LM741 operational amplifier showing voltage offset. (Reprinted with permission of National Semiconductor Corporation.)

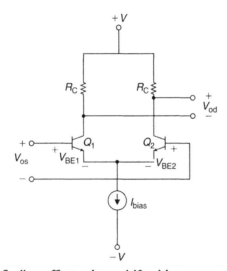

Figure 3-16: Circuit for finding offset voltage drift with temperature in a bipolar op-amp.

Voltage Offset Drift with Temperature

Another important design issue is voltage offset drift with temperature. Referring to **Figure 3-16**, we see a differential input stage typical of that in op-amps. We note that the voltage offset is:

$$V_{os} = V_{BE1} - V_{BE2} \qquad [3\text{-}6]$$

This voltage is the differential voltage that occurs when the output differential voltage is zero. For a transistor under forward bias there is the exponential relationship between collector current and base-emitter voltage, which for transistor Q_1 is:

$$I_{C1} = I_S \left(e^{\frac{qV_{BE1}}{kT}} - 1 \right) \approx I_S e^{\frac{qV_{BE1}}{kT}} \qquad [3\text{-}7]$$

where I_s is the reverse saturation current of the transistor. We now find the equation for the base-emitter voltage of bipolar transistor Q_1:

$$V_{BE1} = \frac{kT}{q} \ln\left(\frac{I_{C1}}{I_S}\right)$$ [3-8]

The temperature coefficient of a single V_{BE} is found by:

$$\frac{dV_{BE1}}{dT} = \frac{k}{q} \ln\left(\frac{I_{C1}}{I_S}\right) - \frac{kT}{qI_S}\frac{dI_S}{dT} = \frac{V_{BE1}}{T} - \frac{kT}{qI_S}\frac{dI_S}{dT}$$ [3-9]

The second term in this expression is the leakage current temperature coefficient. We can now find the total voltage offset drift of the differential amplifier as:

$$\frac{dV_{os}}{dT} \approx \frac{dV_{BE1}}{dT} - \frac{dV_{BE1}}{dT} = \left(\frac{V_{BE1} - V_{BE2}}{T}\right) - \frac{kT}{qI_{S1}} + \frac{kT}{qI_{S2}}$$ [3-10]

In matched transistors, $I_{s1} = I_{s2}$ and the component of voltage offset drift due to leakage current drift cancels, leaving us with:

$$\frac{dV_{os}}{dT} \approx \frac{V_{BE1} - V_{BE2}}{T} \approx \frac{V_{os}}{T}$$ [3-11]

Using this expression, let's predict the voltage offset drift for a 741 op-amp with a maximum 4-mV offset at room temperature:

$$\frac{dV_{os}}{dT} \approx \frac{4\,mV}{300\,K} \approx 13\frac{\mu V}{K}$$ [3-12]

This value is consistent with the datasheet value from the 741 datasheet (**Figure 3-17**).

Input Bias and Input Offset Current

The input differential amplifier of an op-amp requires base current (in the case of a bipolar-input op-amp). This input current is specified on an op-amp datasheet as *input bias* current, as shown in **Figure 3-18**. The bias currents are small but finite. Datasheets also specify *input offset* current, which is the guaranteed maximum difference between I_{B+} and I_{B-}.

Average Input Offset Voltage Drift				15							µV/°C

Figure 3-17: Datasheet excerpt from the National Semiconductor LM741 operational amplifier showing voltage offset drift with temperature. (Reprinted with permission of National Semiconductor Corporation.)

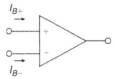

Figure 3-18: Operational amplifier showing small but finite input bias currents I_{B+} and I_{B-}.

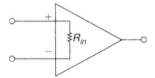

Figure 3-19: Operational amplifier showing differential input resistance.

Differential Input Resistance

There is a differential-mode input resistance that can be modeled as a large-valued resistance across the op-amp input terminals (**Figure 3-19**).

Slew Rate

We have seen before that a small capacitor is needed to tailor the frequency response of the operational-amplifier open-loop characteristic. By pole splitting, we create a dominant low-frequency pole. Another effect of this feedback capacitance is slew-rate limiting. There is a finite amount of current available from the input stage to supply this feedback capacitance. Hence, the output voltage time rate of change is limited by the capacitance and the current available to charge this capacitance, as:

$$\frac{dv}{dt} = \frac{I}{C} \qquad [3\text{-}13]$$

Slew-rate limiting manifests itself when we attempt to switch large amplitude signals. Consider the voltage follower of **Figure 3-20**, where we are switching a 0 to 10 V signal.

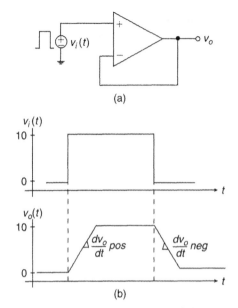

Figure 3-20: Circuit that shows effects of slew-rate limiting.

We note that the output shows slew-rate limiting on the positive-going and negative-going edges. Note that the slew rates in the two directions aren't necessarily the same.

Output Resistance and Capacitive Loading

All operational amplifiers have a finite output resistance (**Figure 3-21**). As we'll see, this finite output resistance can have a significant effect on closed-loop stability in driving capacitive loads.

Example 3-2: Op-amp driving capacitive load

Let's consider a typical[6] operational amplifier open-loop transfer function:

$$a(s) \approx \frac{10^5}{(0.01s + 1)(10^{-7}s + 1)} \qquad [3\text{-}14]$$

---

[6] As pointed out by Bob Pease, we don't exactly know what the DC open-loop gain is or where the low-frequency pole is. In an op-amp, however, we do know fairly accurately what the gain-bandwidth product is. Ignoring the high-frequency pole, the op-amp open-loop transfer function can be expressed as:

$$a(s) \approx \frac{a_o}{\tau s + 1}$$

where a_o is the DC gain and τ is the time constant associated with the low frequency pole. So, although we don't exactly know what a_o or τ are, we do know what the ratio a_o/τ is, since a_o/τ is approximately the gain-bandwidth product of the op-amp.

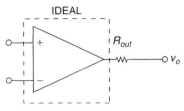

Figure 3-21: Operational amplifier showing output resistance.

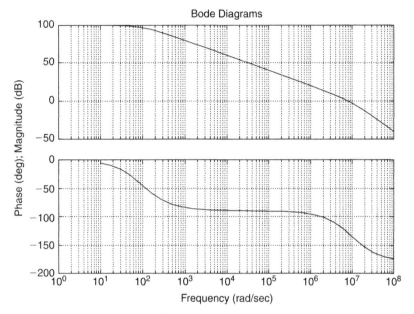

Figure 3-22: Plot of $a(s)$ for typical op-amp.

The open-loop transfer function $a(s)$ of this op-amp is plotted in **Figure 3-22**, where we see the low-frequency pole at 100 radians/second and high-frequency pole at 10^7 radians/second. This type of transfer function is typical of many commercially available op-amps.

We'll now see what happens if we use this op-amp to drive a capacitive load in a follower configuration (**Figure 3-23a**). A model of the op-amp showing the op-amp output resistance is shown in **Figure 3-23b**. In this example we'll assume that the output resistance of the op-amp is 100 Ω.

The block diagram of this capacitively-loaded amplifier is shown in **Figure 3-24**.

When we plot the magnitude and phase of the loop transmission (**Figure 3-25**) we note that there is poor phase margin of 22°. The resultant step response of the closed-loop system (**Figure 3-26**) shows significant overshoot, as expected, with this low value of

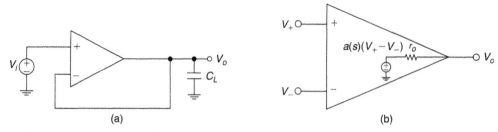

Figure 3-23: Op-amp follower driving capacitive load. (a) Circuit. (b) Model showing op-amp output resistance r_o.

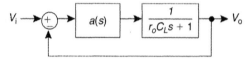

Figure 3-24: Capacitively loaded op-amp follower block diagram.

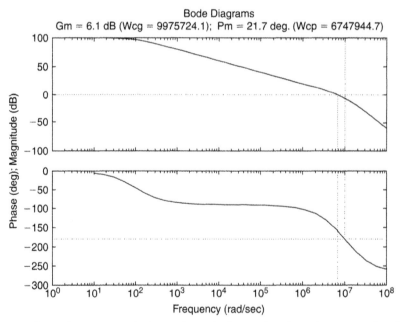

Figure 3-25: Plot of loop transmission gain and phase, showing phase margin of op-amp when loaded with 1000 pF of 21.7°.

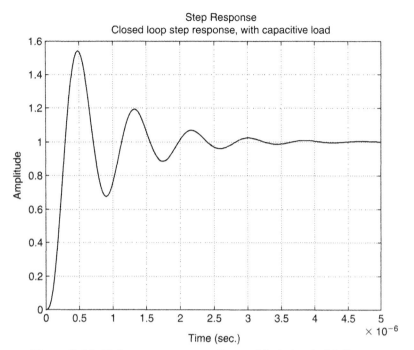

Figure 3-26: Unit step response of capacitively loaded follower.

phase margin. The low phase margin is due to the low-pass filter inside the feedback loop, attributable to the op-amp output resistance interacting with the load capacitor.

References

A wealth of information is provided in the following references. Some of the author's favorites are the references by Bob Widlar, one of the original architects of operational amplifiers. The Solomon and Gray and Meyer references are also excellent overviews of op-amp technology.

Allen, P.E., "Slew-induced distortion in operational amplifiers," *IEEE Journal of Solid-State Circuits*, vol. 12, no. 1, February 1977, pp. 39–44.

Bowers, D.F. and Wurcer, S.A. (1999). "Recent developments in bipolar operational amplifiers," *Proceedings of the 1999 Bipolar/BiCMOS Circuits and Technology Meeting*, September 26–28, pp. 38–45.

Brown, J.L., "Differential amplifiers that reject common-mode currents," *IEEE Journal of Solid-State Circuits*, vol. 6, no. 6, December 1971, pp. 385–391.

Chuang, C.T., "Analysis of the settling behavior of an operational amplifier," *IEEE Journal of Solid-State Circuits*, vol. 17, no. 1, February 1982, pp. 74–80.

Comer, D.T. and Comer, D.J., "A new amplifier circuit with both practical and tutorial value," *IEEE Transactions on Education*, vol. 43, no. 1, February 2000, p. 25.

Erdi, G., "Common-mode rejection of monolithic operational amplifiers," *IEEE Journal of Solid-State Circuits*, vol. 5, no. 6, December 1970, pp. 365–367.

Gray, P. and Meyer, R., "Recent advances in monolithic operational amplifier design," *IEEE Transactions on Circuits and Systems*, vol. 21, no. 3, May 1974, pp. 317–327.

Hearn, W.E., "Fast-slewing monolithic operational amplifier," *IEEE Journal of Solid-State Circuits*, vol. 6, no. 1, February 1971, pp. 20–24.

Huijsing, J.H. and Tol, F., "Monolithic operational amplifier design with improved HF behaviour," *IEEE Journal of Solid-State Circuits*, vol. 11, no. 2, April 1976, pp. 323–328.

Ruediger, V.G. and Hosticka, B.J., "The response of 741 op amps to very short pulses," *IEEE Journal of Solid-State Circuits*, vol. 15, no. 5, October 1980, pp. 908–910.

Solomon, J.E., "The monolithic op amp: a tutorial study," *IEEE Journal of Solid-State Circuits*, vol. 9, no. 6, December 1974, pp. 314–332.

Solomon, J.E., "A tribute to Bob Widlar," *IEEE Journal of Solid State Circuits*, vol. 26, no. 8, August 1991, pp. 1087–1089.

Soundararajan, K. and Ramakrishna, K., "Characteristics of nonideal operational amplifiers," *IEEE Transactions on Circuits and Systems*, vol. 21, no. 1, January 1974, pp. 69–75.

Treleaven, D. and Trofimenkoff, F. (1971). "Modeling operational amplifiers for computer-aided circuit analysis," *IEEE Transactions on Circuits and Systems*, vol. 18, no. 1, pp. 205–207.

Widlar, R.J. (1979). "A new breed of linear ICs runs at 1-volt levels," *Electronics*, March 29, pp. 115–119.

———, "DC error reduction in bipolar op amps," *1980 IEEE Solid State Circuits Conference*, vol. 23, February 1980, pp. 204–205.

———, "Design techniques for monolithic operational amplifiers," *IEEE Journal of Solid-State Circuits*, vol. 4, no. 4, August 1969, pp. 184–191.

———, "Low-voltage techniques [for micropower operational amplifiers]," *IEEE Journal of Solid-State Circuits*, vol. 13, no. 6, December 1978, pp. 838–846.

————, "Some circuit design techniques for linear integrated circuits," *IEEE Transactions on Circuit Theory*, vol. CT-12, no. 4, December 1965, pp. 586–590.

Widlar, R.J., Dobkin, R. and Yamatake, M., "New op amp ideas,"*National Semiconductor Application Note 211*, December 1978.

Widlar, R.J. and Yamatake M. "A 150W op amp". *1985 IEEE Digest of Technical Papers, Solid State Circuits Conference*, 27, February 1985, pp. 140–141.

Wooley, B.A. and Pederson, D.O., "A computer-aided evaluation of the 741 amplifier," *IEEE Journal of Solid-State Circuits*, vol. 6, no. 6, December 1971, pp. 357–366.

Yang, H.C. and Allstot, D.J., "Considerations for fast-settling operational amplifiers," *IEEE Transactions on Circuits and Systems*, vol. 37, no. 3, March 1990, pp. 326–334.

CHAPTER 4

Finding the Perfect Op-Amp for Your Perfect Circuit

Bonnie Baker

"Perfect op-amp"? I just want one that is good enough, and doesn't cost much. /rap

The operational amplifier's operation and circuits are easy to find in the books in your local university library. The amplifier operation and circuit descriptions found in these reference books take you through computational algorithms that theoretically will provide the solutions to your analog amplifier design woes. If there were a perfect amplifier on the market today, the designs found in these books would indeed be easy to implement successfully. But there isn't a perfect amplifier—yet. Throughout the history of analog system design, circuits have required special care in key areas in order to ensure success. As luck would have it, a little common sense and bench sense will pull you out of most of your amplifier design disasters.

In an ideal world, the perfect amplifier would look like the one described in **Figure 4-1**.

The input stage design of this perfect amplifier would use devices whose inputs ($IN+$ and $IN-$) extend all the way to the power supply rails. Some single-supply amplifiers are able to do this with some distortion, but the perfect amplifier would be distortion-free. As a matter of fact, it would be nice if the inputs operated beyond the rails. If this were the case, the common-mode range would go beyond the rails as well.

Additionally, the inputs would not source or sink current—that is, they would have zero-input bias current. This allows source impedances to the amplifier to be infinite. This implies no common-mode or differential-mode input capacitance. Since voltage errors across the two inputs are usually gained by closed-loop circuit configurations around the amplifier, any DC voltage error (offset voltage) or AC error (noise) would be zero. The absence of these errors removes all your calibration worries!

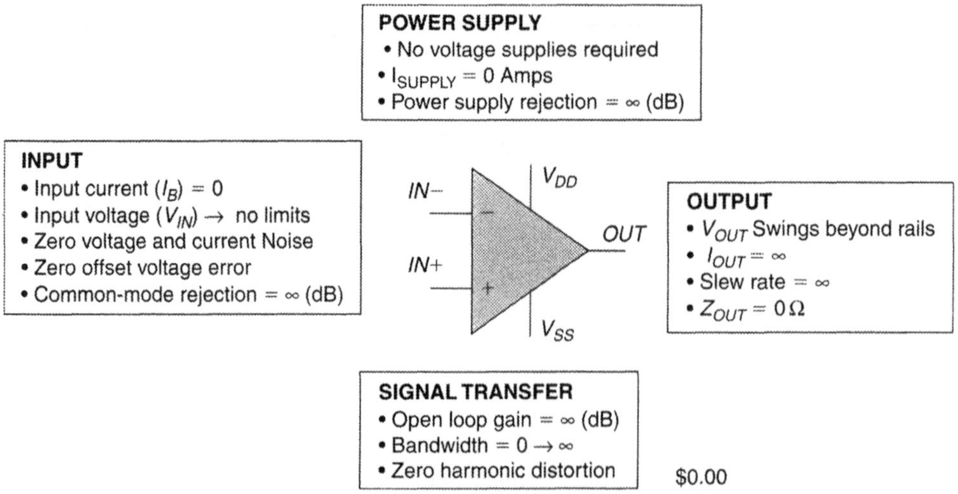

POWER SUPPLY
- No voltage supplies required
- I_{SUPPLY} = 0 Amps
- Power supply rejection = ∞ (dB)

INPUT
- Input current (I_B) = 0
- Input voltage (V_{IN}) → no limits
- Zero voltage and current Noise
- Zero offset voltage error
- Common-mode rejection = ∞ (dB)

IN–

V_{DD}

IN+

OUT

V_{SS}

OUTPUT
- V_{OUT} Swings beyond rails
- I_{OUT} = ∞
- Slew rate = ∞
- Z_{OUT} = 0 Ω

SIGNAL TRANSFER
- Open loop gain = ∞ (dB)
- Bandwidth = 0 → ∞
- Zero harmonic distortion

$0.00

Figure 4-1: A perfect amplifier has an infinite input impedance, open-loop gain, power supply rejection ratio, common-mode rejection ratio, bandwidth, slew rate, and output current. It also has zero offset voltage, input noise, output impedance, power dissipation, and most important, zero cost.

As for the power supply requirements of this ideal amplifier, there would be none. As you know, industry trends are always working on requests for lower supply voltages, and consequently, lower power consumption from active components. The ideal amplifier wouldn't need a voltage supply across V_{DD} and V_{SS} and would have zero power dissipation in its quiescent state.

The output of this amplifier would be capable of really swinging rail to rail or even beyond. This would eliminate the problem of losing bits on the outer rim in the following A/D conversion. The output impedance would be zero at DC as well as over frequency, ensuring that the device connected to the input of the amplifier is perfectly isolated from the external output device. The op-amp would respond to input signals instantaneously—that is, the slew rate would be infinite and it would be able to drive any load (resistive or capacitive) while maintaining an infinite open-loop gain and rail-to-rail output swing performance. Finally, in the frequency domain, the open-loop gain would be infinite at DC as well as over frequency, and the bandwidth of the amplifier would also be infinite. Oh, did I forget price? We would all love to have this ideal amplifier for $0.00.

Welcome to Op-Amp 101! This describes the textbook amplifier.

If I'm able to figure out how to design this amplifier, I guarantee you, I will become a multizillionaire. At this point, you are probably saying "Only in your dreams!" Well, maybe not a multizillionaire, mainly because the profits are $0.00. However, it is certain that I will become a very popular (though still poor) person.

It is interesting to note that many of these design imperfections are used to an advantage by most designers. For example, an amplifier circuit design uses a less than infinite bandwidth to limit the noise and high-speed transients in circuits. An infinite slew rate is not as good as it sounds. The amplifier users enjoy slower signals. This reduces the glitches further down the signal path and simplifies the layout.

So, for today, we know that there isn't an ideal amplifier for all circuit situations. The best we can do with the choices available is to pick the best amplifier for our application circuit and then use it properly.

Choose the Technology Wisely

CMOS and bipolar are the two silicon technologies that single-supply operational amplifiers commonly use. **Figure 4-2** shows the differences between these two operational amplifier technologies. The most important difference between CMOS and bipolar is in the input stage transistors. These transistors have a profound effect on the overall operation of the amplifier.

Because of the difference between the input transistors of these two types of amplifiers, the CMOS amplifier has lower input current noise and higher input impedance. Because of the high input impedance, the input bias current of the CMOS amplifier is much lower. In fact, the electrostatic discharge (ESD) cells at the input of the CMOS amplifier cause the input bias current errors. As will be shown in circuits later in this chapter, we can use this to an advantage for high-impedance sources such as photosensing transimpedance amplifiers.

The CMOS amplifier typically has a higher open-loop gain than bipolar amplifiers. This can minimize gain error in applications where the closed-loop gain is extremely high (60 dB or greater).

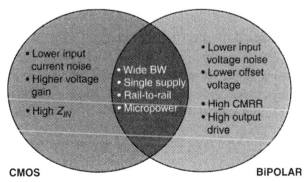

Figure 4-2: The two silicon technologies with which single-supply amplifiers are manufactured are CMOS or bipolar processes. By using the CMOS process, you can manufacture bipolar amplifiers. In these designs, the input transistors are bipolar, and the remaining transistors are CMOS.

In contrast with the CMOS amplifier, the bipolar amplifier usually has lower input-voltage noise, room temperature offset voltage, and offset drift. Bipolar amplifiers are more likely to provide higher output drive. They also exhibit a higher common-mode rejection capability. This is useful if the amplifier is in a buffer configuration. Although these specifications are typically better than the CMOS amplifier counterpart, the input bias current and input current noise are considerably higher.

Single-supply operating conditions are perfect for both CMOS and bipolar amplifiers. With the proper IC design, they are also capable of input and (near) output rail-to-rail operation.

Fundamental Operational Amplifier Circuits

The op-amp is the analog building block that is analogous to the digital gate. By using the op-amp in the design, circuits can be configured to modify the signal in the same fundamental way that the inverter, *AND*, and *OR* gates do in digital circuits. This section of the chapter will show the fundamental circuits using this building block. The list of circuits we discuss includes the voltage follower, noninverting gain, and inverting gain circuits. This discussion is followed by more complex circuits, including a difference amplifier, summing amplifier, and current-to-voltage converter.

Voltage Follower Amplifier

Starting with the most basic op-amp circuit, the buffer amplifier (shown in **Figure 4-3**) is used to drive heavy loads, solve impedance matching problems, or isolate high-power circuits from sensitive, precise circuitry. Usually, heavy loads require an additional specialized amplifier that is capable of supplying the higher output currents that are greater than 20 mA. You will find that the amplifier datasheet has specifications for the magnitude of the amplifier output current, capable of driving higher currents.

Solving impedance matching problems is also a good reason to use a buffer amplifier. This type of problem exists when the signal path has a high-impedance device or resistor that creates an undesirable voltage divider in the circuit. A buffer amplifier breaks up this type of impedance path because of the high-impedance input and low-impedance output of the amplifier.

Another use for a buffer is to keep high thermal changes away from sensitive circuits. In this scenario the buffer follows the sensitive circuit and serves the purpose of driving high-output currents.

The buffer amplifier shown in Figure 4-3 can be implemented with any single-supply, unity gain stable amplifier. In this circuit, as with all amplifier circuits, bypassing the op-amp power with a capacitor is a must. For single-supply amplifiers that operate in

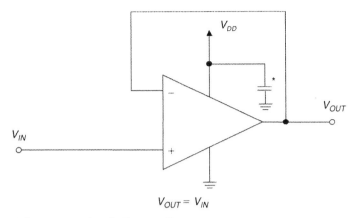

* Bypass capacitor, 1 μF or 0.1 μF

Figure 4-3: A buffer amplifier, also called a voltage follower, is useful when you want to provide a high-current drive stage, match impedances, or electrically isolate signals.

bandwidths from DC to 1 MHz, a 1 μF capacitor is usually appropriate. Sometimes a smaller bypass capacitor is required for amplifiers that have bandwidths of up to the tens to hundreds of megahertz. In these cases, a 0.1 μF capacitor would be appropriate. If the selection of the value of the bypass capacitor is an inappropriate value or placed too far from the power supply pin and not connected to ground directly on the PCB, the op-amp circuit may oscillate. If you are unsure of what the bypass capacitor value should be, refer to the product datasheet for details.

The analog gain of the circuit in Figure 4-3 is +1 V/V. Notice that this circuit has positive overall gain, but the feedback loop is tied from the output of the amplifier to the inverting input. An all too common error is to assume that an op-amp circuit that has a positive gain requires positive feedback. You can configure this amplifier with positive feedback if you connect the noninverting input to the output. I know this sounds unbelievable, but I have had applicants draw buffers with positive feedback during their interviews. If positive feedback is used, the amplifier will most likely drive to either rail at the output.

This amplifier circuit will give good linear performance across the bandwidth of the amplifier. You might be reading this discussion and saying to yourself, "There's that textbook description again." You are right; however, here are the land mines in this type of circuit.

The only restrictions on the signal will occur as a result of a violation of the input common-mode voltage and output swing limits. You need to scrutinize these performance characteristics in your amplifier datasheet and your application's demands on this type of circuit. Oh, by the way, ensure that the bandwidth of the amplifier is at least 100 × higher than the bandwidth of your signal. However, be aware that you need to look at the input and output of the amplifier.

When you use this circuit to drive heavy loads, the specifications of the amplifier must indicate that it is capable of providing the required output currents. Another application where this circuit may be used is to drive capacitive loads. Not every amplifier is capable of driving capacitors without becoming unstable. If an amplifier can drive capacitive loads, the product datasheet will highlight this feature. However, if an amplifier can't drive capacitive loads, the product datasheets will not explicitly say so. This is an instance where features are not in the advertisements or promotions and there is no mention of average performance.

Another use for the buffer amplifier is to solve impedance-matching problems. This would be applicable in a circuit where the analog signal source has relatively high impedance compared to the impedance of the following circuitry. If this occurs, there will be a voltage loss with the signal because of the voltage divider between the source's impedance and the following circuitry's impedance. The buffer amplifier is a perfect solution to the problem. The input impedance of the noninverting input of an amplifier can be as high as 10^{13} Ω for CMOS amplifiers. In addition, the output impedance of this amplifier configuration is usually less than 100 Ω.

Yet another use of this configuration is to separate a heat source from sensitive precision circuitry, as shown in **Figure 4-4**. Imagine that the input circuitry to this buffer amplifier is amplifying a 100 mV signal. This type of amplification is difficult to do with any level of accuracy in the best of situations. Assigning the output current drive to the device that is doing the precision, amplification work can easily disrupt this measurement. An increase in current drive will cause self-heating of the chip, which will induce an offset change. In this circuit, the front-end circuitry makes precision measurements while an analog buffer performs the function of driving a heavy load.

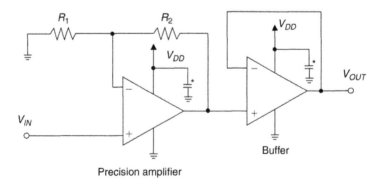

Precision amplifier

* Bypass capacitor, 1 μF or 0.1 μF

Figure 4-4: A buffer amplifier helps achieve load isolation in this circuit. The buffer separates any high-current output requirements from this input amplifier.

Amplifying Analog Signals

The buffer solves many analog signal problems; however, there are instances in circuits where you need to gain a signal. Two fundamental types of amplifier circuits can provide gain. With the first type, the signal gain is positive (or not inverted), as shown in **Figure 4-5**. This type of circuit is useful in single-supply amplifier applications where negative voltages are usually not present, difficult to produce, or just not possible.

The input signal to this circuit is presented to the high-impedance, noninverting input of the op-amp. The gain that the amplifier circuit applies to the signal is equal to:

$$V_{OUT} = (1 + R_2/R_1)V_{IN} \qquad\qquad [4\text{-}1]$$

Typical values for these resistors in single-supply circuits are above $5\,k\Omega$ to $25\,k\Omega$ for R_2. For the input resistor, R_1, restrictions are dependent on the amount of gain desired versus the amount of amplifier noise and input offset voltage as specified in the op-amp product datasheet.

Again, this circuit has some restrictions in terms of the input and output range. The common-mode range of the amplifier restricts the noninverting input. The output swing of the amplifier is also restricted as stated in the product datasheet of the individual amplifier. Most typically, the larger signal at the output of the amplifier causes more signal-clipping errors than the smaller signal at the input. Reducing the gain of this circuit may eliminate undesirable output clipping errors.

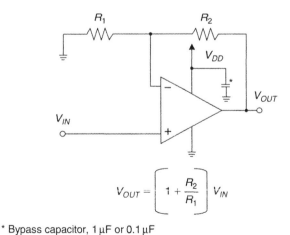

$$V_{OUT} = \left[1 + \frac{R_2}{R_1} \right] V_{IN}$$

* Bypass capacitor, 1 μF or 0.1 μF

Figure 4-5: This is an operational amplifier configured in a noninverting gain circuit. This circuit applies a positive gain to a signal in your circuit. Therefore, you won't need a reference level-shift voltage to keep the output of the amplifier within its operating range.

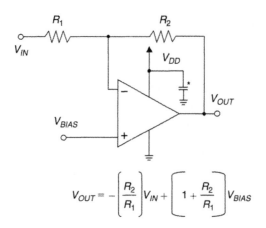

$$V_{OUT} = -\left[\frac{R_2}{R_1}\right]V_{IN} + \left[1 + \frac{R_2}{R_1}\right]V_{BIAS}$$

* Bypass capacitor, 1 µF or 0.1 µF

Figure 4-6: This is an operational amplifier V_{IN} configured in an inverting gain circuit. Single-supply environments usually require V_{BIAS} to ensure the output stays above ground.

Figure 4-6 illustrates an inverting amplifier configuration. This circuit gains and inverts the signal present at the input resistor, R_1. The gain equation for this circuit is:

$$V_{OUT} = -(R_2/R_1)V_{IN} + (1 + R_2/R_1)V_{BIAS} \qquad [4\text{-}2]$$

The ranges for R_1 and R_2 are the same as in the noninverting circuit shown in Figure 4-5.

This circuit has a minor pitfall in single-supply circuits. In single-supply applications, this circuit is easy to misuse. The problem is rooted in the selection of the voltage at V_{BIAS}. You need to select a value for V_{BIAS} so that the output of the amplifier always remains between the supplies.

For example, let R_2 equal 10 kΩ, R_1 equal 1 kΩ, V_{BIAS} equal 0 V, and the voltage at the input resistor, R_1, equal to 100 mV, the output voltage would be -1 V. This would violate the output swing range of the operational amplifier. In reality, the output of the amplifier would try to go as near to ground as possible.

The inclusion of a positive DC voltage at V_{BIAS} in this circuit solves this problem. In the previous example, a voltage of 225 mV applied to V_{BIAS} would level shift the output signal up 2.475 V. This would make the output signal equal (2.475 V $-$ 1 V), or 1.475 V at the output of the amplifier. Typically, you want to make the target average output voltage of the amplifier equal to $V_{DD}/2$.

The Difference Amplifier

The difference amplifier combines the noninverting amplifier and inverting amplifier circuits of Figures 4-5 and 4-6 into a signal block that subtracts two signals. **Figure 4-7**

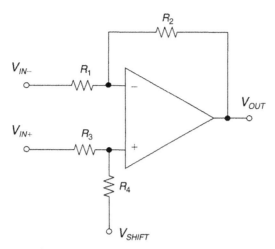

Figure 4-7: This is an operational amplifier circuit configured in a difference amplifier circuit. A difference amplifier implements the subtraction and gain function in one stage.

illustrates an example of the difference amplifier circuit. It illustrates a straightforward implementation of this function. A difference amplifier or op-amp subtractor uses this arrangement of resistors around an amplifier. The DC transfer function of this circuit is equal to:

$$V_{OUT} = V_{IN+} \times R_4(R_1 + R_2)/(R_1 \times (R_3 + R_4)) - V_{IN-} \times (R_2/R_1)$$
$$+ V_{SHIFT} \times R_3(R_1 + R_2)/((R_3 + R_4)R_1) \qquad [4\text{-}3]$$

If R_1/R_2 is equal to R_3/R_4, the closed loop system gain of this circuit equals:

$$V_{OUT} = (V_{IN+} - V_{IN-})(R_2/R_1) + V_{SHIFT} \qquad [4\text{-}4]$$

This circuit configuration will reliably take the difference of two signals as long as the signal source impedances are low. If the signal source impedances are high with respect to R_1, there will be a signal loss due to the voltage divider action between the source and the input resistors to the difference amplifier. Additionally, errors can occur if the two signal source impedances are mismatched. With this circuit, it is possible to have gains equal to or higher than one.

The fact that R_1/R_2 is equal to R_3/R_4 simplifies the mathematics in this system considerably. Since the gain of both signals is equal, the difference amplifier conveniently subtracts the common-mode voltage of the two signals from the system. It is also easy to implement gain by setting the two resistor ratios to be equal or greater than one.

One limitation of this circuit is the lack of flexibility with gain adjustments. If you change the gain dynamically in the application, you must adjust two resistors. In a single-supply environment, a voltage reference centers the output signal between ground and the

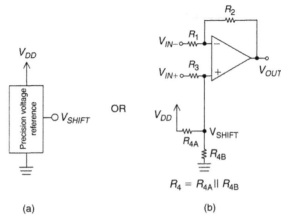

Figure 4-8: A precision voltage reference, (a) or a less expensive solution of replacing R_4 of the voltage divider between the supply, (b) provides the voltage, V_{SHIFT}, of this difference amplifier.

power supply. Figure 4-7 shows this voltage, V_{SHIFT}. The purpose of this reference voltage is to simply shift the output signal into the linear region of the amplifier. A precision, voltage-reference or a resistive network implements the V_{SHIFT} circuit function as shown in **Figure 4-8**.

Summing Amplifier

You can use summing amplifiers to combine multiple signals by addition or subtraction. Since the difference amplifier can only process two signals, it is a subset of the summing amplifier.

The transfer function of this circuit as shown in **Figure 4-9** is:

$$V_{OUT} = (V_1 + V_2 - V_3 - V_4)(R_2/R_1) \qquad [4\text{-}5]$$

You can use any number of inputs on either the inverting or noninverting input sides as long as there is an equal number of both with equivalent resistors. All the inputs to this circuit should be connected to a signal source or (if unused) to ground.

Current-to-Voltage Conversion

If you use a photodetector, feedback resistor, and an operational amplifier in your circuit, you can sense light. This type of circuit converts the output current of a photodetector into a voltage. The single resistor and an optional capacitor are in the feedback loop of the amplifier, as shown in **Figure 4-10**.

In the circuits shown in Figure 4-10, light impinging on the photodetector generates a current. This current flows in the reverse bias direction of the diode. If a CMOS op-amp is used (with low input bias current), the current from the detector (I_{D1}) primarily goes

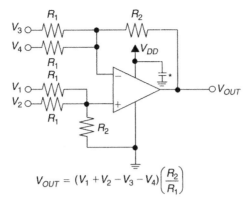

$$V_{OUT} = (V_1 + V_2 - V_3 - V_4)\left(\frac{R_2}{R_1}\right)$$

* Bypass capacitor, 1μF or 0.1μF

Figure 4-9: Operational amplifier configured in a summing amplifier circuit.

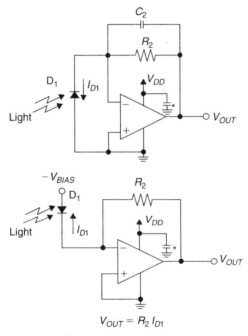

$$V_{OUT} = R_2 I_{D1}$$

* Bypass capacitor, 1μF or 0.1 μF

Figure 4-10: These circuits show how to convert current to voltage by using an amplifier and one resistor. The top light-sensing circuit is appropriate for precision applications. The bottom circuit is appropriate for high-speed applications.

through the feedback resistor, R_2. Additionally, the op-amp input bias current error is low because it is CMOS (typically <200 pA). You would ground the noninverting input of the op-amp, which keeps the entire circuit biased to ground. These two circuits will only work if the common-mode range of the amplifier includes zero and you are not concerned about a zero level of light. If your light source has zero luminance, the output of the single-supply amplifier is unable to go all the way to ground.

The two circuits in Figure 4-10 provide precision sensing from the photodetector (top circuit in the figure) and higher-speed sensing (bottom circuit in the figure). In the top circuit, the voltage across the detector is nearly zero and equal to the offset voltage of the amplifier. With this configuration, current that appears across the resistor, R_2, is primarily a result of the light excitation on the photodetector.

The photosensing circuit at the bottom of the figure works best in a high-speed digital environment. By reverse-biasing the photodetector (which reduces the parasitic capacitance of the diode), this sensing circuit can respond very quickly to digital signals. There is more leakage through the photodetector in this bottom circuit, which causes a higher DC error.

Using These Fundamentals

You can use several amplifiers to build instrumentation amplifiers and floating current sources.

Instrumentation Amplifier

You will find instrumentation amplifiers in a large variety of applications, from medical instrumentation to process control. The instrumentation amplifier is similar to the difference amplifier in that it subtracts one analog signal from another, but it differs in terms of the quality of the input stage. **Figure 4-11** illustrates a classic, three op-amp instrumentation amplifier.

In this circuit, the high-impedance, noninverting inputs of the input amplifiers (A_1, A_2) acquire the two input signals. This is a distinct advantage over the difference amplifier configuration, where source impedances are high or mismatched. The first stage also gains the two incoming signals. One resistor, R_G, adjusts the gain.

Following the first stage of this circuit is a difference amplifier (A_3). The function of this portion of the circuit is to reject the common-mode voltage of the two input signals as well as take the difference between them. The source impedances of the signals into the input of the difference amplifier are low, equivalent, and well controlled.

The reference voltage (V_{REF}) of the difference stage of this instrumentation amplifier is capable of spanning a wide range. Typically, you would connect the voltage reference to half of the supply voltage in a single-supply application. The transfer function of this circuit is:

$$V_{OUT} = (V_1 - V_2)(1 + 2R_2/R_G)(R_4/R_3) + V_{REF} \qquad [4\text{-}6]$$

Figure 4-12 shows a second type of instrumentation amplifier. In this circuit, the two amplifiers serve the functions of load isolation and signal gain. The second amplifier also takes the difference between the two input signals (V_1, V_2).

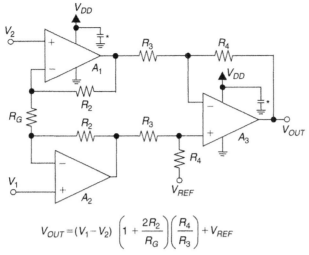

$$V_{OUT} = (V_1 - V_2) \left(1 + \frac{2R_2}{R_G} \right) \left(\frac{R_4}{R_3} \right) + V_{REF}$$

* Bypass capacitor, 1 µF or 0.1 µF

Figure 4-11: You can design an instrumentation amplifier with three amplifiers. The input operational amplifiers (A_1, A_2) provide signal gain. The output operational amplifier converts the signal from the two input amplifiers to a single-ended output with a difference amplifier (A_3).

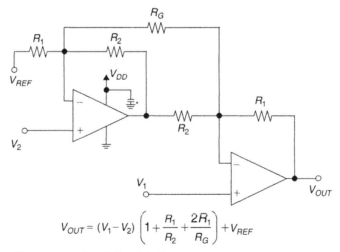

$$V_{OUT} = (V_1 - V_2) \left(1 + \frac{R_1}{R_2} + \frac{2R_1}{R_G} \right) + V_{REF}$$

* Bypass capacitor, 1 µF or 0.1 µF

Figure 4-12: You can design an instrumentation amplifier with two amplifiers. This configuration is best suited for higher gains (gain ⩾3 V/V).

You would connect the circuit reference voltage to the first op-amp in the signal chain. Typically, this voltage is half of the supply voltage in a single-supply environment.

The transfer function of this circuit is:

$$V_{OUT} = (V_1 - V_2)(1 + R_1/R_2 + 2R_1/R_G) + V_{REF} \qquad [4\text{-}7]$$

Floating Current Source

A floating current source (**Figure 4-13**) can come in handy in driving a variable resistance, such as a Resistance Temperature Device (RTD). This particular configuration produces an appropriate 1 mA source for an RTD-type sensor. However, you can change this current reference magnitude to any current.

With this configuration, R_1 reduces the voltage of V_{REF} by the voltage V_{R1}. The voltage applied to the noninverting input of the top op-amp is $V_{REF} - V_{R1}$. This voltage is gained to the amplifier's output by two, to equal $2 \times (V_{REF} - V_{R1})$. Meanwhile, the output for the bottom op-amp (A_2) is presented with the voltage $V_{REF} - 2V_{R1}$. Subtracting the voltage at the output of the top amplifier from the noninverting input of the bottom amplifier gives $2 \times (V_{REF} - V_{R1}) - (V_{REF} - 2V_{R1})$, which equals V_{REF}.

The transfer function of the circuit is:

$$I_{OUT} = V_{REF}/R_2 \qquad [4\text{-}8]$$

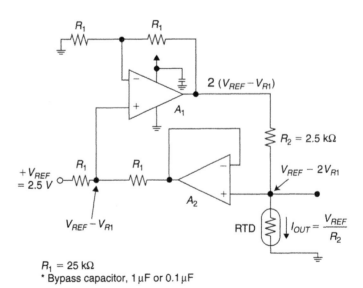

**Figure 4-13: A floating current source uses two operational amplifiers
and a precision voltage reference.**

Amplifier Design Pitfalls

Theoretically, the circuits within this chapter work. Beyond the theory, however, there are few tips that will help get the circuit right the first time. This section lists common problems associated with using an op-amp on a PC board. The following discussion has two categories: general suggestions and single-supply pitfalls.

In General

- Be careful of the supply pins. Don't make them too high per the amplifier specification sheet, and don't make them too low. High supplies will damage the part. In contrast, low supplies won't bias the internal transistors and the amplifier won't work or it might not operate properly.

- Make sure the negative supply is, in fact, tied to a low impedance potential. Additionally, make sure the positive supply is the voltage you expect with respect to the negative supply pin of the op-amp. Placing a voltmeter across the negative and positive supply pins will verify that you have the right relationship between the pins.

- Ground can't be trusted, especially in digital circuits. Plan your grounding scheme carefully. If the circuit has a lot of digital circuitry, consider separate analog and digital grounds and power planes. It is very difficult, if not impossible, to remove digital switching noise from an analog signal.

- Bypass the amplifier power supplies with bypass capacitors as close to the amplifier as possible. Amplifiers usually use a 1 µF or 0.1 µF capacitor. Also, bypass the power supply at the source with a 10 µF capacitor.

- Use short lead lengths to the inputs of the amplifier. If you have a tendency to use the white perf boards for prototyping, be aware that their capacitance and inductance can cause noise and oscillation. There is a good chance that these issues won't be a problem with the PCB implementation of the circuit, except for small twists of fate. The stray capacitance on these boards can complete loops that you had not intended to close. As an example, if you have an op-amp without any feedback capacitor and the feedback resistor is installed between two adjacent strips, the stray white perf board capacitance will prevent an oscillation. Then when you lay this circuit out on a real PCB, you can easily have an oscillation.

- Always include a spot on your PCB for a feedback capacitor across your feedback resistor. Keep this option on your board until you can prove that it isn't needed.

- Avoid using op-amps with very low voltage noise, unless you are sure that the signal source and the feedback networks are very low in impedance. Otherwise, the high op-amp current noise that accompanies the low voltage noise can cause serious noise problems.

- Amplifiers are static sensitive! If damage has occurred, they could fail immediately or exhibit a soft error (like offset voltage or input bias current changes) that will get worse over time.

Single-Supply Rail-to-Rail Amplifiers

- Operational amplifier output drivers are capable of driving a limited amount of current to the load. Check your product datasheet for that number.

- Capacitive loading an amplifier is risky business. Make sure the amplifier can handle any loads you have.

- It is very rare that a single-supply amplifier will truly swing rail to rail. In reality, the output of most of these amplifiers can only come within 50 to 300 mV from each rail. Check the product datasheets of your amplifier.

Moving forward, the next level of troubleshooting amplifier tips is in the frequency domain. You will find with stability issues such attributes as ringing or oscillation.

References

Baker, B.C., "Operational amplifiers: 6 part," first published in *analogZone* (2002, 2003) and reproduced with permission.

Franco, S., *Design with Operational Amplifiers and Analog Integrated Circuits*, McGraw-Hill, New York, 1998.

Frederiksen, T., *Intuitive Operational Amplifiers*, McGraw-Hill, New York, 1988.

Williams, J., *Analog Circuit Design*, Butterworth-Heinemann, Burlington, MA, 1998.

CHAPTER 5

Review of Passive Components and a Case Study in PC Board Layout

Marc Thompson

Here, Marc explains why a resistance and a capacitance are not trivial items. Real resistors and real capacitors have several reactive or lossy components. Is this part of our circuit design work? Yeah. So we have to engineer to get the system working despite component imperfections. Sometimes wise component choice is a very important part of our job./rap

In This Chapter

Here we consider some of the subtleties of passive components, including construction techniques of these devices and parasitic effects. We'll cover some details about resistors, capacitors, and inductors. Then we'll use what we've learned in an illustrative discussion of PC board layout issues.

Resistors

At first blush, a resistor is a resistor is a resistor. However, we'll now delve into some of the subtleties of these devices. The impedance of an ideal resistor is not dependent on operating frequency,[1] and is:

$$Z_{resistor,ideal} = R \qquad\qquad [5\text{-}1]$$

---

[1] Another way to look at this is that the current in an ideal resistor and a voltage across this ideal resistor are in phase.

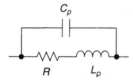

Figure 5-1: Resistor R showing parasitic elements series inductance L_p and parallel capacitance C_p.

A real-world resistor (**Figure 5-1**) includes a parasitic inductance, due to the geometry of the lead length, and a parasitic capacitance across the resistor. The impedance of the resistor, including these parasitic elements, is:[2]

$$Z_{resistor,real}\,(s) = \frac{Ls + R}{LCs^2 + RCs + 1} = \frac{R\left(1 + \dfrac{L}{R}s\right)}{LCs^2 + RCs + 1} \qquad [5\text{-}2]$$

We can put this in "$j\omega$" form by making the substitution $s = j\omega$, resulting in:

$$Z_{resistor,real}\,(j\omega) = \frac{j\omega L + R}{(1 - \omega^2 LC) + j\omega RC} \qquad [5\text{-}3]$$

For a large-valued resistor,[3] the RC time constant dominates. This is because a large resistor will swamp out the value of the parasitic inductance. For a low-valued resistor, the L/R time constant dominates because the resistor effectively shorts out the parasitic capacitance. The magnitude of the impedance of the real-world resistor is:

$$\left|Z_{resistor,real}\right| = \sqrt{\frac{(\omega L)^2 + R^2}{(1 - \omega^2 LC)^2 + (\omega RC)^2}} \qquad [5\text{-}4]$$

In **Figure 5-2**, we see the impedance of a resistor with $R = 1\,\mathrm{M}\Omega$, $C = 0.2\,\mathrm{pF}$, and $L = 10\,\mathrm{nH}$. Note that for this relatively large resistor ($R \gg Z_o = 223\,\Omega$), parasitic capacitive effects dominate; the impedance rolls off at frequencies above approximately $1\,\mathrm{MHz}$.

In **Figure 5-3**, we see the impedance of a low-valued resistor with $R = 10\,\Omega$, $C = 0.2\,\mathrm{pF}$, and $L = 10\,\mathrm{nH}$. Note that parasitic inductive effects dominate; the impedance increases at frequencies above approximately $10\,\mathrm{Mrad/sec}$.

---

[2] One way to sanity-check this result is to consider the limit as $L \to 0$ and $C \to 0$; we want the impedance to be exactly R, and this is indeed the case.

[3] We can see that "large-valued" in this case is large enough so that $RC \gg L/R$, or equivalently $R \gg \sqrt{L/C}$. The term $\sqrt{L/C}$ comes up over and over again in RLC circuits and transmission lines and is called the *characteristic impedance* Z_o of this circuit.

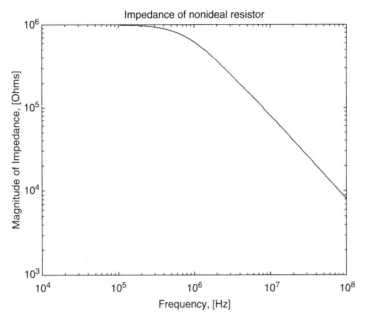

Figure 5-2: Magnitude of impedance of nonideal resistor with $R = 1\,\text{M}\Omega$, $C = 0.2\,\text{pF}$, and $L = 10\,\text{nH}$.

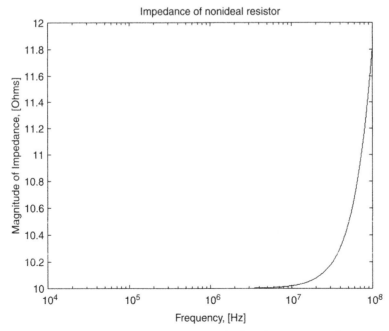

Figure 5-3: Magnitude of impedance of nonideal resistor $R = 10\,\Omega$, $C = 0.2\,\text{pF}$, and $L = 10\,\text{nH}$.

It's difficult to quantify exactly the values of parasitic elements, but in standard through-hole resistors you might expect fractions of a picofarad of parasitic capacitance and a few nanohenries of parasitic inductance.[4] You need to consider such parasitics as series inductance and parallel resistance in high-frequency circuits.

Comments on Surface-Mount Resistors

You can miniaturize your circuit and somewhat reduce the effects of parasitic inductance by using surface-mount resistors. Surface-mount resistors come in a variety of sizes, ranging from 0201 through 0402 and all the way up to 2512. The resistor size numbers indicate the length and width of the resistor. For instance, an 0805 resistor is 0.08″ in length and 0.05″ in width.

There is a trade-off to be made with regard to resistor size and wattage and working voltage rating, as shown in **Table 5-1**. For instance, a longer resistor will have a higher working voltage rating due to a higher voltage breakdown.

Table 5-1: Comparison of surface mount resistor ratings.

Resistor	Size	Typical wattage rating[5]	Typical working voltage rating[6]
0201	0.02″ × 0.01″	50 mW	15 V
0402	0.04″ × 0.02″	50 mW – 62.5 mW	50 V
0603	0.06″ × 0.03″	62.5 mW – 100 mW	75 V
0805	0.08″ × 0.05″	100 mW – 250 mW	100 V
1206	0.12″ × 0.06″	125 mW – 250 mW	200 V
1210	0.12″ × 0.10″	250 mW – 333 mW	200 V
1812	0.18″ × 0.12″	500 mW	200 V
2010	0.20″ × 0.10″	500 mW	200 V
2512	0.25″ × 0.12″	1000 mW	250 V

---

[4] A very rough rule of thumb for the inductance of component leads above a ground plane is 10 nanohenries per centimeter of lead length. So, it behooves you to keep lead lengths short if you want to minimize parasitic inductance. Of course, you can test your resistor using an impedance analyzer (such as the Hewlett-Packard HP4192, now Agilent) and extract the parameters for your device.

[5] There is some variation in wattage rating from manufacturer to manufacturer, so be sure to check the particular device datasheet.

[6] Again, use these numbers for comparison purposes only, and check the specific manufacturer.

Comments on Resistor Types

As the designer, you also have decisions to make regarding the type of resistor you put in your circuit. For instance, do you choose carbon composition, carbon film, metal film, wirewound, or some other type of resistor?

Carbon composition, sometimes called *carbon comp*, resistors are old-style resistors that have been used for years and years in electronics. The main advantage of carbon composition resistors is their ability to withstand high current transient surges. They do have the disadvantage of a high temperature coefficient of resistivity. Remember that the resistance of a resistor varies with temperature and that the resistance can be expressed as:

$$R(T) = R_o(1 + \alpha(T - T_o))$$
[5-5]

- $R(T)$ is resistance at your operating temperature

- R_o is the reference resistance at temperature T_o

- α is the temperature coefficient of resistivity

Carbon comp resistors also have a tendency to drift in value with time, especially if they are overstressed with high currents. Carbon comp resistors have largely been replaced in modern electronics by metal film and carbon film resistors. Film resistors have the advantage of a lower temperature coefficient of resistivity. They are, however, somewhat more susceptible to damage by electrical overloads.

Wirewound resistors are largely used where high-wattage capability is needed. Some of them do suffer, however, from a large series inductance due to the way they are manufactured with wound wires.

A comparison of resistor types is given in **Table 5-2**.

Capacitors

Just as resistors suffer from parasitic components, so do capacitors. A model of a real-world capacitor constructed as a parallel plate filled with a dielectric is shown[7] in **Figure 5-4a**. The resistance R_s is the series resistance of the leads. The parallel plate is filled with

---

[7] In this example, we've ignored the series inductance for simplicity. We're also ignoring the effects of dielectric relaxation, also called *soakage* when referring to capacitors. This effect is described in detail in Bob Pease's article, "Understand capacitor soakage to optimize analog systems," found at www.national. com/rap/Application/0,1570,28,00.html. Dielectric absorption is an issue in high-precision analog circuits such as sample-and-holds and analog filters. Polystyrene, polypropylene, Teflon, and NP0 (C0G) ceramic capacitors have low dielectric absorption.

Table 5-2: Comparison of resistor types.

Resistor	Typical power rating	Temperature coefficient	Comments
Carbon composition	0.25W < 2W	>1000 ppm/°C	Old-style resistors. Typically replaced by carbon film or metal film in new designs. Poor long-term stability and temperature coefficient.
Carbon film	–	Typically −50 ppm/°C to −1000 ppm/°C	–
Metal film	–	Typically +50 ppm/°C to +300 ppm/°C	Low noise.
Wirewound	Typically >5W	Typically +100 ppm/°C	Typically used for high-wattage resistors. Be careful of high parasitic inductance.

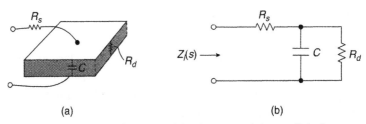

(a)   (b)

Figure 5-4: Capacitor showing parasitic elements. (a) Parallel-plate capacitor filled with dielectric with finite electrical conductivity. This capacitor has series resistance R_s and dielectric resistance R_d that is in parallel with the lumped capacitance C. (b) Electrical model.

a dielectric that has a finite electrical conductivity.[8] This results in a dielectric resistance R_d that is in parallel with the desired capacitance C as shown in **Figure 5-4b**.

The input impedance to the real-world capacitor is:

$$Z_i(s) = R_s + \frac{R_d}{R_d C s + 1} \qquad [5\text{-}6]$$

---

[8] For further information on the lossy capacitor see, e.g., Markus Zahn, *Electromagnetic Field Theory: A Problem-Solving Approach*, Krieger reprint 1987, pp. 184–194. Note that in this initial model we don't include the effects of series inductance. We'll consider this in more detail later.

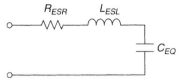

Figure 5-5: Capacitor with equivalent series resistance (R_{ESR}) and equivalent series inductance (L_{ESL}).

We can expand this result to find the real and imaginary parts of the input impedance as follows:

$$
\begin{aligned}
Z_i(j\omega) &= R_s + \frac{R_d}{j\omega R_d C + 1} \\
&= R_s + \frac{R_d(1 - j\omega R_d C)}{1 + \omega^2 R_d^2 C^2} \\
&= \left[R_s + \frac{R_d}{1 + \omega^2 R_d^2 C^2} \right] - j\left[\frac{\omega R_d^2 C}{1 + \omega^2 R_d^2 C^2} \right]
\end{aligned}
\qquad [5\text{-}7]
$$

The first term (the real part) is sometimes called the *equivalent series resistance* of the capacitor, or:

$$
R_{ESR} = \left[R_s + \frac{R_d}{1 + \omega^2 R_d^2 C^2} \right] \qquad [5\text{-}8]
$$

Note that the ESR decreases as frequency increases. The equivalent capacitance is:

$$
C_{eq} = C\left[1 + \frac{1}{\omega^2 R_d^2 C^2} \right] \qquad [5\text{-}9]
$$

A simplified model of the capacitor showing the equivalent series resistance is shown in **Figure 5-5**. To this model we've also added an equivalent series inductance (L_{ESL}). The value of the series inductance depends on the geometry of the internal construction of the capacitor as well as the lead length of the device as it is connected in the circuit. A ballpark rule of thumb for ESL of a capacitance is 10 nanohenries per centimeter of lead length. The ESL of some electrolytic capacitors can be somewhat higher if the capacitor is fabricated with wound foil inside a can.

The dissipation factor (*DF*) is another figure of merit often found on capacitor datasheets. The dissipation factor is given by:

$$
DF = (\omega C)R_{ESR} \qquad [5\text{-}10]
$$

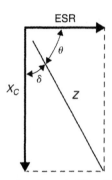

Figure 5-6: Plot illustrating the dissipation factor. The dissipation factor is the tangent of the angle between the impedance of the capacitor X_c and the overall impedance Z.

We note that the dissipation factor is the inverse of the Q of the capacitor. If we plot ESR, capacitive reactance (X_C), and total capacitor impedance (Z) as in **Figure 5-6**, we see that there is a phase angle between the capacitive reactance and the impedance of the capacitor. The dissipation factor is the tangent of this angle, or:

$$DF = \tan(\delta) = \frac{R_{ESR}}{X_C} \qquad [5\text{-}11]$$

The impedance of an ideal capacitor is:

$$Z_{cap,ideal} = \frac{1}{j\omega C} \qquad [5\text{-}12]$$

For a real-world capacitor (ignoring dielectric loss) the impedance is:

$$Z_{cap,real} = \frac{1}{j\omega C} + R + j\omega L = \frac{(1 - \omega^2 LC) + j\omega RC}{j\omega C} \qquad [5\text{-}13]$$

The magnitude of the impedance is:

$$\left| Z_{cap,real} \right| = \frac{\sqrt{(1 - \omega^2 LC)^2 + (\omega RC)^2}}{\omega C} \qquad [5\text{-}14]$$

An impedance plot of an electrolytic capacitor, comparing the ideal with the actual impedance, is shown in **Figure 5-7** for $C = 100\,\mu F$, $L = 25\,nH$, and $R = 0.01\,\Omega$. We note that at frequencies above 100 kHz the impedance of the capacitor looks inductive.

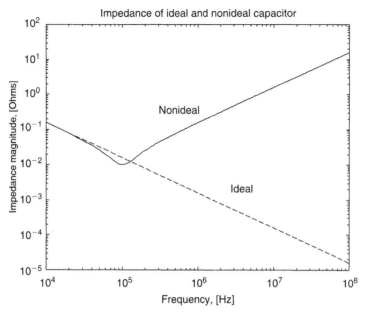

Figure 5-7: Impedance plot of electrolytic capacitor with $C = 100\ \mu F$, $L = 25\ nH$, and $R = 0.01\ \Omega$. The dotted line is the impedance of an ideal 100-µF capacitor.

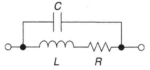

Figure 5-8: Inductor showing parasitic elements of series resistance R and interwinding capacitance C.

Inductors

The impedance of an ideal inductor is:

$$Z_{inductor,ideal} = j\omega L \qquad\qquad [5\text{-}15]$$

For a real-world inductor, the impedance is modified by the resistance of the copper wire[9] and the interwinding capacitance (**Figure 5-8**). This impedance is:

$$Z_{cap,real} = \frac{j\omega L + R}{(1 - \omega^2 LC) + j\omega RC} \qquad\qquad [5\text{-}16]$$

---

[9] This analysis ignores core losses (if any) and other high-frequency effects such as current crowding due to skin effect.

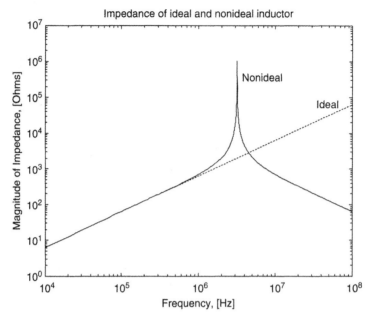

Figure 5-9: Impedance plot of real-world inductor with $L = 100\,\mu H$, $C = 25\,pF$, and $R = 0.1\,\Omega$. The dotted line is the impedance of an ideal 100-μH inductor.

The magnitude of this impedance is:

$$\left| Z_{inductor,real} \right| = \sqrt{\frac{(\omega L)^2 + R^2}{(1 - \omega^2 LC)^2 + (\omega RC)^2}} \qquad [5\text{-}17]$$

A impedance plot of an electrolytic capacitor, comparing the ideal with the actual impedance, is shown in **Figure 5-9** for $L = 100\,\mu H$, $C = 25\,pF$, and $R = 0.1\,\Omega$. Note that the self-resonant frequency of the nonideal inductor (at approximately 3.2 MHz) is clearly shown. Above the self-resonant frequency, the impedance of this inductor is capacitive.

Printed Circuit Board Layout Issues

Printed circuit board (PCB) layout and routing is a task that is sometimes left to the last minute in a design cycle. Doing a good PCB layout requires attention to many details, including:

- Knowledge of where you want the high frequency, high current, or sensitive circuitry to be

- Some information on component limitations

- Information on noise sources

- Real-world constraints, such as PCB form factor and location of connectors and mounting holes

- Other constraints such as PCB design rules mandating minimum trace widths, trace-to-trace spacing, and the like

The following is a discussion of some of these design issues.

Power Supply Bypassing

The need for power supply bypassing from integrated circuits arises from the fact that there is no such thing as a perfect, zero-impedance ground. Consider the model of an integrated circuit IC_1 in **Figure 5-10**, the details of which are unimportant for purposes of this discussion. The *IC* draws DC power from the supply (V_{supply}) through wires or ground and power planes. The series inductance and resistance of the interconnection to the supply are shown.

The integrated circuit IC_1 draws a fast switching current with a high *di/dt* (modeled as current source *i(t)*). The hope in bypassing the *IC* is that proper selection and placement of the bypass capacitor C_B will force transient currents to circulate locally near the integrated circuit and hence voltage transients on the supply lines will be limited. Of course, we want the DC component of the IC_1 current to travel back to the power supply. However, if we send fast current pulses back to the power supply, we will induce voltages on the power supply lines to IC_2 and IC_3 due to distributed resistance and inductance.

The key to selection and placement of bypass capacitor C_B is to choose a capacitor that is sufficiently sized to do the job and to place it in close proximity to the power and ground pins of IC_1. This will minimize the inductance of the bypass path and ensure that the high-frequency switching currents circulate locally near the *IC*.

One way to reduce the impedance of current return paths is to use a ground plane, discussed in the following section.

Ground Planes

A ground is a return path for current. It is desirable that this return path should have as low an impedance as possible, to reduce transient-induced voltage drops and electromagnetic emissions. In the world of two-layer PCBs it's difficult to have a dedicated ground plane,

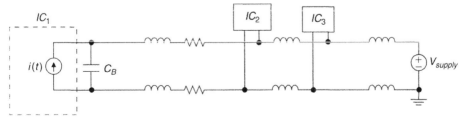

Figure 5-10: Model showing integrated circuit drawing current and its *bypass path*. C_B is the bypass capacitor for IC_1.

since you generally want a couple of PCB layers available for routing signals. In multilayer boards, it's easy to dedicate an unbroken ground plane on an internal PCB layer as ground.

The use of a ground plane helps to reduce the inductance of signal-carrying traces on the PCB. One technique is to have high current and high *di/dt* traces directly above an unbroken ground plane. You can also make the traces wide if you want to reduce the inductance.

PCB Trace Widths

PCB traces must be sized appropriately (both in width and thickness or copper weight[10]) to carry the current that you need without excessive temperature rise (**Table 5-3**). A rule of thumb is that a 10-mil-wide, 1-ounce PCB trace can carry in excess of 500 mA with a

Table 5-3: PC board copper weight vs. thickness.

Copper Weight	Copper Thickness (Inches/Mils)
½ oz.	0.0007" (0.7 mils)
1.0 oz.	0.0014" (1.4 mils)
2.0 oz.	0.0028" (2.8 mils)

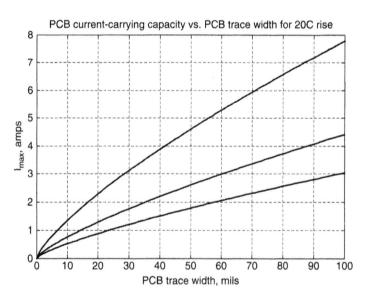

Figure 5-11: Approximate current-carrying capability of 0.5 oz., 1.0 oz., and 2.0 oz. PCB traces with 20°C temperature rise.[11] The upper trace is 2 oz. copper; the lower trace is 0.5 oz. (Note: 1 mil = 0.001")

---

[10] "Copper weight" tells you how thick the PCB trace is. Typical low-power analog boards use ½-ounce or 1-ounce copper. High-power boards may use 2-ounce copper or higher.

[11] From Douglas Brooks, reference at the end of this chapter.

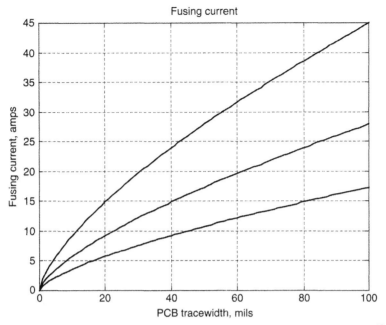

Figure 5-12: Approximate fusing current of 0.5 oz., 1.0 oz., and 2.0 oz. PCB[12] traces.

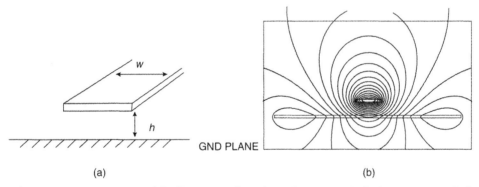

(a)   (b)

**Figure 5-13: 2D FEA model of *w* = 0.01″ PC board trace 0.005″ above a ground plane.
(a) Geometry. (b) 2D FEA model.**

20°C temperature rise above ambient. An estimate of the current-carrying capability for 20°C temperature rise of PCB traces is shown in **Figure 5-11**. The fusing current (**Figure 5-12**) for PC traces is higher.

Approximate Inductance of a PCB Trace Above a Ground Plane

The inductance of a PCB trace above a ground plane can be roughly calculated by assuming a microstrip configuration. For a microstrip line of length l, width w,

---

[12] From Douglas Brooks, "Fusing Current" reference at the end of this chapter.

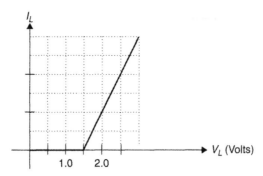

Figure 5-14: Representative laser diode V/I curve showing laser voltage V_L and laser current I_L.

and strip-to-strip spacing $d \ll w$ (**Figure 5-13a**), the inductance is (very roughly[13]):

$$L \approx \mu_o \frac{lh}{w}$$ [5-18]

Using this approximation for a line with $w = 0.01''$ (0.0254 cm) and $h = 0.005''$ (0.0127 cm) we estimate an inductance of 6.3 nanohenries per centimeter of length. A 2D finite element analysis[14] (**Figure 5-13b**) estimates this inductance to be somewhat lower at approximately 3.9 nanohenries per centimeter of length.

Example 5-1: Design case study—high-speed semiconductor laser diode driver

This section considers the design, analysis, and PCB layout of a high-speed switching semiconductor laser diode system, which may be used as a modulated infrared (IR) light source. Direct modulation is a method by which the laser light power output of a semiconductor laser diode is changed by varying the diode current. To use a diode as a high-speed modulated light source, the laser is biased with a small DC current near the *lasing threshold* and a modulation current is superimposed. The light power output of the semiconductor diode is proportional to the laser current in excess of the threshold current. The direct modulation method is used for laser communication, fiber-optic links, industrial applications such as material cutting and in such commercial products as compact-disk players and medical laser printers.

---

[13] Note that this approximation becomes less and less accurate as the trace height h increases above the ground plane, so use this for ballpark estimates only. For more detailed calculations for inductances of all kinds of geometries, see Frederick Grover's excellent reference *Inductance Calculations*, reference given at the end of this chapter. The author gratefully thanks Prof. Dave Perreault from MIT for recommending this book when we were both grad students.

[14] Plots and analysis were done with Finite Element Method Magnetics (FEMM), a finite element package created by Dr. David Meeker at Foster-Miller.

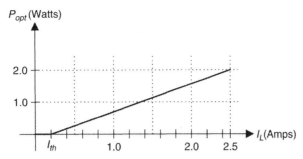

Figure 5-15: Representative high-power laser diode power-current curve. The horizontal axis (I_L) is laser diode current. The vertical axis is optical power output from the laser. Below the lasing threshold I_{th} the diode behaves as an LED.

Electrically, a semiconductor laser behaves like a diode, with a *V*/*I* curve shown in **Figure 5-14**. Since the semiconductor diode is made from gallium arsenide (GaAs) rather than silicon, the voltage "knee" when the diode turns on is approximately 1.5 V.

Under normal operation, the diode is driven by a current source so that the diode current remains constant even if the diode voltage drifts with time and temperature. The optical power output vs. diode current is shown in **Figure 5-15** for a high-power laser diode. For very low currents, the diode does not lase and there is very little optical power. (In fact, for current $<I_{th}$, the laser behaves like an LED and there is some very small amount of optical power emitted.) Once the diode current is increased to a value known as the threshold current (I_{th}), the diode begins lasing, and the optical power output is proportional to the current in excess of the laser threshold current. For a 2W laser diode, the operating laser current is approximately 2.5A, as shown in Figure 5-15. If the laser current is increased further, the laser may be damaged by a process known as *catastrophic optical damage* (COD), where excess heating destroys[15] the laser-emitting area.

Semiconductor laser diodes are inherently fast devices. The intrinsic lasing processes may be modulated at very high rates by variation of the injected current. For representative diodes, the laser power transfer function (optical power output due to current excitation) is flat out to several hundred megahertz or even higher (**Figure 5-16**), depending on the details of the diode construction and the current bias level. The resonance near 10^{10} Hz is due to quantum relaxation processes. Therefore, the high light modulation speed indicated may be achieved in practice if the laser current is changed sufficiently fast. This next leads us to consider how to switch laser current with high current and fast risetimes.

---

[15] This curve is representative of one particular high-power laser diode used by the author; there are other lasers with different power levels and operating currents.

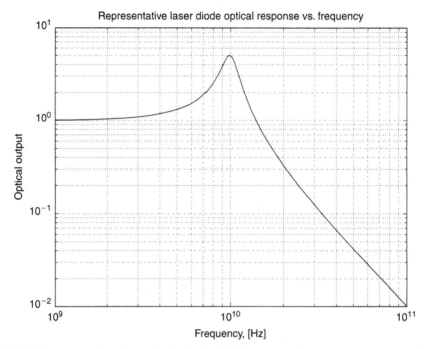

Figure 5-16: Representative laser diode intrinsic light output frequency response with laser resonance at 1 GHz. Horizontal axis—frequency, Hz.

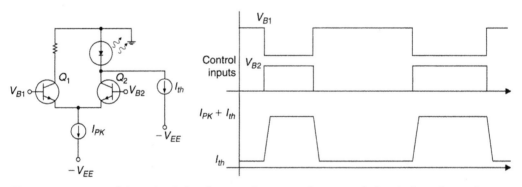

Figure 5-17: Laser driver circuit implemented as an emitter-coupled switch and waveforms. The top traces are the control inputs V_{B1} and V_{B2}. The bottom trace is the laser current.

Driver Implementation

One possible circuit topology suitable for driving a laser diode is shown in **Figure 5-17**. The laser is fed by two DC current sources, I_{BIAS} and I_{th}, corresponding to a laser PEAK current and THRESHOLD currents. When V_{B1} is LOW and V_{B2} is HIGH, Q_1 is OFF and Q_2 is ON, the total current in the laser diode is $I_{PK} + I_{th}$. The resistor in the collector of Q_1 dissipates power so Q_1 will not be damaged.

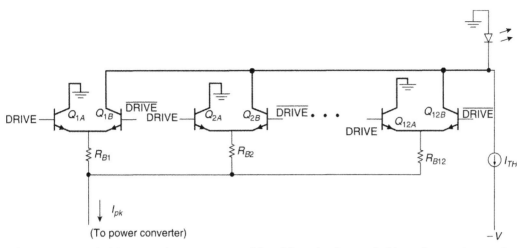

Figure 5-18: Switching transistor array capable of fast risetime switching of up to 2.5A. The array comprises 12 pairs of high-speed switching transistors. The critical high-speed and high-current switching paths are shown in bold.

The author was responsible for the design of a semiconductor diode laser modulator capable of delivering 2.5A pulses to a low impedance load with risetime and falltime of less than 20 ns. The purpose of the circuit board was to drive semiconductor diode lasers for high-speed printing.[16]

There were several design challenges inherent in this design. First, the laser signal is a high-current, fast-risetime set of current pulses with any repetition rate from DC up to 10 MHz, with any duty cycle. This means that extreme care must be taken to ensure low-inductance path from the switching elements on the PCB to the laser. The design of the switch on the PCB presents interesting thermal problems as well.

A simplified schematic of the switching transistor array is shown in **Figure 5-18**. To provide low-inductance paths as well as good thermal management, the fast switch was broken up into a dozen smaller emitter-coupled switches, each pair implemented with a pair of 2N2222 transistors. Note that in emitter coupled pair Q_{1A} and Q_{1B} only one transistor is on at a time; when DRIVE is HIGH and $\overline{\text{DRIVE}}$ is LOW, Q_{1A} is ON and Q_{1B} is OFF (and hence the laser is off, and idling at the threshold current I_{th}). When DRIVE is LOW and $\overline{\text{DRIVE}}$ is HIGH, Q_{1B} is ON and Q_{1A} is OFF and the total laser current is $I_{th} + I_{PK}$.

The critical high-speed and high-current switching paths are highlighted[17] in bold. Each of the transistor arrays switches up to a maximum of more than 200 mA. It's mandatory

---

[16] For more details on the design, see Marc Thompson and Martin Schlecht, "High-power laser diode driver based on power converter technology," *IEEE Transactions on Power Electronics*, vol. 12, no. 1, Jan. 1997, pp. 46–52, and U.S. Patent 5,444,728 (issued 8/22/95).

[17] When doing a PCB layout, the traces in bold are good candidates to be implemented as wide traces over an unbroken ground plane, to reduce parasitic inductance.

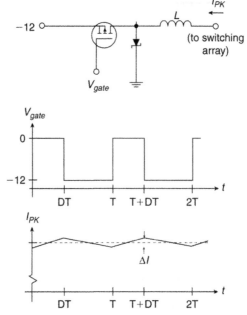

Figure 5-19: Simplified schematic of the DC/DC converter.

to keep the interconnection inductance between transistors and to the laser diode low in order that the transistor arrays can switch as fast as they are capable. Remember from previous chapters that emitter-coupled switches are inherently fast, provided that you provide sufficient base drive capability.[18]

Resistors R_{B1}, R_{B2}, and up to R_{B12} are low-valued *ballast* resistors and ensure that the transistor pairs share the current among them equally.[19]

Another design challenge was that the current rating of the -12 V power supply was only 1A, and we want to deliver 2.5A to the lasers. Therefore, a DC/DC converter was needed to step down the voltage and step up the current. A simplified implementation of this concept is shown in **Figure 5-19**. A detailed discussion of the DC/DC converter is beyond the scope of this book, but this circuit steps *down* the voltage and steps *up* the current. Hence we draw less current from the -12 V power supplies than is delivered to the lasers. We note that this is a high-speed switching circuit; hence we need to take special care in

---

[18] The switching speed of the emitter-coupled pair for signal transistors can be a few nanoseconds. Of course, this assumes that we have a good PC board layout so that parasitic inductances don't slow things down significantly.

[19] The transistors in this design aren't matched, and we want each transistor pair to shoulder an equal value of the load. The emitter ballast resistors, on the order of 1 Ω, force sharing between transistor pairs.

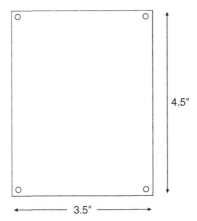

Figure 5-20: Form factor of PC board, 3.5″ × 4.5″, showing mounting holes.

the layout of the MOS FET and diode. The power MOSFET is switched on and off at a high frequency[20] with a variable duty cycle to regulate the current, as shown.

In this design, the form factor of the PCB was mandated to be 3.5″ × 4.5″, since this design was replacing a pre-existing design and the PCB had to be backward-compatible. The form factor and mounting holes are shown in **Figure 5-20**.

The connector locations (**Figure 5-21**) were also set prior to the layout. The connectors are as follows (clockwise from bottom left):

- *Power.* +12 V @ 200 mA; −12 V @ 1A, and 2 ground pins.

- *Laser diode connection.* A microstrip cable was soldered directly to the PCB to provide a low-impedance path to the lasers.

- *Trigger signal.* This is a TTL-level signal that turns the laser diode ON and OFF. When the trigger signal is high, the laser is ON. Repetition rates for the TTL signal is from DC up to 10 MHz.

- *Monitor.* This connector is used to buffer and amplify a photodiode signal used to monitor the optical power output of the laser.

- *Shutdown.* Another TTL-level signal that is used to completely shut down the laser.

Next, the real estate for the various PC board traces was allocated as in **Figure 5-22**. We note that the high-current and high-speed circuitry is segregated from the low-level

---

[20] The switching frequency was 1 MHz. A gate driver circuit (not shown here) supplies the 1 MHz pulsating V_{gate} signal with sufficient current drive to charge and discharge the MOSFET nonlinear gate-source and gate-drain capacitances in tens of nanoseconds. The fast turn-on and -off of the MOSFET ensures that the MOSFET switching losses are low.

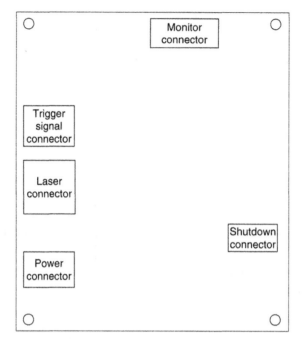

Figure 5-21: PCB connector locations.

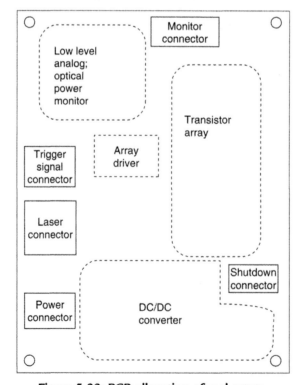

Figure 5-22: PCB allocation of real estate.

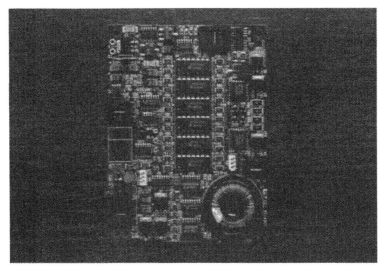

Figure 5-23: PCB showing top side (component side) final layout. Connection to laser diode is not shown.

analog instrumentation circuitry. Furthermore, the PCB was multilayered, ensuring that an unbroken ground plane could be used under the high-speed circuitry. A breakdown of the PCB layers is as follows:

- Top layer: Analog signals

- Internal layer #1: GND

- Internal layer #2: -12V

- Bottom layer: Analog signals, $+12$V

An internal layer was dedicated to -12V since there were significant switching currents drawn from the DC/DC converter.

Figure 5-23 presents a photograph of the resultant PCB. **Figure 5-24** contains an oscilloscope photograph of the laser light output.[21] We note that the laser is switching 2W peak-peak, corresponding to a switched current of 2.5A peak-peak. The risetime and falltime is less than 20 ns.

---

[21] The light output was measured using an extremely fast photodetector. Since the laser is an inherently fast device, the light output shape is representative of the shape of the current pulses to the laser.

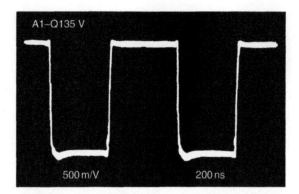

Figure 5-24: Scope photograph showing risetime and falltime of laser optical power. Horizontal: 200 ns per division. Vertical: 2W full scale. The resultant laser light 10–90% risetime and falltime is less than 20 ns.

References

Bartoli, M., Reatti, A. and Kazimierczuk, M., "High-frequency models of ferrite core inductors," *International Conference on Industrial Electronics, Control and Instrumentation (IECON '94)*, 1994, vol. 3, September 5–9, pp. 1670–1675.

Brooks, D., "Fusing currents—When traces melt without a trace," available at www. ultracad.com, printed in *Printed Circuit Design*, vol. 15, no. 12, December 1998, pp. 53.

Cao, Y., Groves, R., Huang, X., Zamdmer, N., Plouchart, J., Wachnik, R., King, T. and Hu, C., "Frequency-independent equivalent-circuit model for on-chip spiral inductors," *IEEE Journal of Solid-State Circuits*, vol. 38, no. 3, March 2003, pp. 419–426.

Demurie, S.N. and DeMey, G., "Parasitic capacitance effects of planar resistors," *IEEE Transactions on Components, Packaging and Manufacturing Technology, Part A*, vol. 12, no. 3, September 1989, pp. 348–351.

Dolan, J.E. and Bolton, H.R. (1991). "Capacitor ESR measurement technique," *Eighth IEEE Pulsed Power Conference 1991*, June 16–19, pp. 228–231.

Franco, S. (1996). "Polypropylene capacitors for snubber applications," *Proceedings of the Thirty-First IAS Annual Meeting (IAS '96)*, 6–10, October, pp. 1337–1342.

Galbraith, J., "Reliable precision wirewound resistor design," *IRE Transactions on Components Parts*, vol. 3, no. 3, December 1956, pp. 116–119.

Grover, F.W., *Inductance Calculations: Working Formulas and Tables*, New York: Dover Publications, Inc., 1946.

Jutty, M.K., Swaminathan, V. and Kazimierczuk, M.K., "Frequency characteristics of ferrite core inductors," (1993). *Proceedings of the Electrical Electronics Insulation Conference and Electrical Manufacturing & Coil Winding Conference,* 1993, October 4–7, pp. 369–372.

Madou, A. and Martens, L., "Electrical behavior of decoupling capacitors embedded in multilayered PCBs," *IEEE Transactions on Electromagnetic Compatibility,* vol. 43, no. 4, November 2001, pp. 549–566.

Manka, W., "Alternative Methods for Determining Chip Inductor Parameters," *IEEE Transactions on Parts, Hybrids, and Packaging,* vol. 13, no. 4, December 1977, pp. 378–385.

Massarini, A. and Kazimierczuk, M.K., "Self-capacitance of inductors," *IEEE Transactions on Power Electronics,* vol. 12, no. 4, July 1997, pp. 671–676.

Naishadharn, K., "Experimental equivalent-circuit modeling of SMD inductors for printed circuit applications," *IEEE Transactions on Electromagnetic Compatibility,* vol. 43, no. 4, November 2001, pp. 557–565.

Neugebauer, T.C., Phinney, J.W. and Perreault, D.J., "Filters and components with inductance cancellation," *IEEE Transactions on Industry Applications,* vol. 40, no. 2, March–April 2004, pp. 483–491.

Reed, E.K. (1994). "Tantalum chip capacitor reliability in high surge and ripple current applications," *1994 Electronic Components and Technology Conference,* 1994, May 1–4, pp. 861–868.

Sakabe, Y., Hayashi, M., Ozaki, T. and Canner, J.P., "High frequency measurement of multilayer ceramic capacitors," *IEEE Transactions on Components, Packaging and Manufacturing Technology, Part B: Advanced Packaging,* vol. 19, no. 1, February 1996.

Sarjeant, W.J., Zirnheld, J. and MacDougall, F.W., "Capacitors," *IEEE Transactions on Plasma Science,* vol. 26, no. 5, October 1998, pp. 1368–1392.

Smith, L.D. and Hockanson, D. (2001). "Distributed SPICE circuit model for ceramic capacitors," *Proceedings of the 2001 Electronic Components and Technology Conference,* May 29–June 1, 2001, pp. 523–528.

Stroud, J. (1990). "Equivalent series resistance-the fourth parameter for tantalum capacitors," *Proceedings of the 1990 Electronic Components and Technology Conference,* May 20–23, 1990, pp. 1009–1012.

Thompson, M. and Schlecht, M. "High-power laser diode driver based on power converter technology," *IEEE Transactions on Power Electronics,* vol. 12, no. 1, January 1997, pp. 46–52.

Ulrich, R.K., Brown, W.D., Ang, S.S., Barlow, F.D., Elshabini, A., Lenihan, T.G., Naseem, H.A., Nelms, D.M., Parkerson, J., Schaper, L.W. and Morcan, G., "Getting aggressive with passive devices," *IEEE Circuits and Devices Magazine*, vol. 16, no. 5, September 2000, pp. 16–25.

Venkataramanan, G. (1998). "Characterization of capacitors for power circuit decoupling applications," *Industry Applications Conference, 1998*, vol. 2, October 12–15, pp. 1142–1148.

Wadell, B.C., "Modeling circuit parasitics 1," *IEEE Instrumentation & Measurement Magazine*, vol. 1, no. 1, March 1998, pp. 31–33.

———, "Modeling circuit parasitics 2," *IEEE Instrumentation & Measurement Magazine*, vol. 1, no. 2, June 1998, pp. 6–8.

———, "Modeling circuit parasitics 3," *IEEE Instrumentation & Measurement Magazine*, vol. 1, no. 3, September 1998, pp. 28–31.

———, "Modeling circuit parasitics 4," *IEEE Instrumentation & Measurement Magazine*, vol. 1, no. 4, December 1998, pp. 36–38.

Yu, Q. and Holmes, T.W., "A study on stray capacitance modeling of inductors by using the finite element method," *IEEE Transactions on Electromagnetic Compatibility*, vol. 43, no. 1, February 2001, pp. 88–93.

CHAPTER 6

Analog Lowpass Filters

Steve Winder

Everybody needs a lowpass filter occasionally. You need to get some nasty high-frequency signal attenuated, and if you are going to feed it to an ADC, you might need an antialiasing filter. Depending on whether your troublesome frequency is near your primary signal or farther out, you might be able to get away with one or two capacitors—or you might need a well-tuned filter with two or four or more stages. Which filter is best for you? Steve Winder has good advice. Fortunately, it only takes a few op-amps and a few passive components to confirm how much filtering you need. Furthermore, there are free computer-based programs to help you choose and design your filter. /rap

In This Chapter

This chapter describes how to design active or passive lowpass filters to almost any desired specification. Formulae and examples of how to use them are given for the denormalization of component values.

A Quick Introduction to Analog Filters

You will sometimes need a *filter* that will attenuate some unwanted signals while it passes your desired signal straight through. Filters can be made of inductors and capacitors ("passive" filters made of L and C) or resistors and capacitors, usually with an "active" amplifier such as a transistor or op-amp. Filtering can also be done with digital filters (digital signal processing, or DSP) or with switched capacitors, which we will not discuss, since those are specialties.

There are *lowpass filters* to reject high-frequency noise and *highpass filters* to reject DC and low frequencies (such as 60 Hz and power-line harmonics). These are useful to prevent "aliasing" when the signal is going to be sampled by an analog-to-digital converter. Similarly, *bandpass filters* reject low and high frequencies to pass only frequencies in a middle region, as in telephony, when audio signals are being sent to an ADC. Conversely, *bandstop* or *notch filters* reject frequencies in a narrow range. In each case, there is a passband, where wanted signals get only minimum attenuation, and a stopband, where unwanted signals get attenuated a lot. In between is a range (the *skirt*) where the gain rolls off rapidly. It is not a surprise that to get a fast rolloff, 12 dB per octave or faster, and a large attenuation in the stop-band could require a filter of more complexity.

There are several types of filters. Bessel and Butterworth filters have a smooth decay in the lower stopband. Chebyshev (or Tchebyshev) filters have amplitude ripple in the passband, but a smooth, well-behaved rolloff going into the stopband. Inverse Chebyshev filters have a smooth passband but ripples in the stopband. Cauer filters have ripples in both the passband and the stopband.

Some of these are optimized for the frequency domain, with the flattest passband followed by a steep rolloff. However, most have ringing and slow settling in the time domain. The various kinds of Chebyshev filters have steeper skirts (better performance in the frequency domain) but poor response in the time domain. Also, the signal's delay through the filter could have various amounts of delay as a function of frequency. The Bessel filter is best behaved for phase delay and has the smoothest response for steps. It is the most tolerant of component tolerances. Other filters with sharper rolloff usually have poorer group delay.

The order of a filter may have to change as your requirements change. If you can afford three capacitors and two inductors, that will be a fifth-order filter. If you need better performance with steeper skirts or flatter passbands, you would have to add more *L*s and *C*s. Similarly, you can get a good fifth-order filter from two op-amps, five capacitors, and five resistors, and if you want better performance, be prepared to spend more parts—and more power. In this book, we will concentrate primarily on Butterworth and Chebyshev filters made with active devices. A complete discussion of all the other types goes beyond the scope of this book. See the last publication in the References section, which is recommended for a full consideration of all kinds of filters.

In the past, a good passive filter even as fast as 5 kHz (or faster) might require *L*s and *C*s. Now that fast amplifiers are cheap and abundant, active filters can extend up into the dozens and hundreds (and thousands) of kHz. There is a moderate range where you can make a filter either way, active or passive, so you have some choice. Just remember that passive filters usually take more space than active ones, and active filters just about always take more power than passive filters. That's a trade-off you have to plan for.

Passive Filters

Passive filters are the simplest to design from the normalized model. The model itself is a lowpass design, although normalized for a passband that extends from DC to 1 rad/s and is terminated with a 1 Ω load resistance. Denormalization for a higher load impedance requires component values to be scaled to have a higher impedance. The impedance of an inductor is proportional to its inductance, but the impedance of a capacitor is inversely proportional to its capacitance. Thus, if the load resistance is a more practical 50 Ω, inductance values are increased fifty-fold and capacitance values are reduced fifty-fold (to increase their impedance).

As an example, let's see how the component values change with a fifth-order Butterworth filter. In **Figure 6-1** is the normalized lowpass model.

Normalization and Denormalization

Because it is customary for filter designers to show a filter normalized to 1 radian/s, or 0.159 Hz, they show a filter with 1 ohm resistors, capacitors scaled in Farads, and inductors in Henries. Most people have never seen precision capacitors in Farads nor inductors in Henries. You don't have to be nervous about this concept! Refer to Figure 6-1.

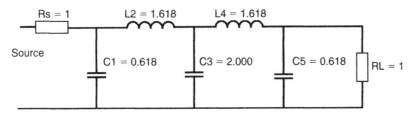

Figure 6-1: Fifth-order Butterworth normalized model.

The first thing you have to do is get reasonable impedance values. Scale the impedances by a factor of m, 50 or 600 or 1000 or 100,000, as appropriate. This should get you into the ballpark of reasonable impedance values. Resistors and inductors (if any) increase by m, and capacitors decrease by the factor m. See **Figure 6-2**.

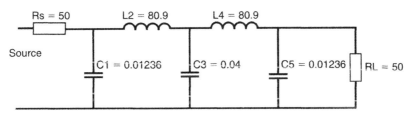

Figure 6-2: Fifth-order Butterworth—Impedance scaled to 50 ohms.

The next thing to do to design a practical filter is to take this filter of the right shape and define the frequency f_{-3DB} that you want. Then take the normalized circuit and divide all the L and C values by $(2\pi f)$. This is pretty obvious. See **Figure 6-3**. Now the capacitors and inductors should not be bigger than your head.

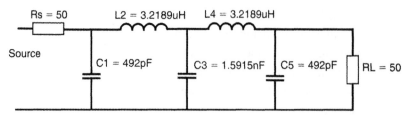

Figure 6-3: Fifth-order low pass filter—Frequency scaled to 4 MHz.

Note that at any time you can scale the impedance by any other factor such as 2.2 or 1/2.2 or *n*. If you have a large stock of capacitors of a certain value, you can use them in a good filter. Most people have a wide selection of resistors, so it is easy to select or buy some resistors to go with the capacitors you have. For example, often you can see a proposed denormalized design with 1 kilohm resistors, "$C1 = 1.414\,\mu F$ and $C2 = 0.707\,\mu F$". This looks absurd! However, it will be easy to convert this to:

$$R = 71.5\,k,\ C1 = 0.02\ \mu F,\ C2 = 0.01\ \mu F\ \dots \qquad\qquad [6\text{-}1]$$

Have you got the idea? And for the first breadboard, you can even make up the $0.02\,\mu F$ from two $0.01\,\mu Fs$.

Poles and Zeros

A proper analysis of analog filters requires a thorough study of the poles and zeros in the S-plane. A complete knowledge of the desired filter response of various types of filters requires this information. Here we give only the lightest possible treatment of poles and zeros. These days, it's easy to choose a very good set of filters without literally studying the poles and zeros by using this book or any of a number of fairly user-friendly computer programs, as found in Chapter 15.

Suffice it to say, the number and location of the poles in the S-plane is important to getting a good rolloff. But you can get good filter results *if* you rely on an expert who did understand the poles and zeros. This book can help provide that expertise.
A complete analysis of the relationship of poles and zeros to the frequency response or the time response is beyond the scope of this chapter. I recommend you see Chapter 3 on poles and zeros in the last publication in the References section. In addition, some free computer programs can help you design filters of almost any desired performance.

Active Lowpass Filters

Active filters are designed using pole and zero locations, which are determined from the frequency response's transfer function. This is not possible in passive filter designs

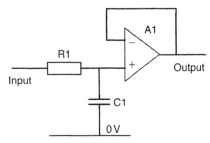

Figure 6-4: First-order active filter.

because all the components interact with each other. However, in active filters the op-amp, the "active" part of the circuit, buffers one stage from the next, so there is no interaction. Each stage can therefore be designed to provide the frequency response of one pair of complex poles or a single real pole, or sometimes both. When all the stages are connected in series, the desired overall response is produced.

Now that I have set the scene, I will describe some active filter designs and see how the pole and zero locations are used to find component values.

First-Order Filter Section

The first-order section is a simple structure comprising a lowpass RC network followed by a buffer, as shown in **Figure 6-4**. The buffer serves to provide a high input impedance so that the voltage at the connection node of the RC network is transferred to the buffer's output without being loaded by following stages. A simple RC network on its own would be loaded by following stages and therefore would not have the expected frequency response.

The first-order section is an all-pole network because it cannot produce zeros in its frequency response. In fact, the first-order section has one real pole at $-\sigma$.

Letting $R1$ equal $1\,\Omega$ in the normalized lowpass model, calculation of $C1$ is simple:

$$C1 = \frac{1}{\sigma}$$
[6-2]

where σ is the pole position on the negative real axis of the S-plane.

Sallen-Key Lowpass Filters

The Sallen-Key filter provides a second-order all-pole response and is a simple active lowpass design. It can be used for Bessel, Butterworth, or Chebyshev responses. High-order filters can be produced by cascading second-order sections. Odd-order filters can be produced using a series of second-order sections and then adding a first-order section at the end.

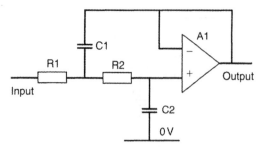

Figure 6-5: Sallen-Key lowpass filter (second order).

The Sallen-Key filter uses an amplifier (which may be connected as a unity gain buffer) with a network of resistors and capacitors at the input. Capacitive feedback from the output is also used, and this can give rise to peaking in the frequency response. Peaking is required in second-order circuits where the Q is greater than unity and occurs due to phase shifts around the feedback loop. If the Q is large, say $Q = 10$, for example, the amplifier is providing a gain of 10 that restricts its bandwidth to 0.1 of the gain-bandwidth product. The diagram in **Figure 6-5** shows the circuit.

By letting $R1$ and $R2$ equal $1\,\Omega$ in the normalized design, the values of $C1$ and $C2$ can easily be calculated.

$$ C1 = \frac{2Q}{\omega_n} = \frac{1}{\sigma} \quad \text{and} \quad C2 = \frac{1}{2\omega_n Q} = \frac{\sigma}{\sigma^2 + \omega^2} \qquad [6\text{-}3] $$

In the case of Butterworth filters, $\omega_n = 1$ and $C2 = \sigma$, that is, the reciprocal of $C1$.

Sallen-Key Rolloff Deficiencies

These Sallen-Key low-pass filters, as shown in Figure 6-5, often show a weakness because their 12-dB per octave rolloff does not continue above 50, 100, or 200 kHz. This is usually related to the rising output impedance of the op-amp at high frequencies. The path through $R1$ and $C1$ forces current through $C1$ into the (finite) output impedance of the op-amp. There are several ways to minimize this problem. One way is to raise the impedance level of all the Rs and Cs by a good factor, such as 10 or 30, so that the value of $R1$ (in Figure 6-5) rises up near 100 k. The problem is thus greatly decreased. If you *just* want to have a perfect response, you might provide a buffer on the output of the amplifier.

Alternatively, if you go into the resistor $R1$ in Figure 6-5, you can break it into two series resistances, $R1a$ and $R1b$, such as $R1a = 10k$, $R1b = 91k$. Add a capacitance Co from the juncture of $R1a$ and $R1b$ to ground, with a time-constant comparable to the rollup of the indicated response, similar to the $C5$ in **Figures 6-6(b) or 6-6(c)**. This can make big (though crude) improvements in the filter response. Even a simple 0.01 µF cap can make a big improvement.

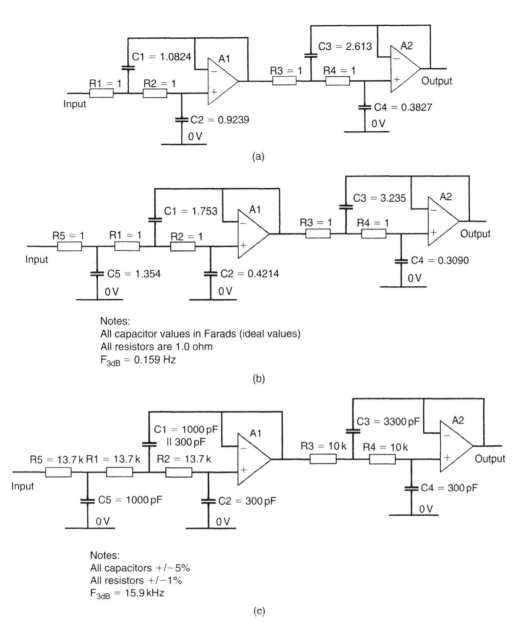

Figure 6-6: (a) Fourth-order filter (normalized). (b) Fifth-order filter (normalized). (c) Fifth-order filter (denormalized).

Components and Tolerances

If you plan to build a simple Sallen-Key filter with 10% capacitors, you might build one and find that its shape is just right—but its *f-3dB* frequency could be 9 or 11% high—or similarly, too low. Or, if the capacitors were mismatched in a different way, the *f-3dB* might be just right but the peaking might be too high or too low. Is that okay with you?

You can buy capacitors with tolerances at 20%, 10%, 5%, 3%, 2%, or 1%. Digikey will sell you any of those for reasonable prices. Of course, you can buy special-order capacitors of any values or of any desired tolerances, down to $+/- 0.01\%$, as long as money is no object. Or you could take some $1\,\mu F +/- 5\%$ capacitors and trim them all up to $1.05\,\mu F$ by adding external trim capacitors. In some cases, you can just measure all those $1\,\mu F$ capacitors, and grade them into 1% bins. Take them out of the bins in pairs or quads, and you get some nice matched pairs. What is your preference?

Temperature Coefficient, or Tempco

If you are planning to build filters that work well over a wide temperature range, you have to put in some *serious* engineering and some *serious* shopping, as well as some *serious* testing, to be sure you will get good results. Computer simulation is not necessarily good enough.

Resistors often have a $+/- 100$ ppm/°C tempco but are available with 50, 25, 10, or better if you don't mind the price and availability. Also, tight-tolerance Rs are available as wirewounds or precision films.

Capacitors such as polyester have about $+300$ ppm/°C. Polystyrene and polypropylene have about -120 to -190 ppm/°C. C0G and NP0 ceramic capacitors have $0 +/-30$ ppm/°C. Silver mica has $+50 +/-50$ ppm/°C.

Inductor datasheets show tempcos of . . . well, makers don't talk about that. You will have to ask. Tempcos are not always linear at extreme temperatures.

Op-amps often have a gain-bandwidth (GBW) product with a tempco of $+/-1000$ ppm/°C, but plus *or* minus 3000 or more may happen on some types. Does the data sheet tell you what that tempco is? Typical? Worst case? Good luck!

In general, FET-input amplifiers have advantages over most bipolar-input op-amps in that their low Ib lets you use higher resistor values without hurting the dc accuracy or noise. And at high frequencies, JFET or MOSFET amplifiers generally have a faster slew rate for better dynamic accuracy.

This is good advice primarily on components for the design of active filters but is also of interest for people designing any kind of analog circuits in general.

Butterworth Filters

To repeat: In the case of Butterworth filters, $\omega_n = 1$ and $C2 = \sigma$, that is, the reciprocal of $C1$.

For example, the first pair of poles of a Butterworth fourth-order filter are $0.9239 \pm j0.3827$. A Sallen-Key filter section that has the same pole locations has $C1 = 1.0824$ and $C2 = 0.9239$.

The second filter section capacitors will number in sequence, being $C3$ and $C4$ and calculated from the same formula by substituting for $C1$ and $C2$, respectively. With poles at $0.3827 \pm j0.9239$, this filter section has capacitor values of $C3 = 2.613$ and $C4 = 0.3827$. The diagram in Figure 6-6(a) illustrates the whole circuit.

Some books recommend that a third amplifier be used to provide a fifth pole, but using the values in **Table 6-1**, it is easy to make a five-pole filter using just two op-amps. The values for Figure 6-6(b) have been specified for 0.159 Hz and large Farad capacitors, and the value for $C5$ is taken from column $C3$ in the table. These values are then de-normalized for a 15.9 kHz filter in Figure 6-6(c). This active filter can be compared to the fifth-order passive filter in Figure 6-3.

Table 6-1: Butterworth active lowpass capacitor values.

Order (n)	C_1	C_2	C_3
2	1.414	0.7071	
3	3.546	0.2024	1.392
4	1.082	0.9241	
	2.613	0.3825	
5	1.753	0.4214	1.354
	3.235	0.3090	
6	1.035	0.9660	
	1.414	0.7071	
	3.863	0.2588	
7	1.531	0.4885	1.336
	1.604	0.6235	
	4.493	0.2225	
8	1.020	0.9809	
	1.202	0.8313	
	1.800	0.5557	
	5.125	0.1950	
9	1.455	0.5170	1.327
	1.305	0.7661	
	2.000	0.5000	
	5.758	0.1736	
10	1.012	0.9874	
	1.122	0.8908	
	1.414	0.7071	
	2.202	0.4540	
	6.390	0.1563	

(Refer to the circuits of Figures 6-5 and 6-6, normalized to 1 ohm at $f_{3dB} = 0.159$ Hz (capacitors in Farads). *Reprinted from *Electronics*, McGraw-Hill, Inc. August 18, 1969.)

Appendix B contains tables of component values for other filters such as Bessel, 1 dB Chebyshev, 0.1 dB Chebyshev, and 0.01 dB Chebyshev.

The Sallen-Key lowpass filter is good if the requirements are not too demanding, with section Q factors below 50. In particular the gain-bandwidth product of the op-amps can limit the filter's cutoff frequency. I previously described this phenomenon in a magazine article (in the second reference) in which I showed that the cutoff frequency limit was given by the empirical expressions:

$$\text{Butterworth passband frequency limit} = \frac{\text{Gain-Bandwidth Product}}{(\text{filter order})^2}$$
$$\text{Chebyshev (1dB) passband frequency limit} = \frac{\text{Gain-Bandwidth Product}}{(\text{filter order})^{3.2}}$$

[6-4]

As an example of how these formulae are used, consider a fifth-order filter using amplifiers with a 1 MHz gain-bandwidth product. If the filter is to have a Butterworth response, its maximum passband frequency is 1 MHz/25 = 40 kHz. If, instead, a 1 dB Chebyshev response is wanted, the maximum passband frequency is limited to 1 MHz/172.5 = 5.8 kHz.

These frequency limits are for a maximum error in the passband of 2 dB. If no error is acceptable, the frequency limit will be much lower. Although the frequency limit can be raised by using an amplifier having a greater gain-bandwidth product, it can lead to instability. Usually, amplifiers with a high gain-bandwidth product have a minimum gain for stability. For example, the OP37 amplifier has a gain-bandwidth product of 63 MHz but at a minimum gain of five. However, modern op-amps are capable of going much faster than that, and many are stable at a gain of 1.

Denormalizing Sallen-Key Filter Designs

In active filter designs, the resistor values used should all be in the range 1 kΩ to 100 kΩ where possible. If resistor values are lower than 1 kΩ, there may be a problem with loading of op-amp stage outputs. Loading can cause distortion and increases the supply current. If resistor values are much higher than 100 kΩ, there may be problems with noise pickup. High-impedance circuits can capacitively couple with external electric fields. These unwanted signals can then interfere with the wanted signal, unless care is taken with good layout, shielding, or guarding. Impedances higher than a megohm are often used in an effort to keep down the capacitor size. Thermal noise voltage generated by the circuit's resistors increases in proportion to the square root of their resistance, but this is often tolerable if the bandwidth is finite. Scaling of impedances continues as mentioned previously, to provide reasonable values of R and C at the frequencies of interest.

State Variable Lowpass Filters

This circuit design has a lower sensitivity to the op-amp's gain-bandwidth product limitation, and section Q factors of up to 200 are possible. It does, however, need three op-amps, as shown in **Figure 6-7**.

Note that the output is in phase with the input (subject to phase shifts due to the filter's response). The output could have been taken from A3 but would have been inverted.

The equations for this filter allow the arbitrary choice of capacitor, C.

$$R1 = \frac{1}{2\sigma C}$$

$$R2 = R3 = R4 = \frac{1}{C\sqrt{\sigma^2 + \omega^2}} = \frac{1}{\omega_n C} \qquad [6\text{-}5]$$

A circuit gain of greater than unity can be achieved if the value of $R4$ is reduced. Dividing the value of $R4$ given in the last equation by a factor K gives the circuit a gain. The gain is equal to K.

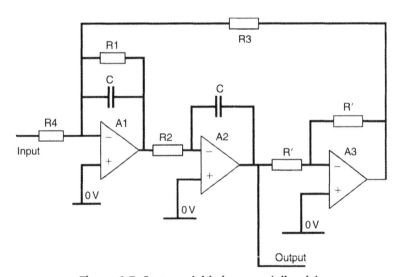

Figure 6-7: State variable lowpass (all-pole).

Cauer and Inverse Chebyshev Active Filters

To design a Cauer or Inverse Chebyshev filter, a different circuit topology is required. The Cauer response has zeros outside the passband, so a notch circuit is required. This can be achieved using a circuit that is an extension of the state variable filter and is known as a *biquad*. This circuit is illustrated in **Figure 6-8**.

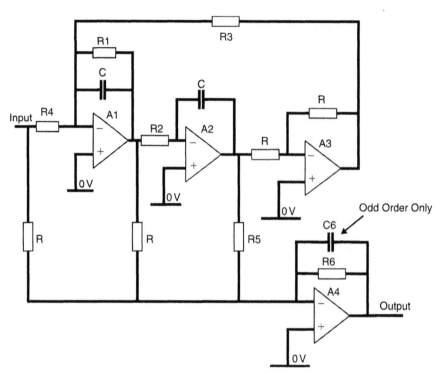

Figure 6-8: The biquad filter.

The following equations give component values for the active biquad filter. As in the case of the state variable, the value of C can be chosen as any suitable value, then resistor values calculated from the equations. First compute the section's frequency from the pole location:

$$\omega_n = \sqrt{\sigma^2 + \omega^2}$$

$$R1 = R4 = \frac{1}{2\sigma C}$$

$$R2 = R3 = \frac{1}{\omega_n C}$$

$$R5 = \frac{2\sigma\omega_n R}{\omega_Z^2 - \omega_n^2}$$

[6-6]

ω_Z = the normalized zero frequency.

$$R6 = \left(\frac{\omega_n}{\omega_Z}\right)^2 \cdot AR$$

The gain at DC and low frequencies is represented by A in the equation. The resistors labeled R can be any arbitrary value; a typical value may be in the range $5\,k\Omega$ to $100\,k\Omega$, say $10\,k\Omega$. Odd-order filter sections can be implemented by adding a capacitor across $R6$. The value of this capacitor is given by the equation below:

$$C6 = \frac{1}{\sigma R6} \qquad\qquad [6\text{-}7]$$

where σ is the value of the pole on the S-plane negative real axis.

Denormalizing State Variable or Biquad Designs

I have shown that the normalized component values used in passive filters and in Sallen-Key active filters can be scaled for different frequencies. However, the simplest approach with state variable and biquad filters is to start by frequency scaling the poles (and zeros in the biquad case). Scaling pole and zero locations is easy: Simply multiply them by the frequency scaling factor, $2\pi F_c$. The frequency scaled pole and zero locations can then be used in the design equations for state variable and biquad filters.

Frequency scaling pole and zero locations can be visualized by considering the S-plane diagram. Frequency scaling moves the poles outward on a line that extends from the S-plane origin. To picture this, think of a pole at, say, $s = -0.75 + j1.2$ in a normalized response. If this is scaled for a frequency of $10\,Hz$, the scaling factor is $2\pi F_c = 62.83\,rad/s$, and the pole moves to $-47.12 + j75.396$. This is shown in the diagram of **Figure 6-9** (not to scale).

Each pole has a certain natural frequency (ω_n) and a certain magnifying factor (Q). The Q depends on the angle of the line from the S-plane origin to the pole location. As the

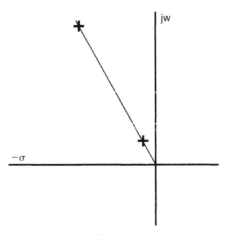

Figure 6-9: Frequency scaling of pole location in S-plane.

pole-zero diagram is scaled for a higher cutoff frequency, the pole moves along the line from the S-plane origin to the pole location. This means that the value of Q remains unchanged as the pole location is scaled for frequency. The natural frequency ω_n is dependent on the σ coordinate (real part), and this changes in proportion to the scaling of the diagram.

Zeros are located on the imaginary axis, so scaling is simple. They are moved along this axis in proportion to the scaling frequency.

Choose a capacitor value and then use the equations given here to find the resistor values. If the resistor values are very small or very large, select a new capacitor value and try again. Again, aim to keep the resistor values between $1\,k\Omega$ and $100\,k\Omega$. Here is an example for a biquad filter.

For example, design a second-order biquad filter based on an Inverse Chebyshev design. The filter should have a passband of $1\,kHz$ and a $30\,dB$ stop-band attenuation. For a $3\,dB$ passband attenuation at $1\,rad/s$, the zero is at 5.71025 and the poles are at $0.70658 \pm j0.72929$.

To scale these for a $1\,kHz$ passband, multiply the pole and zero locations by the frequency scaling factor $2\pi F_c = 6283\,rad/s$. Hence $F_z = 35{,}877.5\,rad/s$. The scaled poles are located at $4439.44 \pm j4582.13$ ($\sigma = 4439.44$ and $\omega = 4582.13$). The natural frequency of this pair of poles is given by:

$$\omega_n = \sqrt{\sigma^2 + \omega^2} = 6380 \text{ rad/s.} \tag{6-8}$$

Component values can now be found by choosing an arbitrary value capacitor, C. Let $C = 100pF$.

$$R1 = R4 = \frac{1}{2\sigma C} = \frac{1}{2 \cdot 4439.44 \cdot 10^{-7}} = 1.126\,k\Omega$$

$$R2 = R3 = \frac{1}{\omega_n C} = \frac{1}{6380 \cdot 10^{-7}} = 1.567\,k\Omega \tag{6-9}$$

$$R5 = \frac{2\sigma\omega_n R}{\omega_z^2 - \omega_n^2} = \frac{56647254 \cdot R}{1.2465 \cdot 10^9} = 0.0454454\,R$$

Letting $R = 10\,k\Omega$ gives $R5 = 454\,\Omega$. This is too low, so let $R = 33\,k\Omega$. Now $R5 = 1500\,\Omega$.

ω_z = the denormalized zero frequency of $35{,}877.5\,rad/s$. Let gain $A = 1$.

$$R6 = \left(\frac{\omega_n}{\omega_z}\right)^2 \cdot AR = \left(\frac{6380}{35{,}877.5}\right)^2 \cdot 33k\Omega. \text{ Hence } R6 = 1\,k\Omega. \tag{6-10}$$

Frequency-Dependent Negative Resistance Filters

Frequency-dependent negative resistance (FDNR) circuits can be used to make an active filter based on a passive ladder filter design. In applications where an elliptical lowpass filter is required and an active filter is possible, FDNR filters can be used as an alternative to a biquad filter. For example, a third-order elliptic lowpass filter requires a biquad design with four op-amps, ten resistors, and three capacitors. The same design using an FDNR requires two op-amps, eight resistors, and four capacitors. An obvious advantage is the reduction of op-amps from four down to two. Halving the number of op-amps required for the filter halves the supply current, assuming that the same type of op-amp would be required in both circuits.

However, there is a catch. For the circuit to work as specified, the source impedance should be zero. This can be compensated for by simply reducing the value of a series resistor in the design (more on this later). The greater problem is the output load. The load must be high impedance for the circuit to work properly. Of course, in multistage filters such as a seventh-order elliptic filter, a biquad design would require three biquad stages connected in series (twelve op-amps). A similar FDNR filter would require six op-amps, seven including a buffer at the output.

The most significant advantage of doubly-terminated lossless ladder circuits is the low sensitivity to component tolerances. However, inductors are bulky and are difficult to obtain. Low-value inductors for radio applications are reasonably easy to find, but audio frequency applications require much larger values. High-value inductors often have to be specially wound to obtain the required inductance.

Replacing the inductors and capacitors by resistors and FDNRs gives the same low sensitivity to component tolerances. If there are two signal paths in a system that must be closely matched in terms of amplitude and phase, an FDNR filter is the better choice. For all these reasons, there is some advantage in using the FDNR for "all-pole" designs, such as Butterworth or Chebyshev. So now I have convinced you, I hope, that in some applications, FDNR filters are a "good thing." But what are FDNRs?

The schematic symbol for an FDNR looks like a capacitor with four plates instead of the usual two and is assigned a letter D. The FDNR is also known as a *D-element*. An FDNR is an active circuit that behaves like an unusual capacitor. In a lowpass RC circuit, the voltage drop across the shunt capacitor falls with increasing frequency. Beyond the passband, doubling the frequency halves the voltage across the capacitor. In a lowpass RD circuit, in which the FDNR has replaced the capacitor, the voltage drop across the FDNR falls at double the rate. Thus, above the passband, doubling the frequency quarters the output signal amplitude.

In decibel terms, a signal applied to an RC network has a rate of fall of 6 dB/octave (a first-order filter). The same signal applied to an RD network has a rate of fall of 12 dB/octave of a capacitor. This double rate of fall is the reason for the four plates in the D-element symbol, rather than the two in a capacitor symbol. The circuit of an FDNR is given in **Figure 6-10**.

In a simple approach where all resistors are equal to 1 Ω and all capacitors are equal to 1 F, the circuit behaves like a negative resistance of −1 Ω. The equation for the negative resistance is:

$$D = \frac{R2 \cdot R4 \cdot C1 \cdot C3}{R5} \qquad\qquad [6\text{-}11]$$

If $C = C3 = 1F$ and $R4 = R5$, the negative resistance equals $R2$.

Now I have shown what an FDNR looks like. How do you use it? Transformation of the passive components is needed. FDNR elements are used to replace the capacitors in passive lowpass filters. Resistors are used to replace the inductors. This allows the filter size to be reduced, and a miniature hybrid circuit is possible. The design begins with a conventional double-terminated lowpass LC filter design, in the T configuration. This

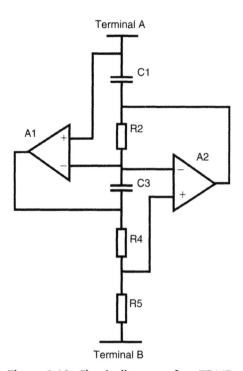

Figure 6-10: Circuit diagram of an FDNR.

has resistors (for the source and load), series and shunt inductors, and shunt capacitors. **Figure 6-11** shows a normalized elliptic lowpass LC filter.

To convert the passive design into an FDNR design, the resistors are replaced by capacitors, the inductors are replaced by resistors, and the capacitors are replaced by FDNRs. If the source and load resistor are $1\,\Omega$, these are replaced by capacitors of $1\,$F. Generally, the capacitor value is $1/R$, so if the load was $0.2\,\Omega$, the capacitor would be $5\,$F.

Inductors are replaced by resistors. A $1\,$H inductor becomes a $1\,\Omega$ resistor. Generally, $R = L$, so a $1.1395\,$H inductor would be replaced by a $1.1395\,\Omega$ resistor.

Capacitors are replaced by FDNRs. In an FDNR, the resistors are normalized to $1\,\Omega$ and the capacitors are normalized to $1\,$F to replace a $1\,$F capacitor. If the normalized capacitor is not $1\,$F, the value of $R2$ (in Figure 6-10) is scaled in proportion. Generally, $R1 = C$. Thus a $1.0844\,$F capacitor is replaced by an FDNR that has $R2 = 1.0844\,\Omega$.

The conversion process is displayed in **Figure 6-12**.

Applying these simple rules to the normalized lowpass design given in Figure 6-11 gives the FDNR equivalent design, illustrated in **Figures 6-13 and 6-14**.

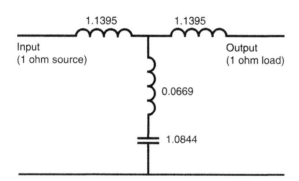

Figure 6-11: Circuit of normalized lowpass LC filter.

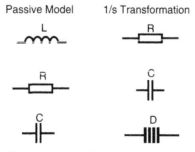

Figure 6-12: 1/S transformation.

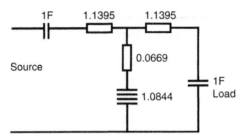

Figure 6-13: Lowpass filter with D-element.

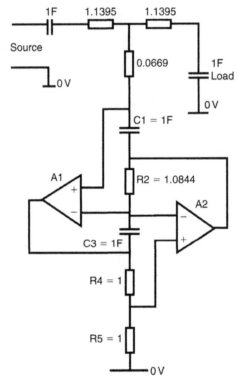

Figure 6-14: Normalized lowpass FDNR filter.

Denormalization of FDNR Filters

Now let's apply frequency scaling to obtain practical component values. I will now design a third-order filter that has a passband of 15 kHz. The normalized design has a passband of 1 rad/s, so the frequency scaling factor is $2\pi F$. The frequency scaling factor is 94,247.78 in this case. All capacitor values must now be divided by 94,247.78, which makes each one equal to 10.6103 μF. This value is a little too large and must be reduced to a more convenient value. Let's divide the capacitor value by 1061.03 so that

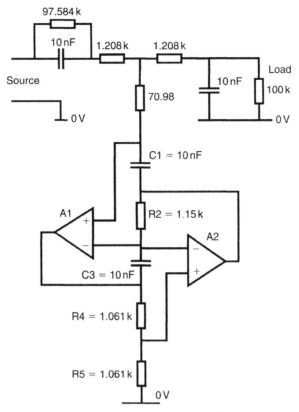

Figure 6-15: FDNR lowpass filter.

all capacitors in the circuit are now 10 nF. Each resistor must now be multiplied by this scaling factor. Resistor values of 1.061 kΩ are now required for *R4* and *R5*.

Before redrawing the filter, the value of *R2* in the FDNR circuit must be defined. If the normalized capacitor is not 1 F, the value of *R2* is given by 1.061 kΩ multiplied by the normalized capacitor value. If, for example, the capacitor in the passive filter has a value of 1.0844 F, the value of *R2* in the FDNR will be 1.061 k × 1.084 = 1.15 kΩ.

Finally, a DC path from the source to the load must be allowed. This will give 6 dB insertion loss, the same as a terminated lossless ladder filter. The output load should be a high value, compared with the other series components; a value of 100 kΩ is often used. The input capacitor must be bypassed by a resistor that has a value less than 100 kΩ. The bypass resistor value should be 100 kΩ minus the sum of other series resistors. Suppose the other series resistors (replacing series inductors in the passive filter) sum to 2.416 kΩ; the bypass resistor should have a value of (100 − 2.416) kΩ, or 97.584 kΩ.

Figure 6-15 gives the circuit diagram of the final FDNR lowpass filter.

An important point is that the common rail of the filter should be connected to the 0 V rail of the supply. The op-amp should then be powered from positive and negative supply rails.

References

Stephenson, F.W., *RC Active Filter Design Handbook*, John Wiley & Sons, New York, 1985.

Williams, A., and F.J. Taylor, *Electronics Filter Design Handbook*, New York, McGraw-Hill, 2006.

Winder, S., "The real choice for active filters," *Electronics World and Wireless World*, September 1993.

Winder, S., *Analog and Digital Filter Design*, Elsevier-Newnes, Amsterdam, 2002.

CHAPTER 7

Highpass Filters

Steve Winder

> *I don't use a real high-pass filter every year, but I do use some audio coupling capacitors, and I do use sample-and-difference circuits in my testing. And that's a form of high-pass filter, so we will explain that. Because, as Steve Winder points out here, it's important to use a good "poly" cap, either polystyrene or polypropylene, not polyester. /rap*

In This Chapter

This chapter describes how to design an analog active or passive highpass filter with almost any desired specification. Examples for most types of highpass filter are given. Formulae will be presented for the denormalization of component values given in previously presented tables.

Passive Filters

Passive highpass filters are designed using the normalized lowpass model. The model is normalized for a passband that extends from DC to 1 rad/s and is terminated with a 1 Ω load resistance. The first part of the process is to carry out the conversion to a highpass model; this can then be scaled for the desired load impedance and cutoff frequency. The highpass model has a passband that extends from 1 rad/s to infinity (in theory, at least). In practice, parasitic components exist to reduce the upper frequency response. These parasitic components are, for example, capacitance between wires in an inductor's windings or inductance in the leads of a capacitor.

Converting the lowpass model into a highpass equivalent is not too demanding in all-pole filters, like Butterworth or Chebyshev types. The process requires replacing each inductor in the lowpass model by a capacitor. Similarly, each capacitor in the lowpass model has to be replaced by an inductor.

In Cauer or Inverse Chebyshev filters there are series or parallel resonant LC networks. For these components, replacing inductors in the lowpass model with capacitors and replacing capacitors in the lowpass model with inductors would appear to give no change. The net result is a series or parallel resonant circuit as before. However, when each component is replaced by one with an opposite reactance, the replacement will have a value that is the reciprocal of its value in the lowpass model. Thus, the inductance value will be the reciprocal of the capacitance value that it replaced. Also, the capacitance value will be the reciprocal of the inductance value that it replaced. The LC network will then resonate at the reciprocal of its lowpass frequency.

Figures 7-1 and **7-2a** show the component-replacing process for a simple all-pole filter. More complex filters, such as Cauer, are described further later in the chapter.

Converting this into a highpass model gives the result in **Figure 7-2a**.

This is not a minimum inductor design any longer. However, a circuit with an entirely equal response is given in **Figure 7-2b** in which shunt inductors have been replaced by series capacitors of the same value. Also, shunt inductors replace series capacitors of the same value.

Denormalization of the highpass model for higher-load impedance requires component values to be scaled to have higher impedance. This is an identical process to that of denormalizing a lowpass filter. The impedance of an inductor is proportional to its inductance, but the impedance of a capacitor is inversely proportional to its capacitance. Thus, if the load resistance is a more practical $600\,\Omega$, inductance values are increased 600-fold and capacitance values are reduced 600-fold.

As an example, let's see how the component values change in the fifth-order Butterworth highpass model given in Figure 7-2a. Now let's scale these values so that the source and load are terminated in $600\,\Omega$. By multiplying the inductance values by 600 and dividing the capacitance values by 600, the result shown in **Figure 7-3** is obtained.

The component values are in Henries and Farads. As found with the lowpass denormalization, these are not very practical values. The cutoff frequency is still 1 rad/s, so the next step is to frequency-scale the design.

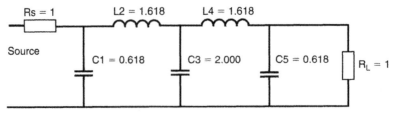

Figure 7-1: Normalized fifth-order Butterworth lowpass model.

How do the values change when the cutoff frequency is scaled? In exactly the same way that lowpass values change: by reducing both capacitance and inductance values by 2π times the cutoff frequency (in Hertz). Inductance values can be reduced because their impedance is proportional to frequency. To maintain the same impedance at a higher frequency requires less inductance. Capacitor values can also be reduced because a capacitor's impedance is inversely proportional to the frequency; to have the same impedance at a higher frequency requires less capacitance.

Since the normalized model has a 1 rad/s cutoff frequency, the scaling factor is $2\pi F_c$ to convert the frequency into Hertz. Let's design a highpass filter with a 100 kHz cutoff frequency and 600 Ω termination. The frequency scaling factor is $2\pi \cdot 100 \cdot 10^3 = 628.32 \cdot 10^3$. In other words, the cutoff frequency required is $628.32 \cdot 10^3$ rad/s. All the inductor and capacitor values in the fifth-order high-pass filter shown in **Figure 7-3**

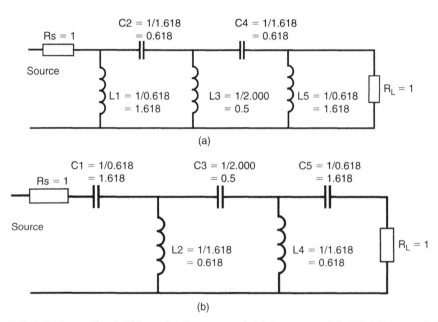

Figure 7-2: (a) Normalized fifth-order Butterworth highpass model. (b) Minimum inductor fifth-order Butterworth highpass model.

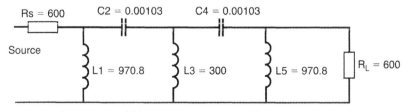

Figure 7-3: Fifth-order Butterworth impedance scaled.

(which has already been scaled for a 600 Ω source and load) must be divided by the frequency-scaling factor. The result is shown in **Figure 7-4**.

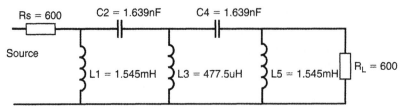

Figure 7-4: Fifth-order highpass filter frequency scaled to 100 kHz.

Active Highpass Filters

Active filters use pole and zero locations from the frequency response's transfer function. (Tables of poles and zero values can be found in Chapter 3 of the last publication cited in the References section.) The op-amp, the "active" part of the circuit, buffers one stage from the next so that there is no interaction. Each stage can therefore be designed to provide the frequency response of one pair of complex poles or a single real pole, or sometimes both. When all the stages are connected in series, the overall response is that which is desired.

A lowpass to highpass translation is required to find the highpass normalized pole and zero locations. Normalized lowpass response pole and zero locations are used as a starting point in the following formulae:

$$\sigma_{HP} = \frac{\sigma}{\sigma^2 + \omega^2}$$
$$\omega_{HP} = \frac{\omega}{\sigma^2 + \omega^2}$$

[7-1]

For a real pole at σ, the imaginary component is zero ($\omega = 0$ in the above equation). Simplifying the equation gives $\sigma_{HP} = 1/\sigma$, which means that the highpass pole is located at the reciprocal of the pole location in the lowpass prototype. Similarly, for a zero on the (imaginary) frequency axis, the real component is zero, so $\sigma = 0$ in the above equation. Simplifying the equation gives $\omega_{ZHP} = 1/\omega_z$, which means that the highpass zero is located at the reciprocal of the zero location in the lowpass prototype.

So, what does the S-plane diagram look like now? In Chapter 6 an example of a fourth-order lowpass filter was given. This had a Butterworth response, with poles on a unit circle at $-0.9239 \pm j0.03827$ and $-0.3827 \pm j0.9239$. Since the poles are on a unit circle, the denominators in the equations are equal to one. Therefore the poles are in the same place, as shown in **Figure 7-5**.

The difference is that the zeros that were at infinity in the lowpass design have now moved to the S-plane origin. In other words, the filter does not pass DC.

Scaling normalized highpass pole and zero locations is easy: Simply multiply them by $2\pi F_c$. The zeros stay at the origin, but the poles move outward away from the origin, just as they did in the case of lowpass filters. This concept is shown in **Figure 7-6**.

Important factors related to the pole locations are ω_n and Q. The values of these for the highpass filter are found in the same way as for the lowpass filter. The method is repeated here. The natural frequency ω_n is dependent on σ, and this changes in proportion to the scaling of the diagram. The origin to pole distance is equal to ω_n. The value of Q is given by the distance from the pole to the origin divided by twice the real coordinate. Thus Q depends on the ratio of ω/σ. As the pole-zero diagram is scaled for a higher cutoff frequency, the value of Q remains unchanged.

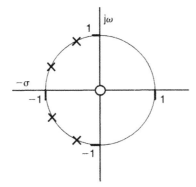

Figure 7-5: Fourth-order Butterworth highpass pole locations.

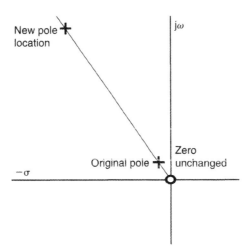

Figure 7-6: Frequency scaling of pole location in S-plane.

Now that I have set the scene, let's take a look at some basic highpass active filter designs and see how the pole and zero locations are used to find component values. I shall return to the S-plane later when discussing active Cauer and Inverse Chebyshev filters; these types both have zeroes in the stopband.

First-Order Filter Section

The first-order section is a simple structure comprising a highpass RC network followed by a buffer, as shown in **Figure 7-7**. The buffer serves to provide a high input impedance so that the voltage at the connection node of the RC network is transferred to the buffer's output and prevents the RC network from being loaded by following stages. A simple RC network on its own would not have the expected frequency response if additional resistance were added in parallel with the shunt resistor.

The first-order section is called an *all-pole network* because zeros cannot be placed on the frequency axis in its frequency response. In fact, the first-order highpass section has one real pole at $-1/\sigma$.

Letting $C1$ equal 1 Farad in the normalized highpass model enables simple calculation of $R1$:

$$RI = \frac{1}{\sigma_{HP}} = \sigma_{LP}, \qquad\qquad [7\text{-}2]$$

where σ is the pole position on the negative real axis of the S-plane. As the cutoff frequency increases, the highpass pole σ_{HP} moves further from the origin. The denormalization process requires the value of σ_{HP} to be multiplied by $2\pi F_c$; hence the normalized value of $R'1$ must be divided by the frequency-scaling factor. Thus, for a given capacitor value, the resistor value must decrease to raise the cutoff frequency.

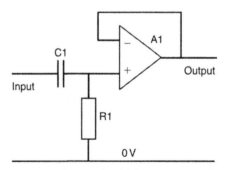

Figure 7-7: First-order highpass active filter.

Does this make sense? Well, intuitively, you might be able to see that by reducing the value of R the potential at the node between C and R will be lower at a given frequency. Increasing this frequency lowers the capacitor's reactance and restores the potential to what it was at the original frequency. In other words, to maintain a certain potential (for example, the 3 dB point of 0.7071 volts) at a higher frequency requires a reduction in the value of R.

Sample-and-Difference Circuit

The basic C-R circuit of Figure 7-7 is often used in testing. A new device under test (DUT) is powered up and its output is fed to a highpass filter so that it can be tested for noise or for other changes in its error signal. To pass low frequencies with full fidelity and accuracy, $C1$ is chosen as a relatively large value such as $0.3\,\mu F$, and $R1$ is chosen as a high value such as 10 megohms (or it may be omitted entirely). The op-amp $A1$ is always a FET-input type, with JFETs or MOSFETs. To get good throughput in the testing, this test must be started in a short time, even though a large DC bias may have been added at the signal input.

How can this DC be rejected quickly? The solution is to add a shorting switch across $R1$. This can be an ordinary push-button switch in series with 100 ohms for manual tests, or a 100-ohm JFET can be used to short out $R1$ and discharge the $C1$. After a short time, perhaps 0.1 to 5 seconds, this switch is turned off and the test is then performed. For this to work properly, the capacitor must have low dielectric absorption, or "soakage." As mentioned in the previous chapter, this capacitor must be polypropylene or polystyrene. (Polyester would give gross long settling tails.)

Sallen-Key Highpass Filter

The Sallen-Key filter produces a second-order all-pole response and is a simple active highpass design. It can be used for Bessel, Butterworth, or Chebyshev responses. Cascading second-order sections can produce high-order filters. Odd-order filters can be produced by using a series of second-order sections and then adding a first-order section at the end.

The Sallen-Key filter uses an amplifier (which may be connected as a unity gain buffer) with a network of resistors and capacitors at the input. Resistive feedback from the output is also used, and this can give rise to peaking in the frequency response. Peaking is required in second-order circuits where the Q is greater than unity and occurs due to phase shifts around the feedback loop. If the Q is large, say $Q = 15$, the amplifier is providing a gain of 15, which restricts its bandwidth to 0.0666 of the gain-bandwidth product. The diagram in **Figure 7-8** shows the circuit.

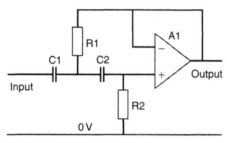

Figure 7-8: Sallen-Key highpass filter (second order).

Using Lowpass Pole to Find Component Values

By letting $C1$ and $C2$ equal 1 F in the normalized design, the values of $R1$ and $R2$ can easily be calculated from the lowpass pole locations.

$$R1 = \frac{\omega_{nLP}}{2Q_{LP}} = \sigma_{LP} \quad \text{and} \quad R2 = 2\omega_{nLP}Q_{LP} = \frac{\sigma_{LP}^2 + \omega_{LP}^2}{\sigma_{LP}} \qquad [7\text{-}3]$$

Lowpass pole positions have been used because they are readily available in tables. Thus it is not necessary to convert to highpass pole positions first. Note that in the case of Butterworth filters, $\omega_n = 1$ (for highpass and lowpass).

For example, assume that the locations of the first pair of lowpass poles of a Butterworth fourth-order filter is $0.9239 \pm j0.3827$. A Sallen-Key filter section, having the same pole locations, has resistor values $R1 = 0.9239$ and $R2 = 1.0824$. As previously stated, to use the simplified equations, the normalized highpass has capacitor values of 1 Farad.

The numbering of resistors in the next filter section follows the number sequence, and resistors are labeled $R3$ and $R4$. The values of $R3$ and $R4$ can be calculated from the same equations that were used to find $R1$ and $R2$. Substitute $R3$ for $R1$ and $R4$ for $R2$. With poles at $0.3827 \pm j0.9239$, this filter section has resistor values of $R3 = 0.3827$ and $R4 = 2.613$.

The diagram in **Figure 7-9** illustrates the whole circuit. Notice the similarity of these resistor values to the capacitor values in **Figure 6-6a**.

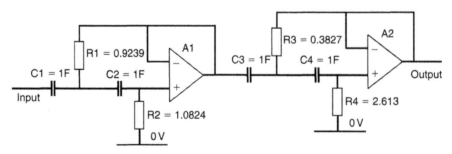

Figure 7-9: Fourth-order filter.

Using Highpass Poles to Find Component Values

If you want to design a Sallen-Key highpass filter from its highpass pole positions, the following equations should be used:

$$R1 = \sigma_{LP} = \frac{\sigma_{HP}}{\sigma_{HP}{}^2 + \omega_{HP}{}^2} = \frac{1}{2Q_{HP} \cdot \omega_{nHP}}$$

$$R2 = \frac{\sigma_{LP}{}^2 + \omega_{LP}{}^2}{\sigma_{LP}} = \frac{1}{\sigma_{HP}} = \frac{2Q_{HP}}{\omega_{nHP}}$$

[7-4]

The relationship between the equations using highpass pole locations and those previously presented using lowpass pole locations can be seen. Note that for both resistors, the equations have a frequency-dependent factor in the denominator. Frequency scaling can therefore be achieved by dividing the normalized highpass resistor values by $2\pi Fc$.

Operational Amplifier Requirements

Sallen-Key highpass filters are good if the requirements are not too demanding, with section Q factors below 50. As with lowpass designs, the gain-bandwidth product of the op-amps can limit the filter's cutoff frequency. The lowpass cutoff frequency limit was given by the empirical expressions:

$$\text{Butterworth passband frequency limit} = \frac{\text{Gain-Bandwidth Product}}{(\text{filter order})^2}.$$

$$\text{Chebyshev (1dB) passband frequency limit} = \frac{\text{Gain-Bandwidth Product}}{(\text{filter order})^{3.2}}$$

[7-5]

These equations can also be used for highpass filters by letting the passband frequency limit equal the highest frequency to be passed (i.e., do not use the $-3\,\text{dB}$ cutoff frequency). Remember that if several amplifiers are cascaded, the gain-bandwidth product of each one has to be higher than what is required overall. This is because each one contributes to high frequency rolloff as the gain-bandwidth frequency is approached.

The passband frequency limit for a given amplifier gain-bandwidth product is for a maximum of 2 dB amplitude error in the passband. A lower passband frequency limit must be set if no amplitude error is acceptable. Although using an amplifier having a greater gain-bandwidth product can raise the passband frequency limit, it can lead to instability. Amplifiers that have a high gain-bandwidth product are often unstable in a unity-gain configuration. However, it is possible to stabilize a unity-gain follower using noise-gain techniques (series R-C dampers), and it is also possible to buy very fast amplifiers, as indicated in Chapter 12. These current-feedback amplifiers can be very fast but require a feedback resistor rather than a wire from the output back to the input.

Denormalizing Sallen-Key or First-Order Designs

In active filter designs, the resistor values used should all be in the range $1\,k\Omega$ to $100\,k\Omega$ where possible. If resistor values are lower than $1\,k\Omega$ there may be a problem with loading of op-amp stage outputs. Remember, as stated in the previous chapter, loading can cause distortion and increase the power supply current. If resistor values are much higher than $100\,k\Omega$ there could be problems with noise pickup. High-impedance circuits capacitively couple with the electric field from other circuits. This coupling could cause the pickup of noise and other unwanted signals, which can interfere with the wanted signal. Also, thermal noise voltage increases in proportion to the $\sqrt{}$resistance.

The normalized highpass active filter model uses 1 F capacitors between the filter input and the op-amp input. The normalized design is based on a cutoff frequency of 1 rad/s. Denormalization is quite simple: (1) scale the impedance; (2) scale for frequency by denormalizing the capacitance value.

Impedance scaling is simply dividing the input capacitor(s) value to give suitable input impedance. The input impedance of an active filter will tend toward $1\,\Omega$ as the frequency approaches the normalized cutoff frequency of 1 rad/s, since the series capacitor $C = 1\,F$ and its reactance is $X_C = 1/\omega C$. The input impedance will therefore change with frequency. To reduce this effect, capacitors with a reactance of about 100 times the desired filter input impedance could be used. A separate terminating resistor could then be used to provide the correct load impedance at all frequencies.

Scaling the resistor values can now be carried out using the following equation:

$$R = \frac{R'}{2\pi F_c C} \qquad [7\text{-}6]$$

where R' is the normalized value calculated earlier and C is the denormalized value chosen to give a suitable input impedance.

For example, suppose you want a second-order Butterworth filter using a highpass Sallen-Key design with an input impedance of $600\,\Omega$ and a cutoff frequency $F_c = 4\,kHz$. The normalized lowpass poles are located at $0.7071 \pm j0.7071$.

Scaling the capacitor for a $60\,k\Omega$ reactance at $4\,kHz$, gives:

$$X_c = 60,000 = \frac{1}{2\pi F_c C}. \qquad [7\text{-}7]$$

Thus:

$$C = \frac{1}{120,000\pi F_c} = 663\,\text{pF}.$$ [7-8]

This is a nonstandard value, so let $C1$ (and $C2$) = 680 pF. A smaller value (higher reactance) could have been used to increase the filter's input impedance.

The normalized resistor values for highpass Sallen-Key designs are related to lowpass pole locations by:

$$R'1 = \sigma_{LP} \quad \text{and} \quad R'2 = \frac{\sigma_{LP}^2 + \omega_{LP}^2}{\sigma_{LP}}.$$ [7-9]

Hence the normalized $R'1 = 0.7071$ and the normalized $R'2 = 1/0.7071 = 1.4142$.

Frequency scaling should now be carried out using:

$$R = \frac{R'}{2\pi F_c C},$$ [7-10]

where R' is the normalized value. Thus $R1 = 41,374\,\Omega$ and $R2 = 82,749\,\Omega$.

State Variable Highpass Filters

The state variable circuit is actually a simple type of biquad. It provides a second-order stage suitable for use in all-pole filter designs. The state variable circuit has a lower sensitivity to the op-amp's gain-bandwidth product limitation, and section Q factors of up to 200 are possible. The penalty for having this good performance is that it needs three op-amps and associated passive components, as shown in **Figure 7-10**.

The equations for this filter require the use of normalized highpass pole locations. They allow the arbitrary choice of capacitor, C.

$$
\begin{aligned}
R1 = R4 &= \frac{1}{2\sigma_{HP}C} \\
R2 = R3 &= \frac{1}{C\sqrt{\sigma^2 + \omega^2}} = \frac{1}{\omega_{nHP}C} \\
R5 &= \frac{2\sigma_{HP}R}{\omega_{nHP}}
\end{aligned}
$$ [7-11]

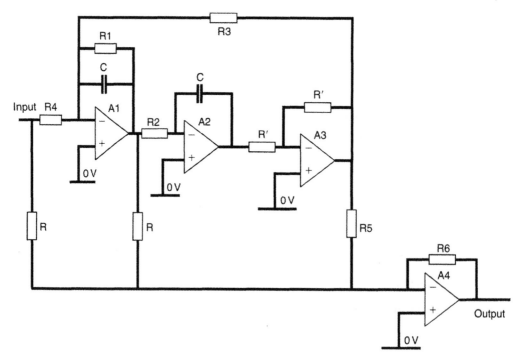

Figure 7-10: State variable highpass (all-pole).

The value of $R6$ determines the gain: $R6 = KR$, where $K =$ gain. The value of R' is arbitrary, but a typical value could be $10\,\text{k}\Omega$.

An odd-order filter is made from second-order sections connected in series, followed by a first-order section. The second-order sections are all as just described, requiring four op-amps for each pole pair. The first-order section is usually added at the end of the second-order sections and comprises a CR network followed by an op-amp.

Cauer and Inverse Chebyshev Active Filters

Let's return to the S-plane and take a look at how the Inverse Chebyshev pole-zero diagram changes when going from lowpass to highpass response. Normalized pole and zero positions of 20 dB Inverse Chebyshev with 3 dB cutoff frequency are given in **Table 7-1**.

These poles and zeros are plotted in **Figure 7-11**.

Using the lowpass-to-highpass conversion equations given earlier:

$$\sigma_{HP} = \frac{\sigma}{\sigma^2 + \omega^2}$$

$$\omega_{HP} = \frac{\omega}{\sigma^2 + \omega^2}$$

[7-12]

Table 7-1: Pole and zero locations of seventh-order Inverse
Chebyshev lowpass response.

Filter Order	Real Part	Imaginary Part
7	0.09360	1.01680
	0.37271	1.15880
	1.13417	1.35424
	2.47872	

Order	Zero 1	Zero 2	Zero 3
7	1.12060	1.39737	2.51797

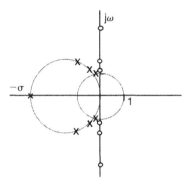

Figure 7-11: Seventh-order Inverse Chebyshev lowpass pole zero plot.

where σ and ω are the real and imaginary parts of the lowpass response. Applying these to the first pair of poles gives:

$$\sigma_{1HP} = \frac{0.0936}{0.0936^2 + 1.0168^2} = 0.089772$$

$$\omega_{1HP} = \frac{1.0168}{0.0936^2 + 1.0168^2} = 0.975214$$

[7-13]

This process can be repeated to find the other highpass pole locations, as shown in **Table 7-2**.

Table 7-2: Highpass pole locations for 20 dB Inverse Chebyshev response.

Pole Number	Real Part	Imaginary Part
1	0.089772	0.975214
2	0.251537	0.782059
3	0.363480	0.434010
4	0.403434	0

The highpass zero locations are the reciprocal of the lowpass locations, as shown in **Table 7-3**.

Table 7-3: Highpass zero locations for 20 dB Inverse Chebyshev response.

Pole Number	Pole Location
1	0.89238
2	0.71563
3	0.39715

When these are all put together, they form the highpass pole-zero diagram shown in **Figure 7-12**.

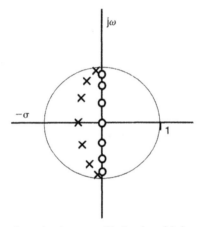

Figure 7-12: Seventh-order Inverse Chebyshev highpass pole-zero plot.

In converting from a lowpass to a highpass Inverse Chebyshev response, the poles and zeros in the S-plane have moved inside the unit circle (unlike those in the Butterworth case). This is because the pole positions are now in a similar position to a lowpass Cauer response. The zero positions are now inside the unit circle.

To design a Cauer or Inverse Chebyshev filter, a different circuit topology is required. The Cauer response has zeros outside the passband, so a notch-generating circuit is needed. This can be achieved using a circuit that is an extension of the state variable filter and is known as a biquad. This circuit, illustrated in **Figure 7-13**, is exactly the same as the state variable circuit previously given for all-pole highpass filters, except that different component values are required.

Also note that in the highpass biquad, *R5* is connected to a different node from that used in the lowpass biquad.

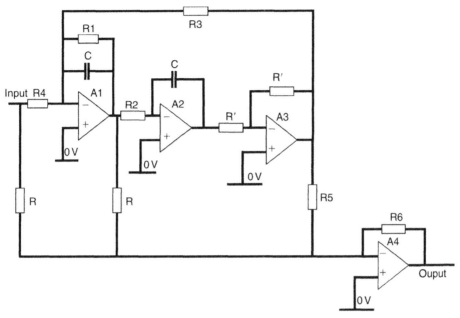

Figure 7-13: The biquad filter.

The following equations give component values. As in the case of the state variable, the value of *C* can be chosen as any suitable value, then resistor values calculated from the equations. First compute the section's frequency from the pole location:

$$\omega_{nHP} = \sqrt{\sigma_{HP}^2 + \omega_{HP}^2}$$

$$R1 = R4 = \frac{1}{2\sigma_{HP}C}$$

$$R2 = R3 = \frac{1}{\omega_{nHP}C}$$

$$R5 = \frac{2\sigma_{HP}\omega_{nHP}R}{\omega_{nHP}^2 - \omega_{ZHP}^2}$$

$$\omega_z = \text{the normalized zero frequency}$$

$$R6 = AR$$

[7-14]

The gain in the passband is denoted by the symbol A. The resistors labeled R and R' can be any arbitrary value; a typical value may be in the range $1\,k\Omega$ to $100\,k\Omega$, say $10\,k\Omega$. The resistors labeled R have an effect on the input impedance of the filter section.

Denormalizing State Variable or Biquad Designs

The simplest approach with state variable and biquad filters is to scale the poles (and zeros, in the biquad case) before using the design equations. Choose a capacitor value and then use the equations to find the resistor values. If the resistor values are very small or very large, select a new capacitor value and try again. Again, aim to keep the resistor values between $1\,k\Omega$ and $100\,k\Omega$.

Here's an example of denormalizing a biquad highpass filter design. Design a second-order Inverse Chebyshev filter that gives $40\,dB$ stopband attenuation. This uses normalized lowpass pole locations $0.70705 \pm j0.71416$ and a zero location 10.04963. The passband cutoff frequency is $1\,kHz$. Let $R = 10\,k\Omega$ and let the reactance of $C \approx 10\,k\Omega$ at $1\,kHz$ ($C \approx 1/[2\pi F_C \cdot 10^3]$). Thus, using the nearest $E6$ preferred value, $C = 150\,nF$.

First compute the pole locations from the normalized lowpass values:

$$\sigma_{HP} = \frac{\sigma}{\sigma^2 + \omega^2}$$
$$\omega_{HP} = \frac{\omega}{\sigma^2 + \omega^2} \cdot \qquad [7\text{-}15]$$

Thus, $\sigma_{HP} = 0.700102$ and $\omega_{HP} = 0.707142$.

The frequency scaling factor at $1\,kHz$ is $2\pi F_C = 6283\,rad/s$. Multiplying the pole locations by this factor gives:

$$\sigma_{HP} = 0.700102 \cdot 6283 = 4398.87$$

$$\text{and } \omega_{HP} = 0.707142 \cdot 6283 = 4443.10$$

Now compute the section's frequency from the normalized highpass pole location:

$$\omega_{nHP} = \sqrt{\sigma_{HP}^2 + \omega_{HP}^2} = 0.995084 \cdot 6283 = 6252.29 \qquad [7\text{-}16]$$

Now compute the highpass zero location, using $\omega_{ZHP} = 1/\omega_Z$ and the normalized lowpass zero location of 10.04963, which gives the highpass zero at 0.099506. Multiplying this by the frequency scaling factor of 6283 gives $\omega_{ZHP} = 625.197\,rad/s$.

Component values can now be found after substituting capacitor value, $C = 150$ nF and $R = 10$ kΩ.

$$R1 = R4 = \frac{1}{2\sigma_{HP}C} = 757\,\Omega \ (750\,\Omega \text{ is the nearest E24 series preferred value})$$

$$R2 = R3 = \frac{1}{\omega_{nHP}C} = 1500\,\Omega \ (\text{a standard value})$$

$$R5 = \frac{2\sigma_{HP}\omega_{nHP}R}{\omega_{nHP}^2 - \omega_{ZHP}^2} = 14{,}213\,\Omega \ (14.3 \text{ k}\Omega \text{ is the nearest 1\% tolerance value})$$

$$\omega_z = \text{ the normalized zero frequency.}$$

$$R6 = AR$$

[7-17]

Let the gain at DC be unity, $A = 1$. Let the resistors $R = R' = 10$ kΩ. High-precision resistors are sometimes necessary to achieve values close to those calculated.

Gyrator Filters

Gyrators are related to the FDNR circuits described in Chapter 6 and are used to replace inductors. The gyrator uses two op-amps, four resistors, and a capacitor. The gyrator can be smaller than the inductor it replaces, especially if surface-mount components are used. Other advantages of using a gyrator instead of an inductor are that using suitable components can reduce temperature effects and that the component value can be adjusted easily.

The gyrator has the same structure as the FDNR: two op-amps connected to a chain of passive elements. The gyrator only has one capacitor instead of the two used in the FDNR. All remaining passive components are resistors. The gyrator has a capacitor in place of the fourth element instead of in place of the first and third element.

A circuit diagram for the gyrator is given in **Figure 7-14**.

The gyrator behaves like a shunt inductor whose value is given by:

$$L = \frac{C \cdot R1 \cdot R3 \cdot R5}{R2}$$

[7-18]

If $C1$, $R1$, $R3$, and $R2$ are all normalized to unity, then $L = R5$. If all resistors in the gyrator circuit are equal to R, $L = R^2C$.

Suppose you want to design a highpass filter. First you should obtain the normalized lowpass passive filter component values. You should then convert the design into a

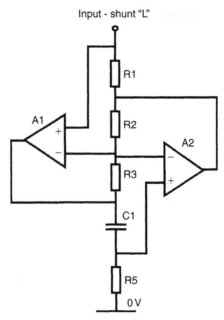

Figure 7-14: Gyrator circuit.

normalized highpass circuit by replacing inductors (that have a value L) with capacitors that have a value of $1/L$.

Also you need to replace capacitors (that have a value of C) with inductors that have a value of $1/C$. The gyrator circuit now replaces the inductor, so $R5$ in the gyrator circuit has a value of $1/C$.

Finally, all component values are normalized. This means that all capacitor values in the final circuit are divided by $Z \cdot 2\pi F_c$ and all resistor values are multiplied by Z.

For example, suppose you want to design a third-order highpass filter using a gyrator. The filter should have a passband cutoff frequency of 10 kHz with input and output impedance of 600 Ω.

A passive filter must be designed first, then the gyrator is used to replace the inductor. The normalized lowpass model has two inductors in series with a central shunt capacitor. The component values are $L1 = 1.4328$; $C2 = 1.5937$; and $L3 = 1.4328$. This is shown in **Figure 7-15.**

The normalized lowpass model is converted into a highpass equivalent by replacing the series inductors by series capacitors; thus $L1$ becomes $C1$, and so on. The capacitor values in the highpass model are the inverse of the inductor values in the lowpass model. In this case, $C1 = 1/1.4328 = 0.697934$. Due to symmetry, $C3 = 0.697934$. The shunt

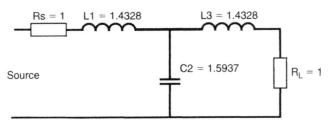

Figure 7-15: Lowpass model.

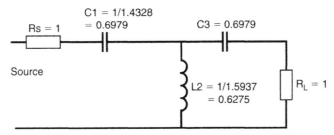

Figure 7-16: Highpass model.

capacitor in the normalized lowpass model becomes a shunt inductor in the highpass model. The value of the shunt inductor is the inverse of the shunt capacitor in the lowpass model, so $C2$ becomes $L2$. The value of $L2 = 1/1.5937 = 0.627471$. This is illustrated in **Figure 7-16**.

To replace $L2$ with a gyrator, the value of $R5$ becomes 0.627471, with $R1 = R2 = R3 = 1\,\Omega$, and $C2$ of the gyrator circuit equals 1 F.

To denormalize the filter, all resistor values must be multiplied by the load impedance of 600 Ω. Resistors $R1$, $R2$, and $R3$ all become 600 Ω. $R5$ becomes 376 Ω. The capacitor values must all be divided by the load impedance and by the cutoff frequency in radians ($2\pi F_c$). Thus, capacitors $C1$ and $C3$ become 18.5133 nF and $C2$ becomes 26.5258 nF. The circuit is given in **Figure 7-17**.

The gyrator resistors all have a low value, which could be a problem for op-amp drive capability. Although most op-amps do have a reasonable output drive performance, low-power devices do not. To overcome this, the resistance values of $R1$, $R2$, $R3$, and $R5$ can be increased, provided that the combined multiplying factor of $R1$, $R3$, and $R5$ is equal to the multiplying factor of $R2$.

Suppose, for example, that $R1$, $R3$, and $R5$ were all multiplied by 2. The value of $R2$ would have to be multiplied by 8 to restore the balance of the equation. The modified component values are then $R1 = R3 = 1.2\,\text{k}\Omega$, $R5 = 752\,\Omega$, and $R2 = 4.8\,\text{k}\Omega$. The value of $C2$ was unchanged for this modification, but it could be reduced so that the value of $R2$

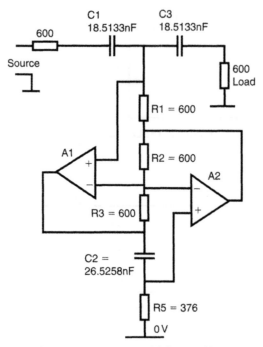

Figure 7-17: Gyrator highpass filter.

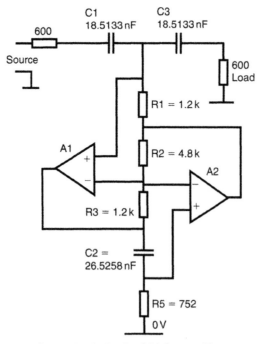

Figure 7-18: Revised highpass filter.

would not have to increase by such a large factor. The highpass filter circuit with revised component values is given in **Figure 7-18**.

The secret is to design the filter as initially described and then modify component values to make them practical. Remember to keep the equation for the gyrator inductance (equivalent to the value of *L2*) balanced. In practical circuits, the value of *C2* would probably have to be produced by two or more capacitors wired in parallel. Standard capacitor values are usually in the *E*6 range, which is coarsely spaced. It is unlikely that the gyrator capacitor would just happen to fall on one of these *E*6 values. Fortunately, it is easier to find resistor values that are close tolerance and finely spaced, so a single component can usually be used.

References

Stephenson, F.W., *RC Active Filter Design Handbook*, Chapter 13, John Wiley & Sons, New York, 1985.

Winder, S., *Analog and Digital Filter Design*, Elsevier-Newnes, Amsterdam, 2002.

CHAPTER 8

Noise: The Three Categories—Device, Conducted, and Emitted

Bonnie Baker

If you haven't bumped into noise problems, perhaps you haven't been looking too closely! Modern solid-state systems almost always have enough noise to see, because even the quietest amplifiers can amplify their noise up to levels that are easy to study. Here Bonnie Baker helps you plan to keep the noise levels down low, despite a whole thicket of (potential) noise generators. /rap

Types of Noise

Are your circuits unstable? Or do they tend to give you different results from one moment to the next? Too much noise is a typical problem confronting many circuit designers. In this chapter, noise is defined as undesirable signals that are present in a circuit. This definition excludes analog nonlinearities, which can produce distortion. Once you evaluate where your noise sources are, eliminating circuit noise may be quite simple. Tools such as filters or lower noise devices provide effective solutions. Your circuit board layout will also be critical.

Figure 8-1 shows three primary types of noise found in analog applications. Each type of noise has its own set of possible solutions. The first noise type is *device noise*. Device noise is the intrinsic noise of the devices in the circuit. Examples of device noise are the thermal noise of a resistor or the shot noise of a transistor. Another type of device noise is the switching noise from a switched mode power supply (inductive based) or a switched capacitor converter (capacitive based).

A second type of noise is *radiated noise*. You will find that radiated noise originates with EMI sources, such as switches or motors. These sources couple radiated noise into the signal. Radiated noise can also transmit from trace to trace on your PCB.

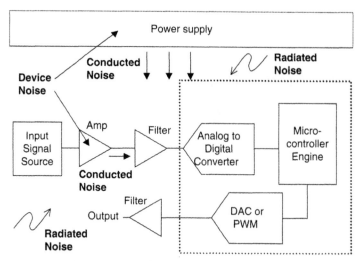

Figure 8-1: An analog circuit can acquire noise from a variety of sources. You can categorize these noise sources into three groups: device noise, radiated noise, and conducted noise.

The third type of noise that affects the performance of analog devices is *conducted noise*. This type of noise already exists in the conductive paths of the circuit, such as the power lines or signal path. Conducted noise mixes with the desired electrical signal. The origin of conducted noise is either device noise or radiated noise. In this chapter we will primarily cover device and conducted noise.

Definitions of Noise Specifications and Terms

Noise in electronics can be random or connected to some circuit generated frequency. If it is a random event over the frequency spectrum, it is void of coherent frequencies. Although you know the input signal to your circuit, noise events are somewhat unpredictable. They occur inside all analog devices, including passive and active devices. If you sample these noise events, they will build a normal distribution over time, as illustrated in **Figure 8-2**. If sampled noise events do not have a normal distribution, it's difficult to make the prediction of peak-to-peak over time. If the noise falls into a normal distribution, you can apply mathematics to bound and describe the apparent noise randomness. We will discuss the type of circuit noise that does not fall into a normal distribution later in this chapter, in the "Power Supply Noise" section.

If the noise samples fall within a normal distribution, repeated samples differ around a central value. The distribution is roughly symmetric around this central value. The distribution produces a curve with its highest occurrence at the center point, tailing off to zero in both directions. Because this distribution is consistent with the Central Limit Theorem, you can use standard calculations such as mean and standard deviation to predict the general magnitude of future occurrences with respect to the normal curve.

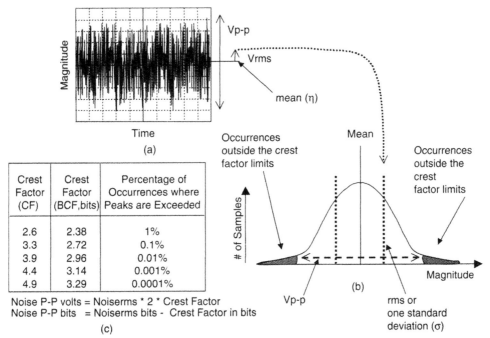

Crest Factor (CF)	Crest Factor (BCF,bits)	Percentage of Occurrences where Peaks are Exceeded
2.6	2.38	1%
3.3	2.72	0.1%
3.9	2.96	0.01%
4.4	3.14	0.001%
4.9	3.29	0.0001%

Noise P-P volts = Noiserms * 2 * Crest Factor
Noise P-P bits = Noiserms bits - Crest Factor in bits

(c)

Figure 8-2: If noise is a random event (a), these "obviously" unpredictable events can be characterized with a normal distribution (b) where you can calculate mathematical equivalents of the mean (η) and standard deviation (σ). If you multiply the standard deviation (equal to rms) by a 2 × crest factor, you can predict the probability of occurrences outside the crest factor limits (b and c).

Each of the three diagrams in Figure 8-2 shows a different representation of the same data. The first diagram, Figure 8-2a, is a time-based oscilloscope picture of a noisy signal. This signal is void of coherent frequency signals, and it seems to be random. This picture of noise might be similar to what you have seen on the bench, particularly if you zoom in using your magnitude/div and time/div knobs.

The noise in Figure 8-2a is sampled and mapped into the magnitude-based graph in Figure 8-2b. In this graph, the number of samples is on the y-axis and the sample magnitude on the x-axis. The time scale from the previous diagram is lost and only the magnitude of each sample is preserved. It is interesting that the samples from Figure 8-2a build a normal distribution in the histogram plot (Figure 8-2b). This is not unusual due to the random nature of noise. With this sampled data, you can calculate the root-mean-square (rms) value (one standard deviation) and mean of the sample set. And if this histogram forms a normal distribution, you can exercise a multiple (called a *crest factor*, shown in Figure 8-2c) to the rms value to establish peak-to-peak noise values for your data. This calculated peak-to-peak value will predict the percentage of future samples that will fall outside those limits.

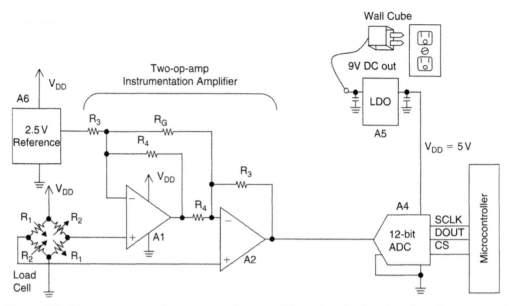

Figure 8-3: The two-op amp instrumentation amplifier gains the DC signal at the output of the load-cell sensor. The 12-bit A/D converter then digitizes that signal.

It is important to know that when you use the noise calculations, they should be treated as best estimates rather than absolutes. This could give you a feeling that you can't count on any of the results that you arrive at, but to the contrary, this is where statistics come into the picture. These statistical estimates provide a degree of confidence that you will consistently get the expected results if, in fact, your sample is a good representation of the population.

Evaluating Noise with a Circuit Example

Take, for example, the circuit in **Figure 8-3** and the test results in **Figure 8-4**.

In Figure 8-3, a load-cell circuit measures the weight applied to the sensor. With a 5 V excitation voltage applied to the high side of the sensor, the full-scale output swing is a $\pm 10\,mV$ differential signal with a 32-ounce maximum excitation. This small differential signal is gained by a two-op-amp instrumentation amplifier (G = 153 V/V). I chose a 12-bit converter to match the required precision of this circuit. Once the converter digitizes the voltage presented at its input, the microcontroller receives the digital code through the SPI port. The microcontroller then uses a lookup table to convert the digital signal from the ADC into load-cell weight.

Figure 8-4 shows how the circuit noise can contaminate a perfectly good A/D converter. In this figure, the 12-bit converter was used to unsuccessfully convert a DC signal. Initially,

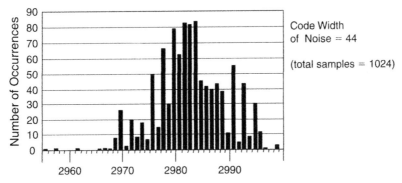

Figure 8-4: A 12-bit A/D converter is used to convert DC signal to (hopefully) one code. If you don't manage the passive (from resistors) and active (from amplifiers) noise well, the converter will give an unrepeatable result over time. The LSB size in this graph is 1.22 mV. The code width of 44 codes equals 53.68 mV.

one might assume that the 12-bit converter is not very good. But it is possible, as is with this circuit, that the active and passive devices in the signal path are noisier than expected. The components I chose for this first circuit were:

$$R_3 = 300 \, \text{k}\Omega, \, R_4 = 100 \, \text{k}\Omega, \, R_G = 4020 \, \Omega, \, (+/-1\%) \qquad \text{[8-1]}$$
A1 and A2 are single-supply op-amps with an input voltage noise density of $29 \, \text{nV}/\sqrt{\text{Hz}}$ (MCP601, Microchip Technology)

Figure 8-4 shows the results of ignoring noise issues in this circuit. The sampling rate was 10 ksps for this data. There were 1024 samples taken for the histogram. I took care to ensure that the input signal was DC or relatively noiseless. The noise spread of my data is 44 codes wide. I was surprised to find that this code width was repeatable (± 2 codes)! Since this data was taken using a 12-bit A/D converter, which has 4096 possible output code combinations with a 5 V reference, one LSB equals about 1.22 mV. If you translate the code width of 44 into millivolts, this would equal approximately 53.68 mV of noise. You will note that the data in Figure 8-4 doesn't really have a normal distribution. Consequently, I suspected that there is a frequency component embedded in my data.

Then I went back in an effort to reduce the noise so that my converter would provide a single conversion value every time. **Figure 8-5** shows this modified circuit. It is not a coincidence that you find the circuit in Figure 8-5 in Chapter 14 (Figure 14-8). This is a common circuit solution for pressure-sensing applications, for good reason. One of the reasons is that it produces an accurate, low-noise solution.

In this modified circuit I reduced the resistor values (without changing the function of the circuit) and replaced the operational amplifiers with a lower-noise version (MCP6021,

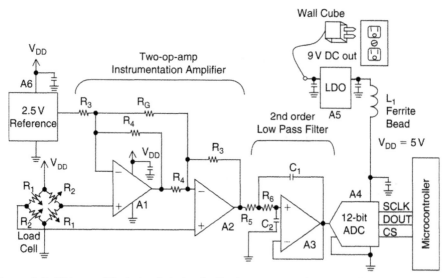

Figure 8-5: This modified circuit is basically the same as the circuit in Figure 8-3. The modifications include the addition of a lowpass, antialiasing filter, bypass capacitors, and an inductive ferrite bead (L_1) on the power supply trace.

8.7 nV/$\sqrt{\text{Hz}}$, Microchip Technology). These modifications attacked my device noise problems. From there I added a ground plane, which is highly recommended in any analog circuit. I did this to address emitted noise problems as well as conducted noise problems. I proved that emitted noise was a problem by moving the board around my lab. Sometimes I would place it closer to the wall cube and other times closer to my fluorescent lights. This would increase the circuit noise in both places. I also found that when the copy machine was working in the cubicle next to me, I had an increase of noise.

I finally tackled my conductive noise problems by inserting an analog lowpass filter after the instrumentation amplifier. I also put a "choke" inductor (L_1) on the power supply trace and installed appropriate bypass capacitors.

The components I chose for this second pass circuit were:

$R_3 = 30\,k\Omega$, $R_4 = 10\,k\Omega$, $R_G = 402\,\Omega$, (+/ − 1%)   [8-2]
10 Hz, lowpass filter − $R_5 = 27.4\,k\Omega$, $R_6 = 196\,k\Omega$, $C_1 = 470\,nF$, $C_2 = 100\,nF$
Single-supply op-amp with input voltage noise density = 8.7 nV/$\sqrt{\text{Hz}}$ @ 10 kHz
Bypass capacitors = 0.1 µF on every active device

This type of attention to detail paid off. As **Figure 8-6** illustrates, the converter is perfectly capable of converting to one bit in a reliable and repeatable manner.

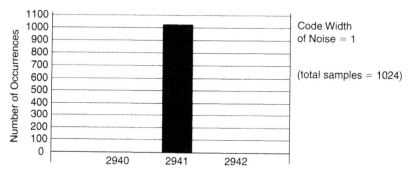

Figure 8-6: This data uses the same 12-bit A/D converter to convert DC signal to one code. In this new circuit, I am reducing the resistor values and using lower-noise operational amplifiers. These two actions reduce the device noise in the circuit. An added ground plane reduces emitted noise. Finally I add a low-pass filter, bypass capacitors, and a power supply inductive filter to reduce the conducted noise.

These results are impressive, but you might be curious to learn how I know that these changes were the right ones. So let's dig into each type of noise in some detail.

Device Noise

Devices come in many flavors, each with their own frequency spectrum of noise. The passive and active devices, all capable of creating noise, can have an impact on your circuit. The passive devices include resistors, capacitors, and inductors. Of these three types of passive devices, the resistor is most notable for circuit noise. This type of noise can be a nuisance. Capacitors generate random noise that is dependent on temperature and Boltzsman's constant, equally kT/C (where k is Boltzman's constant, T is temperature in Celsius, C is capacitance in farads). Capacitors and inductors can also reduce noise in a signal path or on the power supply lines. The capacitor and inductor also create noise if there is some kind of switching signal going across them. So we will discuss these passive devices as they relate to switch-mode power supplies and charge pumps.

Another source of device noise comes from active devices. The typical active devices that will be in your circuits are operational amplifiers, A/D converters, voltage references and power supply chips. Of these devices, the operational amplifier has the largest direct effect on the signal chain and the power supply chips will inject noise through the device's power supply pins, which may eventually enter the signal chain.

Resistor Noise

An ideal resistor creates noise inside the resistive element. This type of noise is commonly called *thermal noise* or *Johnson noise*. The resistive element generates noise due to the agitation of particles in the resistor. You cannot avoid resistor noise by not applying power. This noise will appear across the resistor whether it is powered with a current source, voltage source, or not at all. However, this noise can be filtered and consequently reduced later down the signal path as conducted noise. The ideal resistor will create a predictable noise that is flat across the full frequency spectrum. The ideal resistor rms noise is equal to:

$$R_{NOISE} = \sqrt{(4 \times k \times T \times R \times BW)} \qquad \text{[8-3]}$$

Where k is Boltzman's constant equal to 1.38×10^{-23}

        T is temperature in Kelvin

        R is the resistor in ohms

        BW is the bandwidth of interest

In this calculation, Kelvin is also equal to 298.16°K (25°C) at room temperature. Each degree increase in temperature in Kelvin is equivalent to a 1 degree increase in Celsius.

With this calculation, it is easy to quickly determine whether your resistor is too noisy for your circuit. For instance, the noise of a 1 kΩ resistor at 25°C is approximately 4 nV/$\sqrt{\text{Hz}}$ (rms). If you want to calculate the amount of noise generated by an ideal 1 kΩ resistor across a frequency bandwidth of 1 Hz to 1000 Hz, that noise would be equal to:

$$V_{NR} = \sqrt{(4 \times k \times T \times R \times (BW))}$$
$$V_{NR} = \sqrt{(4 \times 1.38 \times 10^{-23} \times 298.16\,K \times 1000\,\Omega\,(1000\,Hz - 1\,Hz))}$$
$$V_{NR} \sim 126 \text{ nV rms}$$
$$V_{NR} \sim 834 \text{ nV p-p (assumes a crest factor of 3.3; see Figure 8-2)} \qquad \text{[8-4]}$$

This doesn't seem like very much noise. For example, if your system uses a 16-bit A/D converter with a 5 V reference (5 V FSR), the LSB size with this converter is 76.3 µV. The question is: "If I have an input resistance of 1 kΩ to my converter, will it affect my signal accuracy?" The answer is no. Our peak-to-peak estimate is equal to 1% of one LSB. On the other hand, if you are designing with a 20-bit A/D converter (5 V FSR), our peak-to-peak estimate is equal to 17.5% of one LSB. Now, that very well could be a problem!

Table 8-1 gives the ideal resistor noise for a variety of resistors.

The units of these ideal calculations are nV/$\sqrt{\text{Hz}}$ (rms). The values in this table are easy to convert to higher or lower resistors that are not on the chart. This is done by multiplying or dividing the quantity by the square root of the ratio between the table value and the value of interest. For instance, if you want to know what type of noise a 1,000,000 Ω

Table 8-1: All resistors generate noise. This table tabulates the ideal room-
temperature noise that a variety of resistors will generate.

Table of Resistance Noise @ 25°C (298.16°Kelvin)			
Resistance (Ω)	**Noise Density (nV/rt Hz)**	**Resistance (Ω)**	**Noise Density (nV/rt Hz)**
1	0.1283	100	1.283
2	0.1814	200	1.814
3	0.2222	300	2.222
4	0.2566	400	2.566
5	0.2869	500	2.869
6	0.3142	600	3.142
7	0.3394	700	3.394
8	0.3629	800	3.629
9	0.3849	900	3.849
10	0.4057	1000	4.057
20	0.5737	2000	5.737
30	0.7027	3000	7.027
40	0.8114	4000	8.114
50	0.9072	5000	9.072
60	0.9937	6000	9.937
70	1.0734	7000	10.73
80	1.1475	8000	11.48
90	1.2171	9000	12.17
100	1.2829	10000	12.83

resistor will create, you need to multiply the noise of the 10,000 Ω resistor by 10, or $\sqrt{(1,000,000/10,000)}$. If you work with these ideal calculations, you will generally be able to determine whether the resistors in your circuit are causing noise problems.

Real resistors, such as wirewound, film type, or composition resistors, produce noise that is higher than the ideal. Of these three types, the wirewound resistor is the quietest, followed by the film type, with the worst being carbon composition. A good-quality wirewound can produce noise that is nearly ideal. On the other end of the spectrum, the

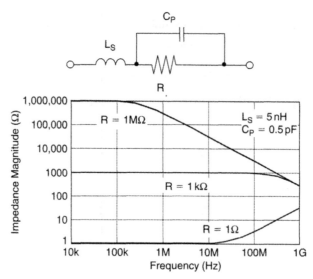

Figure 8-7: The nonideal resistor can be modeled using an ideal resistor, *R*, a parasitic capacitor, *C_P*, and series lead inductance, *L_S*. These parasitic elements attenuate the resistor noise at higher frequencies, particularly with higher-value resistors.

composition resistor will generate noise because of its contacts. This noise is a result of the individual particles in the contact. Contact noise is proportional to the DC current flowing through the resistor. This noise will appear at lower frequencies and looks very similar to the 1/f noise of amplifiers. A composition resistor that has no DC current conducting through it will exhibit near-ideal noise behavior. Film-type resistors also have contact noise but to a lesser degree. This is primarily because the film resistor contact is made of a more homogenous material. Variable resistors, such as analog and digital potentiometers, exhibit the same types of noise as described above, and don't forget the additional wiper resistor noise. All these types of resistors will operate at a lower noise level if you keep them under their power rating.

Figure 8-7 shows the AC model for a real, wirewound resistor. The parasitic capacitor (*C_P*) will have the most impact on the noise behavior (as well as the frequency response). This is because it attenuates higher-frequency noise. This is a nice benefit, but you should still use caution when you use higher-value resistors. They will still have higher low-frequency noise.

Revisiting Our Application Circuit With Better Resistors
Going back to the circuit in Figure 8-3, I have run a SPICE simulation (Figure 8-8) to try to determine whether or not a reduction in resistor values around the instrumentation amplifier will make a difference. I know that I need these resistors, and I know that the

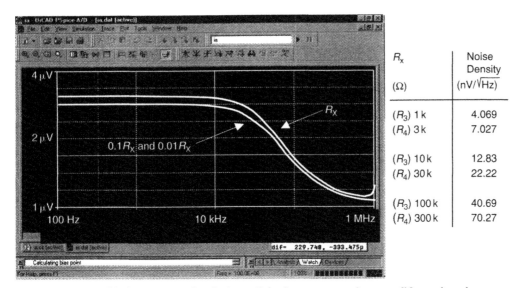

R_x	Noise Density
(Ω)	$(nV/\sqrt{Hz})$
(R_3) 1 k	4.069
(R_4) 3 k	7.027
(R_3) 10 k	12.83
(R_4) 30 k	22.22
(R_3) 100 k	40.69
(R_4) 300 k	70.27

Figure 8-8: This is a SPICE simulation of the instrumentation amplifier using three different sets of resistors. The x-axis is a log scale of frequency in Hertz, and the y-axis is the magnitude of noise divided by the square root of Hertz. The top curve, R_X, has $R_3 = 100\,k\Omega$ and $R_4 = 300\,k\Omega$. This combination of resistors is the noisiest. The curves 0.1 R_X and 0.01 R_X are on top of each other. This is because the amplifiers in the instrumentation amplifier start to dominate the noise response. With these two curves, the circuit noise is lower than the R_X curve.

output of these amplifiers should have at least a $1000\,\Omega$ load. It is easy enough to reduce all these resistors by $10\times$ or $100\times$ without changing the gain of the instrumentation amplifier circuit.

An easy way to determine whether or not I have resistor noise is to run a SPICE simulation (see Figure 10-8). In my simulation, I have made sure that the DC simulation is correct. In particular I am looking for extraordinarily unexplained high voltages, input amplifier nodes that are not where they are supposed to be, and an output voltage on A_2 that is between the supply voltages. After this reality check, I have more faith in the AC analysis of Figure 8-8.

In this figure, I ran the circuit SPICE simulation with the resistor values called out in Figure 8-3. The second and third AC simulation replaced all the resistors with 1/10 of the values and then 1/100. As you can see in this figure, resistors that are $10\times$ lower buy me some reduction in noise. But not as low as I would expect. By lowering the resistors by $10 \times$ I should see an improvement in noise equal to $\sqrt{(1/10)}$, or a 32% reduction. This did not happen. Furthermore, reducing them by $100\times$ buys me nothing. This suggests that my amplifiers are the limiting factors.

Operational Amplifiers

It is important to understand the noise that operational amplifiers create, because almost every analog circuit will have an op-amp somewhere in the circuit. The op-amp noise behavior, over frequency, has a signature that is unmistakable.

If you look for the amplifier noise specification in the typical amplifier datasheet, you will notice that it is a "referred-to-input" specification. The location of this noise source is at the noninverting input of the amplifier. In the specification table, you will typically find input noise and input noise density specifications. The input noise specification will describe the low-frequency noise of the amplifier in terms of bandwidth. You will find this bandwidth in the "conditions" column. 1/f noise is this lower-frequency noise. This is mainly because this part of the curve actually follows the ratio of 1:frequency times a multiple. The transistors in the input stage of the amplifier generate the noise through this frequency band. This is primarily the differential input stage, but it also includes the input stage load transistors.

Input noise density calls out a noise figure that refers to one frequency. For instance, the noise specifications in **Figure 8-9** identify the input voltage noise density at 10 kHz to equal 8.7 nV/√Hz. You measure the input voltage noise density at the specified frequency across a 1 Hz bandwidth. Usually this specification appears in the broadband noise portion of the frequency plot (Figure 8-9). Theoretically, this broadband noise is flat. Assuming that it is

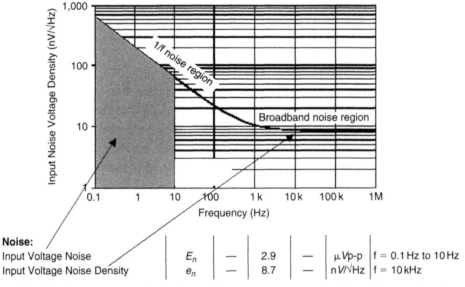

Noise:						
Input Voltage Noise	E_n	—	2.9	—	µ Vp-p	f = 0.1 Hz to 10 Hz
Input Voltage Noise Density	e_n	—	8.7	—	n V/√Hz	f = 10 kHz

Figure 8-9: This is a representation of the noise of an example amplifier. The specifications for the noise performance of the amplifier are in tabular form at the bottom of the figure. These specifications numerically refer to the input noise voltage density versus frequency plot.

flat is a good estimate of the amplifier's behavior. It is also the foundation or baseline of the 1/f noise portion of the curve. The diffused resistors inside the operational amplifier primarily generate the broadband noise. These resistors can be diffused resistors or the source or drain of the transistors in the amplifier.

Further, on in the amplifier datasheet you will find a typical specification graph that will show you the input noise voltage density vs. frequency. Figure 8-9 shows an example of this type of graph. In this example, the input voltage noise specification is equal to the area beneath the input-voltage, noise-density curve between the specified frequencies of 0.1 Hz to 10 Hz. Note that in the table the units for this specification are peak-to-peak. To convert this to an rms value, simply divide the value by 6.6 (industry standard crest factor = 3.3).

You can easily calculate the noise underneath the curve for different input voltage noise bandwidths in the 1/f region. The first order of business in this calculation is to determine the input noise density at 1 Hz. Once you find that value, this simple formula will provide the rms noise under the curve:

$$V_{(1/f)\,:\,f_2 - f_1} = B\sqrt{\ln(f_2/f_1)} \qquad\qquad [8\text{-}5]$$

where B is equal to the input noise density at 1 Hz. As an example, the amount of rms noise produced by the amplifier shown in Figure 8-9 from 0.1 Hz to 1000 Hz is equal to:

$$
\begin{aligned}
V_{(1/f)\,:\,f_2 - f_1} &= B\sqrt{\ln(f_2/f_1)} \\
V_{(1/f)\,:\,f_2 - f_1} &= 200\,\text{nV} \times \sqrt{\ln(1000/0.1)} \qquad\qquad [8\text{-}6] \\
V_{(1/f)\,:\,f_2 - f_1} &= 607\,\text{nV rms or } 4\,\mu\text{V p-p}
\end{aligned}
$$

When you think about noise at these low frequencies you may jump to the conclusion that you should take this formula down to a very low frequency, such as 0.0001 Hz (0.0001 Hz = 1 cycle per 2.8 hours). Be careful when you look at frequencies lower than 0.1 Hz, which is one cycle every 10 seconds. At lower frequencies, it is very possible that other things are changing in your circuit, such as temperature, aging, or component life. If you think of this realistically, low-frequency noise from your amplifier will probably not appear at this sample speed. But changes in your circuit, such as temperature or power supply voltage, could.

The amplifier table of specifications also gives the input noise density value. This specification is always at a higher frequency in the area where the input voltage noise is relatively constant. For this region of the curve, multiplying the square root of the bandwidth and the noise content is defined by multiplying the square root of the bandwidth by the noise density. For example, if the noise of the amplifier is $8.7\,\text{nV}/\sqrt{\text{Hz}}$ @ $10\,\text{kHz}$, the noise from the amplifier across the bandwidth of $1\,\text{kHz}$ to $100\,\text{kHz}$ is equal to:

$$V_{100k-1k} = (\text{Noise Density} \ @10\,\text{kHz}) \times \sqrt{\text{BW}}$$
$$V_{100k-1k} = (8.7\,\text{nV}/\sqrt{\text{Hz}}) \times \sqrt{(100,000 - 1,000)}$$
$$V_{100k-1k} = 2.74\,\mu\text{V rms or } 18.1\,\mu\text{Vp-p}$$

[8-7]

Where BW is equal to the bandwidth of interest.

The challenge from the manufacture is to give you good data so you can work through the impact of their device in your application. So how do you get from the manufacture's graph to a meaningful result in your application circuit? You calculate the area beneath the noise curve and multiply that times the noise gain of the amplifier. Let's go through this process with a real circuit and real component values.

The amplifier in **Figure 8-10** is in a typical inverting gain stage. The input to the circuit is V_{IN} and the output is V_{OUT}. The voltage at V_{SS} is equal to 0 volts or ground, and the voltage at V_{DD} is equal to 5 volts. There is a 2.5 V reference connected to the noninverting input of the amplifier at the V_{REF}.

Figure 8-10 shows the internal capacitors of the amplifier. They will come into play when we start to calculate the gain of the circuit over frequency and look at the noise. C_{CM} is equivalent to the common-mode capacitance of the input stage of the amplifier. For our example, C_{CM} is equal to 6 pF. This capacitance is referenced to ground. C_{DIFF} is equivalent to the differential input capacitance of the amplifier, and you will notice that it appears between the two input terminals. For our calculations we will use $C_{DIFF} = 3\,\text{pF}$.

The parasitic capacitance of the external resistors, R_1 and R_2, are C_{P-R1} and C_{P-R2}, as shown in this diagram. Although you might think that these are insignificant capacitances (at ~0.5 pF), they are worth paying attention to. They could affect the noise gain of the amplifier circuit at higher frequencies. Figure 8-7 shows the frequency effects of this parasitic capacitance.

The noise gain calculation of this amplifier circuit uses the noise source, e_n, as an input signal. This source is graphically inside the amplifier symbol. You will notice that this formula is not the same as the formula for the signal gain.

$$\text{Signal Gain} = V_{OUT}/V_{IN} = -Z_2/Z_1$$
$$\text{Noise Gain} = V_{OUT}/V_{IN} = 1 + Z_2/Z_1$$

[8-8]

Where Z_1 is the equivalent input resistor, capacitor network, and
Z_2 is the equivalent feedback resistor, capacitor network

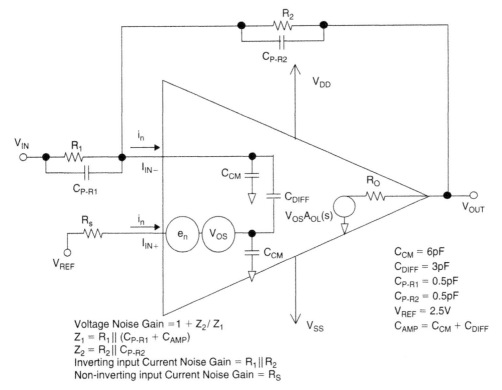

Figure 8-10: This amplifier circuit model is in an inverting gain configuration. This diagram illustrates the pertinent parasitics of the amplifier and resistors along with the calculation for noise gain.

When you are calculating the amount of noise that an amplifier produces, the noise gain equation will provide the correct results. This equation will also provide the correct closed-loop bandwidth of the amplifier circuit.

Figure 8-11 shows the frequency response of this amplifier circuit. The capacitors and resistors surrounding the amplifier, as well as the frequency response of the amplifier, affect the bandwidth of the circuit.

The DC noise gain of this circuit is dependent on the resistors in the circuit. At higher frequencies, the noise gain is dependent on the capacitors. It is possible in many circuits to design the second-corner frequency, f_2, higher than the f_{AOL} crossing. If this is the case, you can ignore the effects of f_2. If the value of R_2 is high ($>100\,k\Omega$), f_2 may come down in frequency, lower than the open-loop gain crossing. To optimize the noise and bandwidth performance of this type of amplifier circuit, the pole, f_2, should occur at or slightly before the point where the noise gain plot intersects the open-loop gain curve of the amplifier. This could require an additional capacitor in parallel with R_2.

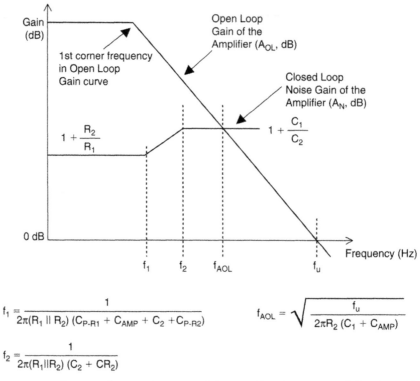

$$f_1 = \frac{1}{2\pi(R_1 \parallel R_2)\,(C_{P\text{-}R1} + C_{AMP} + C_2 + C_{P\text{-}R2})}$$

$$f_{AOL} = \sqrt{\frac{f_u}{2\pi R_2\,(C_1 + C_{AMP})}}$$

$$f_2 = \frac{1}{2\pi(R_1 \parallel R_2)\,(C_2 + C_{R2})}$$

Figure 8-11: The open-loop gain curve of the amplifier is on top of the closed-loop gain curve of an amplifier circuit. With the open-loop amplifier gain curve (A_{OL}), the first corner frequency follows the DC gain. Past this first pole, the gain of the amplifier attenuates at a rate of −20 dB/decade. With the closed-loop noise gain curve (A_N), the poles and zeros of the transfer function are shown along with their corner frequencies. The bandwidth of the noise gain is equal to f_{AOL}.

Once you calculate the gain across the frequency spectrum of the amplifier's bandwidth in this circuit, you can start to determine the circuit's referred-to-output noise. **Figure 8-12** separates the noise into six parts. Five of these noise parts are in the graph, and the sixth is part of the formula in the figure.

In region e_1, the $1/f$ noise of the amplifier is gained by the DC gain of the amplifier circuit. The specifications for amplifier noise are in nanovolts per root hertz. So the analysis is complete when you multiply the average noise over the region by the square root of the bandwidth of that region. For CMOS amplifiers, the $1/f$ region is usually from 0.1 Hz to 100 Hz up to 1000 Hz. Since this noise value is multiplied by the square root of the bandwidth, its contribution is low.

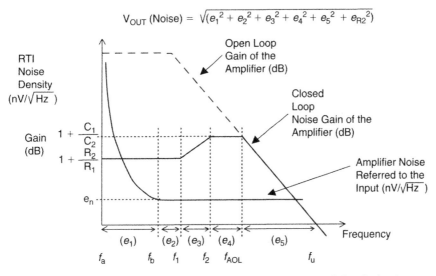

$$V_{OUT} \text{ (Noise)} = \sqrt{(e_1{}^2 + e_2{}^2 + e_3{}^2 + e_4{}^2 + e_5{}^2 + e_{R2}{}^2)}$$

Figure 8-12: The noise of the amplifier is referred to the input of the device because the manufacturer doesn't know which configuration you are going to put your op-amp into. With the referred-to-input values you can calculate your own referred-to-output values, which you can use moving forward in the rest of your circuit.

In the second region, the broadband noise of the amplifier is multiplied by the DC noise gain. Again, the average noise is multiplied by the square root of the bandwidth of that region. The contribution of noise in this region is also relatively low.

The third, fourth, and fifth regions are calculated in the same manner, with each region contributing more to the overall noise of the circuit. The sixth part of the noise equation in Figure 8-12 represents the noise contribution of the feedback resistor, R_2. The noise contribution of this resistor might or might not be significant, depending on the magnitude of the resistor. This calculation will quickly demonstrate where the highest noise contribution is coming from and make it easier to refine the design.

Region e_1:
$$e_1 = (1 + R_2/R_1) \times B \sqrt{(\ln(f_b/f_a))} \qquad [8\text{-}9]$$

Region e_2:
$$e_2 = (1 + R_2/R_1) \times e_n \times \sqrt{(f_2 - f_1)} \qquad [8\text{-}10]$$

Region e_3:
$$e_3 = (1 + R_2/R_1) \times e_n \times (1\,Hz/f_1) \sqrt{(f_2/3 - f_1/3)} \qquad [8\text{-}11]$$

Region e_4:
$$e_4 = (1 + C_1/C_2) \times e_n \times \sqrt{(f_{AOL} - f_2)} \qquad [8\text{-}12]$$

Region e_5:

$$e_5 = (1 + C_1 / C_2) \times e_n \times \sqrt{(\pi / 2(f_u - f_{AOL}))} \qquad [8\text{-}13]$$

Region e_{R2}:

$$e_{R2} = \sqrt{(4 \times K \times T \times R_2 \times (BW))} \qquad [8\text{-}14]$$

Note: In this calculation, C_1 is the parallel combination of the input capacitors, or $C_{P\text{-}R1} \| 2C_{CM} \| C_{DIFF}$. C_2 is the feedback capacitor or C_{P-R2}.

All this being said, SPICE is a useful tool as you verify your noise calculations. The two graphs in Figure 8-13 demonstrate how SPICE can help you understand the noise in your circuit. The graph on the left (**Figure 8-13a**) shows the simulated noise as frequency increases. You will notice that the noise is very low at the lower frequencies. This is because the lower bandwidths are multiplied by the square root of a small number, the bandwidth. As frequency increases, the cumulative noise also increases. You would think that at higher frequencies the increases in noise would be less due to the characteristics of the left-hand graph (Figure 8-13a). As you can see, though, this is not true. The reason is that the bandwidth multiplier (square root of the bandwidth) is larger at higher frequencies.

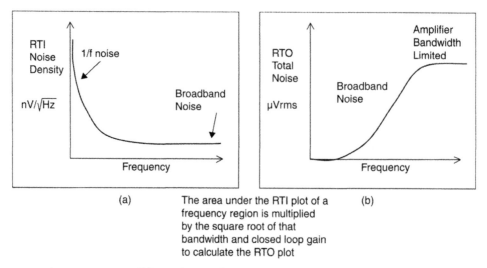

(a)   The area under the RTI plot of a   (b)
frequency region is multiplied
by the square root of that
bandwidth and closed loop gain
to calculate the RTO plot

Figure 8-13: Amplifier noise can be graphically represented as though the noise source is at the input of the amplifier (a), otherwise known as *referred-to-input (RTI)*, or as though it is at the output of the amplifier (b), otherwise known as *referred-to-output (RTO)*.

Going back to Figures 8-3 and 8-8, we concluded that reducing the resistor values was beneficial to a point. The next step is to reduce the amplifier noise. If the resistors are reduced ten times and the amplifiers are changed, the noise code width response in Figure 8-4 is reduced from 44 to 21 codes, p-p. This is not bad, considering we haven't changed the layout, just the devices.

A/D Converter Noise

The most talked-about noise from the A/D converter is *quantization noise*, which is the noise that an A/D converter generates as a consequence of dividing the input signal into discrete "buckets." **Figure 8-14** illustrates this concept. The width of these "buckets" is equal to the LSB size of the converter. The quantization noise of a converter determines the maximum Signal-to-Noise Ratio ($SNR_{IDEAL} = 6.02\ n + 1.76\ dB$). This noise is immediately apparent in the converted signal. If you want more accuracy, you need to change to a converter with a higher number of bits. Just as a caveat: Making this change does not guarantee a better SNR, because the converter might have other noise sources inside, but it is a good start.

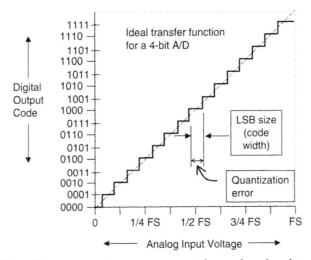

Figure 8-14: The A/D converter does not convert the analog signal to an ideal value. The A/D converter has a discrete number of output conditions and the analog signal has an infinite number of voltage states. This is called the *quantization error*, which accounts for quantization noise.

There is also noise inside the A/D converter that comes from the internal transistors. This type of noise is discussed in Chapter 13 in more detail, in the section "AC Specifications Imply Repeatability." Just to preview the discussion in Chapter 13, AC domain specifications, such as Signal-to-Noise Ratio (SNR), Effective Resolution (ER), Signal-to-(Noise + Distortion) (SINAD), or Effective Number of Bits (ENOB) help you

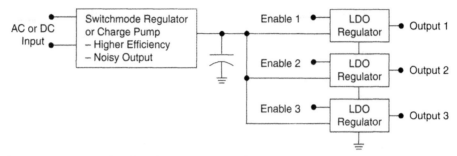

Figure 8-15: Power is taken from an AC (such as the wall) or DC (such as a battery) source and converted down (or up) to the required voltage level for the electronics. This conversion can be done in one stage or multiple stages, using LDO, switch-mode converters, or capacitor charge pumps.

understand how repeatable your A/D converter might be. These specifications do not imply accuracy, only repeatability.

Power Supply Noise

There are three fundamental types of power supply devices that you can use to deliver power to your circuit. They are the *regulator* (also *low dropout regulator,* or *LDO*), the *switched power supply circuit* (SPC), and the *capacitive charge pumps.* Use **Figure 8-15** as an example of how you might want to connect these parts.

Low Dropout Regulators
Figure 8-16 shows a simplified CMOS LDO circuit. In this circuit, the input voltage directly supplies the power for the integrated circuits as well as the output load. Usually the quiescent current of the LDO device is much lower than the output current.

Proper LDO operation requires that the input voltage is DC and higher in magnitude than the output voltage. When an input voltage is applied, the low-drift, bandgap reference voltage is established. The bandgap voltage and an operational amplifier are used to sense the resistor divider voltage at the output of the operational amplifier. This resistor divider establishes the output voltage magnitude. Regardless of input voltage, as long as it remains higher than $V_{DROPOUT} + V_{OUT}$, the output voltage will remain constant. The input source through the p-channel MOSFET, Q_1 provides the output current load. Q_1 can also be other types of transistors, such as a Darlington pair of PNP bipolar transistors or an NPN transistor. The input current is equal to the output current plus the internal current required for bandgap generation, operational amplifier bias and p-channel MOSFET turn ON.

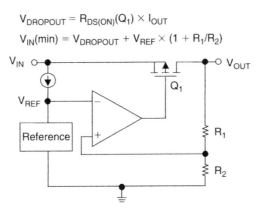

$$V_{DROPOUT} = R_{DS(ON)}(Q_1) \times I_{OUT}$$
$$V_{IN}(min) = V_{DROPOUT} + V_{REF} \times (1 + R_1/R_2)$$

Figure 8-16: This figure illustrates a simplified diagram of an LDO, which uses a MOSFET, Q_1, to provide the output current drive. Q_1 is also the fundamental limiting factor at the output. In other topologies, Q_1 can be a bipolar transistor.

The LDO noise is relatively small as compared to active device's capability to reject that noise on its power supply pin. For instance, the spot noise of a typical LDO at 10 kHz is $1 \mu V/\sqrt{Hz}$ rms. The noise rejection capability of a 10 MHz operational amplifier at that frequency is 70 dB. With this type of performance, the amplifier will reduce the noise from the LDO by 3162 times. Another way to look at this is that the LDO noise that gets into the amplifier (referred-to-input) is equal to approximately 316 pVrms.

In the general scheme of things, this is not critical. However, the noise rejection capability of the LDO is another story. If a SPC is in front of the LDO (as is the case with the circuit in Figure 8-3), it is highly probable that the LDO will not be able to remove that noise from the power supply line. In other words, LDOs are well known for "sharing" a lot of their noisy input with the rest of the circuit.

Switched Mode Power Supplies (SPCs)
Figure 8-17 shows a simplified example of a Buck-SPC circuit.

Several issues have an impact on the amount of noise that this converter generates. The noise magnitude at the output of these types of devices depends on the type of switching mode employed. Generally, if the switched mode power converter is in a pulse width modulation (PWM) mode, the peak-to-peak noise is lower than if the device is in a pulse frequency modulation (PFM) mode. The frequency of the switching noise of the converter is dependent on the converter's oscillator source. There is also an overshoot when a large amount of current is pulled from the device. **Figure 8-18** shows an example of the switching noise from this type of device.

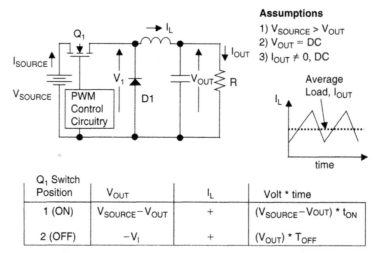

Q_1 Switch Position	V_{OUT}	I_L	Volt * time
1 (ON)	$V_{SOURCE} - V_{OUT}$	+	$(V_{SOURCE} - V_{OUT}) * t_{ON}$
2 (OFF)	$-V_1$	+	$(V_{OUT}) * T_{OFF}$

Figure 8-17: This diagram shows a Buck-SPC with the dynamic evaluation calculated for the device in the continuous mode. This power supply requires an external inductor, making manufacturing more difficult, and is also capable of higher emitted noise.

The device noise from a SPC changes into conducted noise once it is injected into the power supply trace. Figure 8-18 is not representative of the specific kind of noise that these types of devices generate. Every device is different, but there could be switched mode power supply noise in your circuit, and it might have a detrimental effect on the signal.

A second type of noise that comes from this type of device is *radiated noise*. The inductor in the circuit is a storage device and so it emits magnetic noise. The wall cube in Figure 8-5 has this type of power supply. The noise from the wall cube was effectively reduced with a ferrite bead (L_1, per Figure 8-5).

Capacitive Charge Pump

Figure 8-19 shows a simplified circuit example of the charge pump power management system.

Power supply noise is generated at the frequency rate of the internal oscillator of this device. Again, the magnitude and frequency of the noise at the output of this device, which is injected into the power trace, is device dependent. For details, refer to your vendor's datasheet. This device also emits noise due to the switching action through the capacitors.

Minimizing Device Noise

If you are interested in minimizing the device noise in your circuit, you should first inspect the value of the resistance. If possible, make the resistors as low as possible.

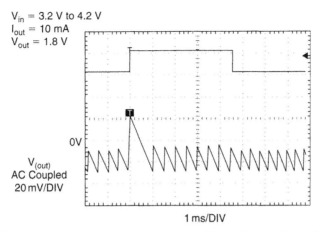

V_{in} = 3.2 V to 4.2 V
I_{out} = 10 mA
V_{out} = 1.8 V

0V

$V_{(out)}$
AC Coupled
20 mV/DIV

1 ms/DIV

Figure 8-18: This figure shows the output voltage of a Buck switch-mode power supply converter with time. The converter is in its frequency mode of operation. The output signal produces a complex 20 mVp-p signal that has a switching frequency of 2000 Hz.

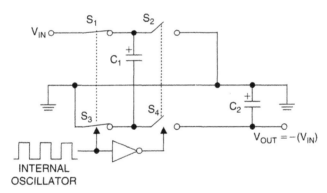

V_{IN}

S_1 S_2

C_1

S_3 S_4 C_2

$V_{OUT} = -(V_{IN})$

INTERNAL
OSCILLATOR

Figure 8-19: This figure shows an ideal switched capacitor charge-pump inverter. The low count in external components makes implementation of this circuit relatively easy.

Some design constraints could prevent you from lowering the circuit resistors as low as you would like. For instance, amplifiers are generally not designed to drive low resistive loads. Another reason you might hesitate is power considerations. If you have resistors spanning the power supply, their power dissipation could be an issue, particularly in battery-powered applications. But with these constraints in mind, lowering the values of your resistors can provide a noise improvement of $\sqrt{F}$, where F is the reduction factor of the resistor.

Active device selection could also be the way to go as you reduce the noise in your circuit. In this chapter, we discussed replacing the amplifiers in the circuit to lower noise amplifiers. This turned out to be critical, mainly because the amplifier noise (along with the resistor noise) was gained by the instrumentation amplifier configuration. Amplifier noise reduction is usually critical as the gain of the amplifier circuit increases. You might have noticed that I did not talk about the voltage reference in Figure 8-3. This is because that reference voltage and its noise goes straight through the instrumentation amplifier without the gain applied. But if you are using a voltage reference that is gained or a reference to a converter that is 16 or more bits, the voltage reference selection could be critical.

Other active devices, such as the A/D converters or D/A converters, should be selected to meet the requirements of your application. If you have a noisy device, you might have to change it out for a higher-bit device.

Of the three power supply devices, the LDO has a relatively low noise output. However, that device passes noise almost directly from its input to output. This is the case with the LDO that is used in the circuit in Figure 8-3. The other two power supply devices do generate switching noise. If you sample this noise, you will find that it is not random like the resistor or operational amplifier. Instead, it has a complex signal that rides on top of the DC output of these devices. The magnitude of this small signal can be from a few millivolts to tens of millivolts. You can try to replace these devices with lower-noise devices, but many times you will have other requirements in your circuit, such as efficiency. If this is the case, a replacement device might not be feasible. However, help is on the way as you read on. In the next section of this chapter we discuss power supply filters for your active devices.

Conducted Noise

The third type of noise that affects the performance of analog devices is conductive |noise. This type of noise already exists in the conductive paths of the circuit, such as the power lines or the signal path. Conducted noise mixes with the desired electrical signal. The best weapon against conducted noise is to go back to the source and implement noise reduction strategies on the offending device or radiating source. However, you'll find that sometimes you know the origin of the conducted noise and you can't further reduce it at the source. For instance, you might need to use a switching power supply because of its improved efficiency compared to an LDO. This power supply device allows you to run your application circuit with lower power dissipation. Now that this noise problem has changed from device noise to conducted noise, there are filtering techniques that you can use to overcome it. The origin of conducted noise is either device noise or radiated noise.

Noise in the Signal Path

Signal path noise can come from a variety of devices. For instance, resistor noise is a likely candidate in your circuit. Amplifier noise is another. Earlier in this chapter, we discussed the techniques that you might employ to reduce noise at the source. If you can't change the components or if there are no lower-noise alternatives, the next tactic is to insert some kind of filter. Your fundamental choices are lowpass, highpass, or bandpass. Of these three types of filters, you will find that the most beneficial one is the lowpass filter.

You will note that the circuit in Figure 8-3 does not have an antialiasing filter. As the data shows, this oversight has caused noise problems in the circuit. When the board has a second-order, 10 Hz, antialiasing filter inserted between the output of the instrumentation amplifier and the input of the A/D converter, the conversion response improves dramatically (refer back to Figure 8-5).

Analog filtering can remove noise superimposed on the analog signal before it reaches the A/D converter. In particular, this includes extraneous noise peaks. Analog-to-digital converters will convert the signal that is present on its input. This signal could include the sensor voltage signal or noise. The antialiasing filter removes the higher frequency noise prior to the conversion. Chapter 6 covers the topic of lowpass filters in detail.

Noise in Your Power Supply Bus

There are several things you can do to reduce noise in your power supply. The recommended strategy is to have a bypass or decoupling capacitor straddling the supply pin to ground of every active device. You will find that with analog devices the product datasheets recommend bypass capacitor values.

Analog devices and digital devices all require these types of capacitors. In both cases, these devices require a capacitor as close to the power supply pin(s) as possible, with a common value for this capacitor of $0.1 \mu F$. A second class of capacitor in the system is required at the power supply source. The value of this capacitor is usually about $10 \mu F$.

Bypass capacitors belong in two locations on the board: one at the power supply ($10 \mu F$ to $100 \mu F$ or both) and one smaller capacitor for every active device (digital and analog, see **Figure 8-20**). The value of the device's bypass capacitor is dependent on the device in question. If the bandwidth of the device is less than or equal to ~1 MHz, a $1 \mu F$ will reduce injected noise dramatically. If the bandwidth of the device is above ~10 MHz, a $0.1 \mu F$ capacitor is probably appropriate. Between these two frequencies, both or either one could be used. Refer to the manufacturer's guidelines for specifics.

Every active device on the board requires a bypass capacitor. It must be placed as close as possible to the power supply pin of the device, as shown in Figure 8-20. If you have two bypass capacitors for one device, the smaller one should be closest to the device pin. Finally, the lead length of the bypass capacitor should be as short as possible.

Bypass or decoupling capacitors and their placement on the board are just common sense for both types of devices but, interesting enough, for different reasons. In the analog layout design, bypass capacitors generally serve the purpose of redirecting high-frequency signals on the power supply that would otherwise enter into the sensitive analog chip through the power supply pin. Generally speaking, these high-frequency signals occur at frequencies beyond the analog device's capability to reject those signals. The possible consequences of not using a bypass capacitor in your analog circuit result in the addition of undue noise to the signal path and, worse yet, oscillation.

For digital devices such as controllers and processors, decoupling capacitors are required but for a different reason. One of the functions of these capacitors is as a "mini" charge reservoir. Frequently in digital circuits, a great deal of current is required to execute the transitions of the changing gates. Because of the switching transient currents that occur on the chip and throughout the circuit board, having additional charge "on call" is advantageous. Not having enough charge locally to execute this switching action could result in a significant change in the power supply voltage. When the voltage change is too large, it will cause the digital signal level to go into the indeterminate state, more than likely resulting in erroneous operation of the state machines in the digital device. The switching current passing through the circuit board traces would cause this change

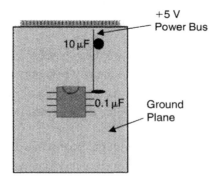

Figure 8-20: If you want to reduce the overall noise in your circuit (as well as enhance stability), the bypass or decoupling capacitors shown in this figure are crucial. The smaller-value capacitor (0.1 μF) is as close to the device power pin as possible. The higher-value capacitor (10 μF) is as close to the power supply source as possible.

in voltage. The circuit board traces have parasitic inductance, and you can calculate the change in voltage results using the formula:

$$V = L\delta I/\delta t \qquad [8\text{-}15]$$

Where V = Voltage change

L = Board trace inductance

δI = Change in current through the trace

δt = The time it takes for the current to change

So, for multiple reasons, it is a good idea to bypass (or decouple) the power supply at the power supply and at the power supply pin of active devices.

Although the manufacturer usually recommends a bypass capacitor value, you can determine the impact of changing the recommended value. The plot on the left in **Figure 8-21** shows the power supply rejection capability of a 12-bit A/D converter. If power supply noise exists at lower frequencies, the converter will attenuate them by ~80 dB, or 10,000 times. But at higher frequencies the converter is less able to reject signals on the supply.

Since this is a 12-bit converter, any signal injection that causes less than a 1/4 LSB error is not noticed. But as the interfering signal from the power supply starts to cause conversion errors, attenuation of that signal is required. In this example, the power supply to the converter is 5 V and it has ±20 mV noise riding on it. Near DC, this noise

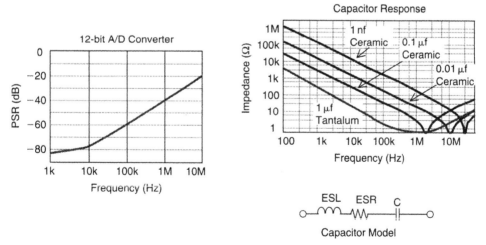

Figure 8-21: Bypass capacitors reduce the noise on the power supply pin of a device at frequencies where the power supply rejection capability of the device is too low to reject noise.

is attenuated 10,000 times by the converter, or to a voltage level of $2\,\mu V$ peak to peak. You would never see this noise at the output of the converter. However, the point where the noise is equal to about $\pm 1/4$ LSB or $\pm 1/4 \times$ (FSR/2^{12}), or $\pm 31\,mV$, the required attenuation for the $\pm 20\,mV$ noise signal is $-36.3\,dB$ or lower. This starts to occur at approximately $2\,MHz$ as you go up in frequency. If the noise signal is not attenuated at a frequency of $2\,MHz$ or higher, the noise will start to make its way into the output code.

An easy solution to this problem is to select a bypass capacitor that passes higher-frequency signals to ground. The graph on the right side of Figure 8-21 shows the frequency response of several capacitors. The frequencies where these curves extend down toward zero are the frequencies that are passed to the ground plane. For the application circuit, where the 12-bit A/D converter of Figure 8-21 is used, the best bypass capacitor would be a $0.1\,\mu F$ ceramic.

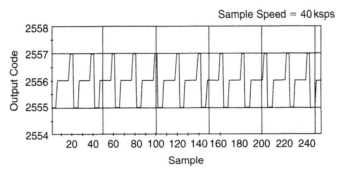

Figure 8-22: Power supply noise from the wall cube in Figure 8-3. The power supply frequency that appears at the output of the A/D converter is approximately 2 kHz.

Sometimes the simple bypass capacitor is not adequate. For instance, in the circuit in Figure 8-3 a power-supply small signal that came from the wall cube appeared at the output of the A/D converter. **Figure 8-22** shows this signal in the time domain.

You cannot filter this low-frequency noise with a bypass capacitor unless you are willing to use an extraordinary high-valued, large, expensive capacitor. This is not recommended. There are other techniques that you can use to filter power supply noise of this sort.

The most common power supply filtering for analog circuits uses a bypass or decouple capacitor (**Figure 8-23a**). This will reject frequencies that are present on the power supply line when the analog device is not able to reject an AC signal with its power supply rejection capability. If lower-frequency noise is still injected into the analog signal path (as illustrated in Figure 8-22), an additional component might be required in series with the device, as shown in Figure 8-23b. The resistor in series with the device that is shown in Figure 8-23b creates a lowpass filter in conjunction with C_2. This lowpass filter can be designed to eliminate lower-frequency noise if the correct resistor/capacitor combination is

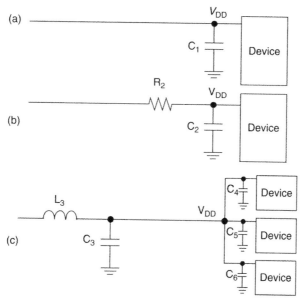

Figure 8-23: You can use these three power supply filters for higher-frequency filtering with (a) a decoupling capacitor, (b) a low pass filter, or (c) a ferrite bead.

chosen. This type of lowpass filter begins to attenuate power supply noise at the frequency $1/(2\pi RC)$.

The negative side of using the strategy is the loss of power supply voltage near DC. This loss is due to the voltage drop across R_2. Consequently, if you are trying to design a low-frequency filter, high-value capacitors are required. This can become a vicious circle.

Figure 8-23c shows an alternative circuit. In this circuit a ferrite bead is in series with the device. For this type of circuit, the ferrite bead can accept DC current while rejecting low-frequency AC signals. At low frequencies, ferrite beads are inductive. This combines with the bypass capacitor to form a lowpass LC filter. At higher frequencies the impedance of these beads is primarily resistive. Since the ferrite bead is an inductor that carries high currents, the DC loss across this element is very low.

Proper selection of your ferrite bead can reduce power supply noise. For example, a switcher that is operating with a 40 mVp-p ripple at ~150 kHz can use a ferrite bead to cut the ripple down to 3 mVp-p. This is can be achieved with a 50 µH ferrite bead and a 100 µF tantalum capacitor (C_3).

Revisiting Our Application Circuit With These Low-Noise Improvements
I always like to bring us back to reality (or better yet, the lab) to prove that things work the way we are theorizing they will. So let's return to the circuit in Figure 8-5. I show this circuit again in **Figure 8-24** so you won't have to leaf through the book pages.

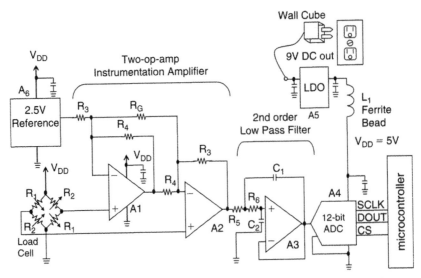

Figure 8-24: Starting from Figure 8-3, this circuit has reduced resistors, lower-noise amplifiers, a second-order lowpass filter, and an inductive choke or ferrite bead on the power supply. You can implement all these tactics in the circuit to reduce analog noise and achieve 12-bit accuracy from the 12-bit ADC.

In this new and improved circuit, I reduced the resistors in the instrumentation amplifier by $10\times$. I then swapped out the amplifiers for lower-noise devices. As I moved on to conducted noise problems, I added a second-order, lowpass filter in the signal path because I know that my instrumentation amplifier circuit had a high gain. Then I added the appropriate bypass capacitors near the active devices. From there I put in my ground plane. This addition made a big difference because the ground trace resistance was reduced and the board now had some shielding from outside interference. Finally, I added a ferrite bead to the power supply trace to remove any lower-frequency noise from the power supply. These noise-reduction activities changed my "~7-bit" converter to a true 12-bit converter.

References

Baker, Bonnie C., "Comparison of noise performance between a fet transimpedance amplifier and a switched integrator," AN-057, Texas Instruments.

Box, Hunter, *Statistics for Experimenters*, John Wiley & Sons, New York, 1978.

Fair-Rite Products Corp., *The Effect of Direct Current on the Inductance of a Ferrite Core.*

Morrison, Ralph, *Noise and Other Interfering Signals*, John Wiley & Sons, New York, 1991.

Ott, Henry W., *Noise Reduction Techniques in Electronic Systems*, John Wiley & Sons, New York, 1976.

Ryan, Scranton, "D-C amplifier noise revisited," *Analog Dialogue* 18-1, 1984.

Smith, Sheingold, "Noise and operational amplifier circuits," *Analog Dialogue* 3-1, 1969.

"Understanding data converter frequency domain specifications," AN-4, Datel.

CHAPTER 9

How to Design Analog Circuits Without a Computer or a Lot of Paper

Richard S. Burwen

Well, I sure do agree with Mr. Burwen, that he has the right attitude for designing analog systems and circuits. I've designed several settling boxes, and we use similar circuits. We even tend to use the same kinds of parts. /rap

Thoughts on Designing a Circuit

When designing a circuit, some people generate pages and pages of notebook paper full of calculations. Others fill their computer screens with elaborate formulas and simulations. For over 40 years I've been designing analog and analog-to-digital circuits. Although I have a small roomful of engineering notebooks, they are mostly full of schematics, prints, printed circuit layouts, and written reports on how the circuits work—and not much in the way of calculations.

Not that I don't calculate; I am a pretty good computer. I design my circuits as I am drawing them and generally do the calculations in my head. Using simple approximations, I can probably come within 5% to 10% of the results that somebody else might generate from a notebook full of calculations.

I have nothing against using computers, calculators, manual computation, or even cookbooks, if that's the way you like to work. Probably I am the last circuit designer in the world to have acquired a computer. Right now I have two AT clones on my desk and a portable on my workbench for taking notes. I love my computers and I hate my computers. I hate them for the three months or so of my life that I wasted on program idiosyncrasies, incompatibilities, disk problems due to copy protection, IO port interferences, computer hangups, and a lot of little problems you have to learn about the hard way. My disks have tens of megabytes of programs I own but don't use.

My computers help me with three things. First is schematic drafting using Futurenet. I don't try to analyze or simulate my circuits, just draw them. I used to be a terrible draftsman, but now my schematics are clear and beautiful—drawn by my plotter. Second, I do a lot of writing using a word processor. That has not speeded up my writing at all. I used to dictate my reports to my secretary and ship them out with just minor revisions. Now I can change my mind as much as I want, and it is less efficient.

The third thing I use my computers for is storing tidbits of information. I use a program called MemoryMate, which is very simple to use. I use one copy of it for my telephone file. I use a second copy of the same program to store my bench notes, usually done on the portable computer. The third and fourth copies of the program store information on my sound recordings and anything else I want to remember. The nice thing about this program is that if you can remember a single word of what you wrote there is no trouble finding it. Usually, just to make sure, I add a few keywords at the top of the page, so if I can think of something related I can find it. Using computer memory is a lot better than stuffing notes into my engineering notebooks, where I think it is too much trouble to go searching for the information.

Once in a while I use an electronic calculator. When I was in college and during my early engineering years I used to pride myself on carrying around the most complex slide rule and a miniature version of it in my shirt pocket. Today I use a pocket calculator, a Sharp 5100. It has a zillion buttons, but what I really like about it is the ten memory buttons that can store intermediate results. I use it mostly for calculating precision resistors and values that have to be closer than 1%.

To give you an idea of how I think about circuits, it might help to know more about my background.

My Background

I have known since I was eight years old that I wanted to be a circuit designer. That's lucky. A lot of kids take thirty years to figure out what they want to do. My dad once sold radios back in the 1920s, and he got me started building a crystal radio. Probably most of you are too young to remember the old galena crystal with its cat whisker. The crystal was a little rock of mineral embedded in a cylinder of lead or solder about 0.4 inches in diameter and 0.2 inches high. To make it receive local AM radio signals you fished around for a sensitive spot on the crystal using a small pivoted arm with a cat whisker. A cat whisker was a little piece of springy wire. When you touched the crystal at the right spot with the tip of the wire and with the right pressure, presto—the first crystal diode or semiconductor rectifier.

With my dad's advice and finding what I could in books, I built at least three different crystal sets. They all received something. I used my bedspring for an antenna and a radiator for ground. Then I improved reception with a long wire antenna in the attic. I tried a lot of different coils, including one wound on a Quaker Oats box and different variable tuning capacitors. The third crystal set was a sleek one that I built into a redwood cigar box, which I sanded, stained, varnished, and polished like a mirror.

After the crystal radios, I graduated to a one-tube radio. From the first vacuum tube set I advanced to two tubes and then three tubes. None of the circuits was exactly like what I found in the books but used the general ideas with my own improvements.

At age eleven I developed a very healthy respect for high voltage. One day when I was visiting a ham radio operator friend in his attic, I managed to get one hand across his telegraph key while holding a headphone in the other hand. There was 400 V DC between them. I couldn't let go, but fortunately he was right beside me and shut off the power quickly. Now I keep one hand in my back pocket.

When I was thirteen I became a ham radio operator, but I was always more interested in building the equipment than in talking. At age fourteen I took the tests and acquired commercial radio telephone and telegraph licenses. While I was still in high school, World War II started. I worked afternoons and Saturdays servicing and aligning HRO shortwave receivers used by the military. At the beginning of the war the government shut down ham radio activities, and at that point, I became interested in high-fidelity music reproduction.

I realized from the beginning that frequency response was one of the primary factors in making good sound, and I have spent a lot of my life designing various kinds of analog filters, equalizers, and tone control systems related to hi-fi. The radio-phonograph I built during high school and early college days used multiple speakers, a complex crossover network, bass and treble controls, and most important, speaker equalization. It was built into large maple and walnut cabinets with transparent doors and colored lights to enhance the music.

After a half year at Harvard College, I entered the U.S. Navy and attended the electronic technician schools. That was a terrific practical education I wish every engineering school would provide. It gave an insight into the practical application of all the theory one gets in college. I often wondered how anybody could get through college physics, mathematics, electrical theory, Maxwell's equations, and such without dropping out from boredom unless he had such a practical course in electronics.

The Navy course consisted of preradio, primary radio, and secondary school, totaling a year. Because I had a lot of practical experience, I was able to take the final exam at the end of my second month in primary school and graduate at the top of the third month graduating class. Students with the best grades usually got their choice of where to go to secondary school, so the Navy sent me to Treasure Island in San Francisco Bay, which was my last choice. By the time I graduated from secondary school, the war was over. The Navy made me a typist for three months, discharged me, and I went back to Harvard College.

When I wanted to take somewhat advanced electronics courses and skip one or two of the more fundamental courses, my adviser, Professor F. B. Hunt, asked, "Do you want to be a scientist or a hi-fi nut?" My answer was "both." I guess I've succeeded at least at the latter.

After finishing college and getting a master's degree at Harvard, I promptly retired. What I did was go down into my workshop in my parents' basement, where I tried to develop a top-notch hi-fi system built into the base of a Webster record changer. My intention was to sell hi-fi systems to wealthy people. It was a great experience because I learned a lot about multiple feedback loops and equalization. I built two complete systems. About the time I finished construction I realized that the people who had the money to buy the systems didn't care about the fidelity, and the people who cared didn't have the money.

Just at that time I got a call from an electronics company and took a job in 1950 designing equipment for one of the first cable television systems. Over the next eleven years I worked for a half-dozen companies designing analog circuits for laboratory instruments, hi-fi, and military equipment. Since 1961 I have been working full time as an analog circuit design consultant. Companies don't hire you for this kind of work unless they are overloaded, in a hurry, or you have a capability they don't have. So, for thirty years I have had a lot of fun and been involved in extremely interesting projects, including medical equipment, hi-fi, space, automotive, TV, analog function modules and ICs, power supplies, laboratory instruments, and lately, switching power amplifiers.

My first love, hobby, and part of my business has been hi-fi. My home system has 169 speakers in one room and 20,000 W. It took 25 years to build so far and will be finished around 2010.

Over the years, only a small percentage of my bread and butter work has been concerned with hi-fi. That doesn't matter; almost anything you can design in analog circuits has some bearing on hi-fi. My sound system uses about 2,000 op-amps. The first op-amps I designed for my hi-fi were discrete potted modules. These modules, somewhat refined, became the first products of Analog Devices. Later I developed more complex signal processing modules, which helped start two high-end audio companies and also companies bearing my own name.

You can see that my practical background has given me some feel for how circuits work. Although I have hardly ever directly used the mathematical theory I gained in college and graduate school, I am firmly convinced that a good mathematical background is an absolute necessity to help you make the best trade-offs in designing a circuit. Most of the tools I use in my designs are pretty simple. Here is how I go about designing analog circuits.

Breaking Down a Circuit

In the beginning there were transistors (what is a tube?). They became very cheap and you could use a lot of them. When the first op-amps were developed, they were expensive and you had to conserve circuits. Now you can buy several op-amps on one chip so cheaply that you can use them as high-quality transistors. That makes it easier to design circuits separated into simple functions.

For example, suppose you want to design a standard phono equalizer. It has a high-frequency rolloff and a bass boost. You can build a single network around one op-amp stage, or you can separate the high- and low-frequency parts of the equalization into separate op-amp circuits, cascaded. Separating the circuits allows you to adjust one time constant without affecting another, and the circuit is easier to calculate.

The first thing to do is to break down a circuit into all its blocks. If each block has a very high input impedance and a near-zero output impedance, one block can feed another without interaction. That's the beauty of using a lot of op-amps. Noise buildup from using a lot of separate circuits can be more of a problem or less of a problem. If you keep all the signals at the highest possible level consistent with not overloading, you will probably generate less noise than trying to perform several functions in a single circuit. The more functions you perform in one circuit, the more interaction there is between them. Usually circuits toward the output of a network have to be higher in impedance than circuits near the input, to reduce loading problems. The higher the impedance, the more noise it generates.

The lowest-noise circuit you can make that performs a lot of different functions usually consists of a number of near-unity gain op-amp circuits. Low gain means less amplification of noise. So I use a lot of op-amp followers in my designs.

If you separate all the functions of a circuit into building blocks that don't interact, then the design job is relatively simple. Each block can be designed independently of the others, provided it can feed the load.

Equivalent Circuits

If you break a circuit apart at any point, it looks like a source feeding a load. The source has an internal impedance, and the load affects the final output. However, if you have broken your circuit into individual op-amp circuits, each with a near-zero output impedance compared with its load, you don't have to worry about the interaction. Within each individual block, you can use Thevenin or Norton equivalents to determine the gain vs. frequency.

There are two equivalents. The source can be thought of as a voltage source having an internal impedance, or the source can be thought of as a current source in parallel with its

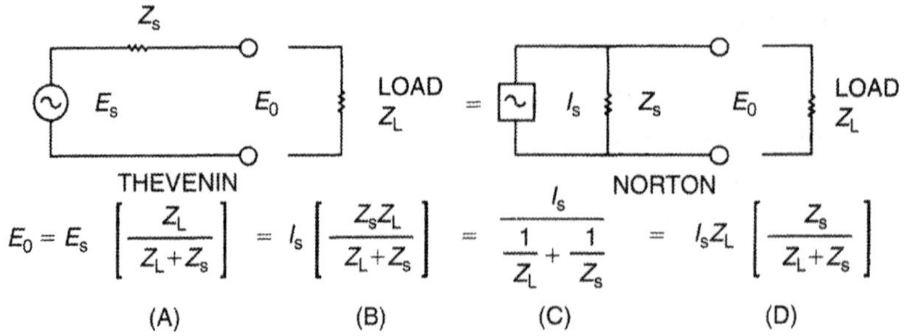

Figure 9-1: Equivalent sources.

own internal impedance (see **Figure 9-1**). If you have a complex network, it is frequently convenient to alternate between voltage and current source equivalents. All you have to know to calculate the gains of these circuits is how to calculate the gain of a two-element voltage divider and how to parallel impedances.

In the case of the voltage source, formula A gives the output voltage as determined by the ratio of the load impedance to the total impedance consisting of the source and the load. In the case of the current source, the source current flows through both the source and the load in parallel. So the output voltage in formulas B and C is just the source current times the parallel impedance.

If the load is open, all the source current flows through its own internal impedance Z_s, producing an output $E_0 = I_sZ_s = E_s$, the equivalent source voltage. If the load is a short circuit, all the source current flows through the load path and none through Z_s, producing zero output voltage. In between, a fraction of the source current flows through the load impedance, producing the output in formula D, which is equal to B.

When more than one source contributes to the output of a linear circuit, you can consider the effect of each source separately. Leave all the impedances connected and short all but one voltage source. Compute the output due to that source using Thevenin equivalents. Next, short that voltage source, turn on the next, and calculate the output. After calculating the outputs due to each source, you can add them all together to get the total. If a source generates current, open it, but leave its source impedance connected while calculating the effect of another source. Use whichever type makes calculation easier.

You can plug the numbers into your calculator or you can make an estimate in your head. The ratio of load to source impedance Z_L/Z_s gives you gain G.

$$G = \frac{E_0}{E_S} = \frac{1}{1 + Z_S/Z_L} \qquad\qquad [9\text{-}1]$$

Table 9-1

Ratio Z_S/Z_L	Gain	Attenuation
0	1	1
1/4	4/5 = 0.8	5/4 = 1.25
1/3	3/4 = 0.75	4/3 = 1.33
1/2	2/3 = 0.667	3/2 = 1.5
1/1	1/2 = 0.5	2
1.5	0.4	2.5
2	0.333	3
3	0.25	4
4	0.2	5
9	0.1	10
100	0.0099	101

It also gives you attenuation $1/G$.

$$\frac{1}{G} = \frac{E_S}{E_0} = 1 + \frac{Z_S}{Z_L}$$
[9-2]

Frequent numbers appear in **Table 9-1**.

Stock Parts Values

Unless your system requires a very precise odd gain in one of its blocks, you don't have to calculate very accurately. You just have to arrive at the nearest stock resistor value. That makes calculation easy.

One-percent resistors are so cheap and inserting them in boards is so expensive that there is no worthwhile savings when using 5% or 10% resistors. Your company can waste a lot of money stocking all the different values. My designs use standard 1% 0.25 W, 100 ppm/°C resistors with twelve values selected from each decade, according to **Table 9-2**. Once in a while I need an accurate, stable resistor, and I select from a very few 0.05%, 0.1-W, 10 ppm/°C values shown in **Table 9-3**.

Similarly, I use a limited number of capacitor values (see **Tables 9-4** and **9-5**).

For precision filtering I use 2%, 50 V polypropylene capacitors having values limited to 0.0033, 0.0047, and 0.01 uF.

Table 9-2

Stock Resistor Values in Ohms, 1%, 0.25 W, 100 ppm/°C					
10.0	100	1000	10.0 k	100 k	1.00M
11.0	110	1100	11.0 k	110 k	1.50M
12.1	121	1210	12.1 k	121 k	2.00M
15.0	150	1500	15.0 k	150 k	3.01M
20.0	200	2000	20.0 k	200 k	4.99M
30.1	301	3010	30.1 k	301 k	
40.2	402	4020	40.2 k	402 k	
49.9	499	4990	49.9 k	499 k	
60.4	604	6040	60.4 k	604 k	
69.8	698	6980	69.8 k	698 k	
80.6	806	8060	80.6 k	806 k	
90.9	909	9090	90.9 k	909 k	

Table 9-3

Stock Precision Resistor Values in Ohms, 0.05%, 0.1 W, 10 ppm/°C
100
4990
10.00 k
49.90 k
100.0 k

Table 9-4

Stock Ceramic Capacitor Values in picofarads, 5%, 50 V, ±30 ppm/°C, 10 pF and Larger		
5		
10	100	1000
15	150	
22	220	
33	330	
47	470	
68	680	

RC Networks

Most engineers have no feel for the relationships among reactance, time constant, and frequency response. They have to plug all the numbers into formulas and see what comes out. It's fairly simple to calculate RC circuits in your head.

Table 9-5

Stock Metallized Film Capacitor Values in microfarads, 5%, 50 V, ±200 ppm/°C			
	0.01	0.1	1.0
0.0015	0.015	0.15	
0.0022	0.022	0.22	
0.0033	0.033	0.33	
0.0047	0.047	0.47	
0.0068	0.068	0.68	

First, let's look at the simple RC lowpass filter in **Figure 9-2**. The filter is a simple voltage divider whose gain is 0.707 or 3 dB down at the frequency where the reactance of the capacitor equals the source resistance R in ohms. The magic numbers are the $-3\,\text{dB}$ frequency, f_0, and the time constant T, simply the product of resistance and capacitance in microseconds.

While capacitive reactance can be calculated from:

$$X_C = \frac{1,000,000}{2\pi f C} = \frac{159,155}{f C} \qquad [9\text{-}3]$$

$$C = \frac{1,000,000}{2\pi f X_C} = \frac{159,155}{f X_C} \qquad [9\text{-}4]$$

where f is frequency in hertz and capacitance is in microfarads, all you have to remember is that $1\,\mu\text{F}$ has a reactance of approximately $160,000\,\Omega$ at 1 Hz. You can figure out everything else from this. For example, $1\,\mu\text{F}$ at 1 kHz has a reactance of $160\,\Omega$. At 1 MHz it has a reactance of $0.16\,\Omega$. And 1 pF at 1 MHz has a reactance of $160,000\,\Omega$; 1 nF $(0.001\,\mu\text{F})$ has a reactance of $160,000\,\Omega$ at 1 kHz.

Suppose you want to design an RC lowpass filter that attenuates $-3\,\text{dB}$ at 1 kHz (the cutoff frequency f_0). Let's start with a resistance of $4990\,\Omega$. This is one of my frequently used stock values and is an appropriate load for an op-amp. We need a capacitive reactance of $4990\,\Omega$ at 1 kHz. What is the capacitor value?

Just divide 160,000 by the frequency in hertz (1000) and then by the number of ohms (5000), as in Equation 9-4. The tough part is getting the decimal point right. Remember the number 160,000 is associated with ohms and microfarads. It also works with kilohms and nanofarads (1 nF = 0.001 μF) or megohms and picofarads. The simple RC lowpass filter works out to need 0.032 μF, actually 0.0318948 μF, for a 1 kHz rolloff. That's close to my stock value of 0.033 μF.

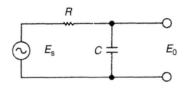

Figure 9-2: RC lowpass filter.

Another way of looking at the simple RC low-pass filter is to associate its time constant $T = RC$ with its cutoff frequency f_0.

$$f_0 = \frac{160,000}{T} = \frac{160,000}{RC} \qquad [9\text{-}5]$$

$$T = \frac{160,000}{f_0} = RC \qquad [9\text{-}6]$$

T is in microseconds and f_0 is in hertz.

A filter having a 1 MΩ resistor and a 1 μF capacitor has a time constant of 1 sec, or 1,000,000 μsec, and cuts off −3 dB at 0.16 Hz. That is the frequency at which 1 μF has a reactance of 1 MΩ and equals the 1 MΩ resistor. You can calculate all other simple RC filters from that point.

If capacitance is in microfarads and resistance is in ohms, the time constant is in microseconds. If you know the time constant you can figure the cutoff frequency, and vice versa. As examples, a time constant of 1 msec produces a cutoff frequency of 160 Hz. And 1 μsec corresponds to 160 kHz. To find the frequency, just divide 160,000 by the time constant in microseconds. To find the time constant, just divide 160,000 by the frequency in hertz.

Once you have done this calculation in your head a few times, you acquire a feel for what time constant goes with what frequency. I know that 100 μsec goes with 1600 Hz, 160 μsec goes with 1 kHz, 300 μsec goes with about 500 Hz, 10 μsec goes with 16,000 Hz, and so on.

The response of the simple RC lowpass filter at any frequency is determined by its frequency ratio.

$$G = E_0/E_S = \frac{-jX_C}{R - jX_C} = \frac{1}{1 + j2\pi fT} \qquad [9\text{-}7]$$

$$|G| = E_0/E_S = \frac{1}{\sqrt{1 + (f/f_0)^2}} \qquad [9\text{-}8]$$

Table 9-6

f/f_0	Decibels
1/7	−0.1
1/2=0.5	−1
1	−3
2	−7
10	−20
100	−40
1000	−60

If you know what the 3 dB cutoff frequency f_0 of a simple RC filter is, you can plot its entire response curve on semi-log graph paper, or you can do it in your head. The curve is universal. You just have to move it to the right frequency. Find the ratio of the frequency of interest f to the 3 dB cutoff frequency f_0 and you can determine the response. In **Table 9-6** you can see that at half the cutoff frequency the response is down 1 dB, and at twice the cutoff frequency it is down 7 dB. At 1/7th of the cutoff frequency it is down 0.1 dB. Well beyond the cutoff frequency, the response goes down at 20 dB/decade.

If the filter is a highpass type instead of a lowpass type, simply interchange f_0 and f in Table 9-6.

Often I need to estimate the transient response of a simple RC filter or find how far the capacitor will charge in a given time for a step input. The step response of the filter is:

$$\frac{E_0}{E_S} = 1 - e^{-t/T} \qquad\qquad [9\text{-}9]$$

Table 9-7 is a table of useful values of output vs. time as a fraction of T, the RC time constant.

You can figure out most of this table in your head if you can remember that the capacitor charges up to 0.63, or 63%, in one time constant. Also, at 2.3 time constants the capacitor charge reaches within 10% of its final value, at 0.9, or 90%.

The exponential curve has the same shape in the next 2.3 time constants but starts at a point 10% away from the final value. Therefore, the value at the end of $2 \times 2.3 = 4.6$ time constants is within 1% of final value, at 0.99, or 99%. Similarly, at $3 \times 2.3 = 6.9$ time constants the capacitor charges to within 0.1% of final value, at 0.999, or 99.9%. At small fractions of a time constant the fractional charge is the same as the fractional time.

Table 9-7

t/T	E_0/E_s
0.001	0.001
0.01	0.01
0.1	0.1
0.2	0.18
0.5	0.39
1	0.63
2.3	0.9
4.6	0.99
6.9	0.999
9.2	0.9999

If the simple RC filter is a highpass type instead of a lowpass type, subtract the above outputs from 1.

Stabilizing a Feedback Loop

The first rule for making a feedback loop stable is to keep it simple. Flat response feedback around a single RC rolloff or an integrator produces a lowpass filter that looks like a single RC rolloff. It goes down at 6 dB/octave.

The single RC rolloff produces a nice exponential step response with no overshoot. If the open-loop response goes down at 6 dB/octave, it has 90° of phase lag. When you close the loop and make a lowpass filter out of it, the phase shift at the −3 dB point is 45°. Anything else in the loop that adds phase shift tends to cause a peak in the frequency response and an overshoot in the step response.

Let's start with a simple active low-pass equalizer, **Figure 9-3**. The gain vs. frequency of this equalizer is Z2/Z1, which is simply the ratio of the feedback impedance Z2 to the input resistor R1. I have chosen the feedback network to be a parallel resistor R2 and capacitor C1, having a time constant of 150 μsec. This active filter has the same frequency response as the simple lowpass filter in Figure 9-2 if the time constants are the same. The difference is the active filter inverts and has a near-zero output impedance, and you can design it to provide DC gain.

The −3 dB cutoff frequency associated with 150 μsec is approximately 1,000 Hz. Remember 160,000/T? You can look at this circuit as an integrator consisting of an input resistor R1 and a feedback capacitor C1. Adding resistor R2 provides an overall DC feedback path to convert the integrator into a lowpass filter.

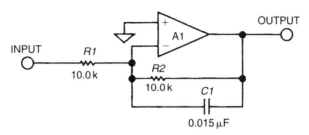

Figure 9-3: Active 6 dB/octave lowpass filter.

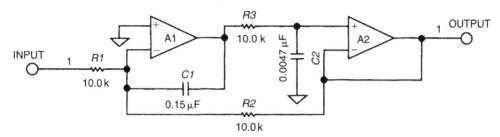

Figure 9-4: Active 12 db/octave lowpass filter.

Another way to look at this circuit is to consider A1 as a current source whose terminals are its output and its negative input. This source has an internal resistance *R2*. Because large feedback keeps the negative op-amp input at near-ground potential, the source current through *R2* is the same as the input current through *R1*. This internal source resistance *R2* is loaded by capacitor *C1*, making the equivalent circuit of the simple RC lowpass filter in Figure 9-2.

Now suppose the DC feedback, instead of coming directly from the output of op-amp A1, comes from a more complex system shown in **Figure 9-4**. Here the integrator A1 is followed by a lowpass filter *R3* and *C2*, buffered by a unity-gain follower amplifier A2. I have chosen the cutoff frequency of this filter at $3 \times 1{,}000$ Hz, or 3 kHz. Its time constant $T = RC = 160{,}000/3000 \, \text{Hz} = 53.3 \, \mu\text{sec} = 0.0047 \, \mu\text{F} \times 10{,}000 \, \Omega$. What happens?

The gain at 1,000 Hz is practically the same as in Figure 9-3, so the loop gain is determined almost entirely by the integrating capacitor *C1* and the feedback resistor *R2*. The main effect of the 3000 Hz lowpass filter is an additional 6 dB/octave rolloff and a contribution to phase shift which results in about 1% overshoot in the step response.

Here is the rule. If you have feedback around an integrator and a 6 dB/octave lowpass filter, you can achieve transient response with only 1% overshoot by making the cutoff frequency of the lowpass filter three times the cutoff frequency of the integrator with DC feedback alone. If the cutoff frequency of the lowpass filter is lower, you get more phase shift and more overshoot. At 1,000 Hz, the lowpass filter contributes a phase lag of 18.4°.

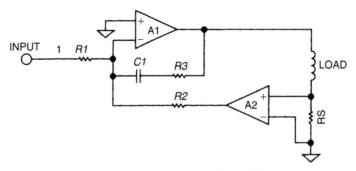

Figure 9-5: Current control amplifier.

Added to the integrator phase lag of 90°, the total open-loop phase shift is 108.4°, less than 110°. That's a nice number.

Remember that 110° total phase shift at the unity-gain frequency gives you beautiful transient response. Unity gain means that if you break the loop at a convenient point and connect a signal generator there, the magnitude, but not the phase, of the signal coming back is the same as that of the signal generator. If there is more than one lowpass filter in the circuit or contributor to phase lag at the unity-gain frequency, you have to add up all the phase shifts. Much below the 3 dB cutoff frequency of a simple RC lowpass, the phase shift is approximately proportional to frequency based on one radian or 57° at the cutoff frequency. At frequencies higher than half the cutoff frequency, the formula is inaccurate. For example, at the cutoff frequency the phase shift is 45°, not 57°.

A filter such as that in Figure 9-4 involves more than a single feedback loop. Follower A2 has its own feedback from output to input. Feeding back through R2 to A1 provides additional feedback around A2 at DC and low frequencies, producing open loop nearly a 12 dB/octave slope and 180° phase shift. A very high-frequency version of this circuit might oscillate if A2 does not have favorable overload and slew rate characteristics. Depending on the method of internal stabilization, some op-amps delay several microseconds in coming out of overload, effectively adding to system phase lag.

Any feedback loop involving integration or lowpass filtering in the forward path may be subject to overload recovery problems. This is because the capacitor involved becomes overcharged when the output of the system saturates. Therefore it is necessary to understand what happens to a feedback loop when various parts are driven into overload. You should know how an op-amp recovers from overload before designing it into your circuit. Sometimes the problem can be avoided by limiting the input signal amplitude.

Another kind of loop you may have to stabilize is one in which the load is inductive and behaves as an integrator. This happens with many magnetic loads such as deflection circuits, magnets, and motor drives. In **Figure 9-5**, a low-resistance shunt measures the current in the load coil, and its output is amplified by A2 to provide feedback. This loop already has nearly 90° phase shift over a wide range of frequencies due to the load. Therefore, the feedback network around A1 has to have a low phase-shift flat response region at high frequencies determined by resistor *R3*. We want maximum feedback at DC for accurate control of the output current, so it uses an integrating capacitor *C1*. Here is my simple way of stabilizing the loop without knowing anything about it:

1. Short out the integrating capacitor *C1* and connect a potentiometer in place of *R3*.

2. Connect an oscilloscope to look at the feedback from A2 and the error signal output of A1.

3. Feed in a small signal square wave and adjust *R3* to the maximum resistance value that gives you a satisfactory amount of overshoot.

4. Connect in a large value integrating capacitor *C1* and then select smaller and smaller values, accepting the smallest that does not seriously degrade the good transient response you just had.

That's it. No simulation, no calculations. A great time saver. This method works for all kinds of feedback systems that can be stabilized by a simple series RC network. If the system has additional contributors to phase lag, you might need to compensate by adding a phase lead network such as a capacitor in series with a resistor across the feedback resistor *R2*. This network can reduce the total phase shift at the unity-gain frequency and thereby reduce overshoot and ringing.

Circuit Impedance

High-impedance circuits are affected by small stray capacitances, and they generate more noise than low-impedance circuits. When using operational amplifiers at frequencies below 100 kHz, my rule is to use circuit impedances in the vicinity of 5000 Ω. Most BIFET op-amps can feed a 5000 Ω load without distorting much at 20 kHz. If you are using an op-amp as a follower, it has maximum feedback. Phase shift caused by stray capacitance can make the circuit ring in the megahertz region.

For example, suppose instead of connecting the output of the follower directly to the input, it goes through a 100 kΩ resistor. Stray capacitance of only 3 pF to ground will make a 500 kHz lowpass filter at the inverting input. Its phase shift at 1.5 MHz is about 71.5°, which will cause many cycles of ringing, perhaps oscillation. If the feedback resistance is only 5000 Ω, the 3 dB point of the stray capacitance lowpass filter is 10 MHz and will not cause ringing if the unity-gain frequency is only 3 MHz.

In some circuits you need a high resistance, at least at low frequencies. If the resistance is more than 5000 Ω in a circuit that produces unity feedback at high frequencies, it is necessary to bypass the output to the negative input with a capacitor. Circuits that have high closed-loop gain can tolerate higher impedances because the unity-loop gain frequency is lower. Remember the 3:1 cutoff frequency rule for low overshoot and less than 110° phase shift. High-frequency op-amps have more than 90° phase shift at high frequencies and can tolerate very little phase shift in the feedback path.

Lower-impedance circuits are less susceptible to noise pickup from other circuits. On a printed circuit board, for example, where two adjacent conductors may have a capacitance between them of about 1 pF per inch, you can estimate how much crosstalk you will get by estimating the ratio of circuit impedance to coupling reactance at the frequency of interest.

New Parts

Before choosing a part you have never used before, it is important to find out its characteristics—not just those on the specification sheet but important characteristics that are unspecified. For example, before using a new op-amp, you should find out if the manufacturer really designed it to be stable at unity gain, or if it is on the verge of oscillation. When you overdrive either input, does the op-amp go into phase reversal? Can it be driven by an op-amp of the same type without phase reversal but not by a different type that delivers more voltage? Does it produce a lot of popcorn noise that could bother your system? Does it delay in coming out of overload? If you don't see the characteristic in which you are interested on the specification sheet or covered in application notes, you should assume that the part performs poorly in that respect.

Many of the more serious troubles I have encountered in making my circuit designs work resulted from incomplete knowledge of parts I was using for the first time. For example, in more than one instance I have been burned by parasitic effects in an integrated circuit. One section of the chip that is supposed to be unrelated to another section affects the other when you feed current into one of the pins driving it below ground or above the supply rail. It really pays to make a few crucial experiments before designing in any part with which you are not completely familiar.

Breadboarding

If the circuit involved is closely similar to circuits I have used before and the operating frequencies are not too high, I usually skip the breadboarding phase and go straight to a printed circuit layout. Parts of the circuit that involve tricky feedback loops or unfamiliar parts or are susceptible to wiring inductance or capacitance need to be tested.

Breadboard circuits should be carefully constructed with attention paid to grounding, shielding, and lead lengths. Use a ground-plane board. You can waste a lot of expensive engineering time finding troubles in a breadboard circuit that has been just thrown together.

Some engineers prefer computer simulation. That's okay, but the one big advantage of the experimental method is that the results agree with experiment.

Testing

I can't believe it—a technician turns on one of my circuits for the first time, feeds in an input signal, and expects the correct signal to appear at the output. I don't have that much confidence. When I test a circuit I break it into its blocks and check DC voltages, gain, frequency response, and other important characteristics of every single part of each block. It is important to know that every component you design into a circuit is really serving its purpose. If blocks of the circuit cannot be easily separated from others or if test signals cannot be injected, you can measure the output signal from one block used as an input signal to the next block and then see that the next block does its job relative to that signal. Once the individual sections of a circuit are working, I check groups of blocks and finally the whole system. Even if the system seems to deliver the correct signal all the time, that does not mean that every intermediate part of the circuit is really functioning correctly, optimally, or reliably.

How Much to Learn

As a consultant I have had the opportunity to work in many fields of electronics. Many times I have been surprised at how soon new circuit knowledge gained in one field became useful in an entirely different area, sometimes within a week. Efficient circuit design comprises building on what others and especially you have done before, with a bit of innovation, but not too much. I have a stock of circuits in my computer, such as common mode rejection amplifiers, output followers, crystal oscillators, and triangular wave generators, but I rarely use any circuit exactly as I did before. They keep evolving with new parts and characteristics adapted to new requirements.

Once in a while you need to take on a project involving circuits, parts, and ideas entirely new to you. Pioneering usually is not a way to make money directly. You run into too many unforeseen problems. However, it gives you knowledge which, if applied over and over again with small improvements to other projects, really puts you ahead.

Settling Time Tester

I needed a production test instrument to measure the settling time of a power amplifier used to drive the gradient coils in magnetic resonance imaging machines. In this application the output current to a load coil has to follow an input pulse and settle to within 0.1% of final value within 1.3 msec. This settling requirement applies both at the top of the pulse and following the pulse. Pulses can be either positive or negative.

To avoid overloading the amplifier, the input pulse must have a controlled slope, typically lasting 1 msec, on the leading and trailing edges. For an accurate settling test, the top of the pulse has to be extremely flat and free of noise.

In addition to generating the pulse, the instrument has to provide a means of filtering out 81 kHz noise and magnifying the top of the pulse without distortion caused by poor overload recovery of an oscilloscope. I decided to build an analog signal generator and error amplifier using op-amps and some HCMOS logic.

The tester consists of two sections. A wave form generator delivers the slow rising and falling pulse to the amplifier and a synchronizing signal to an oscilloscope. An error amplifier then processes the amplifier's current monitor signal for viewing on an oscilloscope. Processing consists of filtering out 80 kHz noise, offsetting the top of the pulse to zero, and then amplifying and clipping the error.

The block diagram, **Figure 9-6**, shows the organization of the system. The upper set of blocks is the wave form generator, and the lower set of blocks is the error amplifier. The wave form generator starts with a pulse generator block that delivers −3.3 V to +3.3 V pulses, selectable in polarity, and adjustable in width and frequency. An integrator that saturates and recovers quickly slopes the leading and trailing edges and increases the pulse size to ±13 V.

After the integrator, two different clipping circuits select portions of the signal. One passes the portion of the integrator output signal from 0 to +10.5 V, while the other

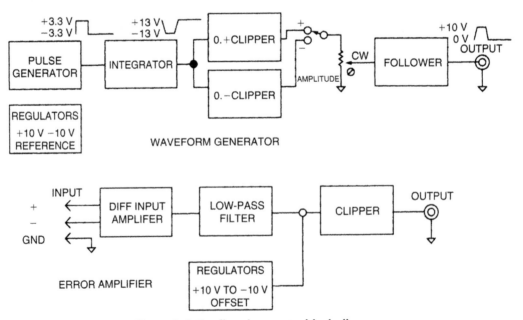

Figure 9-6: Settling time tester block diagram.

passes the negative portion of the signal from 0 to -10.5 V. After selecting the output of one or the other clipper, the operator adjusts the amplitude of the signal using a 10-turn potentiometer, and the output goes through a follower amplifier to the power amplifier under test. The diagram shows the wave forms at important points.

The error amplifier system uses a differential input buffer to get rid of ground voltage noise at the input connections. Lowpass filtering at 45 kHz attenuates 81 kHz and higher frequency noise. Then coarse and fine offset potentiometers adjust the top of the pulse to 0 V. The resulting signal is amplified 10 $\times$ in a fast recovery amplifier, which clips the output at ± 1 V. An oscilloscope connected to the output will display a range of ± 100 mV referred to the top of the pulse. You can clearly see 0.1% of a 5-V signal as 1 cm deflection at 50 mV/cm.

Now let's run through the schematic in **Figure 9-7** so you can get an idea of my thinking and how little calculation was necessary. First, I needed an oscillator adjustable from at least 4 Hz to 50 Hz. This oscillator consists of the simplest possible circuit, an HCMOS Schmitt trigger inverter *U1A* with negative feedback via a lowpass filter consisting of trimpot *R1* and capacitor *C2*. The output of the HCMOS chip swings from 0 to $+5$ V, and its input triggers at typically $+2$ V and $+3$ V. This means that every time the output swings, the capacitor charges 3 V/5 V $= 60\%$. That takes about 1 time constant. Using a 200 msec network makes each half cycle last 200 msec, producing a 2.5 Hz oscillator. The 15-turn trimpot has more than a 20:1 adjustment range, so there is no problem getting to 50 Hz. I wouldn't use this circuit in a production instrument because the threshold levels of a Schmitt trigger logic device, such as *U1*, could vary widely from manufacturer to manufacturer and possibly from batch to batch. To construct two instruments, this was no problem.

A highpass RC network and Schmitt trigger inverter next convert the square wave to narrower pulses ranging from 3 msec to 30 msec in width for duty factor adjustment. This network consisting of potentiometer *R2* and capacitor *C3* converts the square wave to exponentially decaying pulses offset toward the $+5$ V supply. Resistor *R3*, which has three times the potentiometer resistance, keeps the load impedance high when the wave form is large enough to be clipped by the input diodes of *U1B*. The small bypass to ground *C4* prevents *U1B* from false pulsing on noise picked up from other circuits.

Two more gates, *U1D* and *U1E*, buffer the output to provide a synchronizing signal for the oscilloscope. Selector switch *S1A* selects either the narrow positive pulse output of *U1B* or an inverted signal from *U1C*. Next an inverting MOSFET driver switch *U2* raises the pulse level from 0 to 5 V to 0 to $+10.00$ V set by a precision reference voltage.

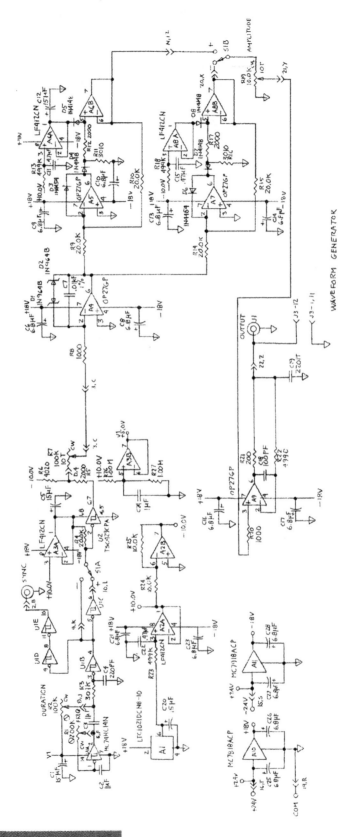

Figure 9-7: Complete schematic of the settling time tester.

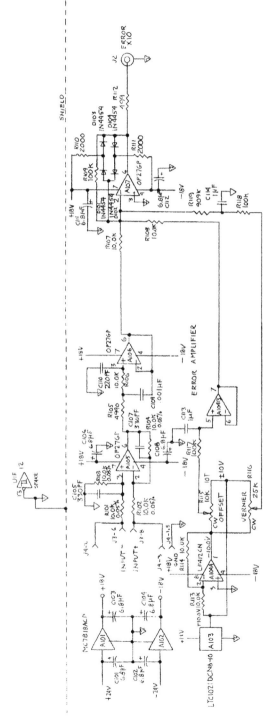

Figure 9-7: Continued

Gate *U2* feeds an integrating operational amplifier *A4*, which produces the leading and trailing edge slopes. To achieve equal slopes a divider, *R5* and *R6* connected to a precision -10.00 V reference, offsets the pulse output of U2 to ±3.3 V. I could have solved a pair of equations to determine the division ratio required for equal positive and negative swings. It was easier to try two or three different ratios in my head to converge on 2/1 for *R6/R5*.

Calculating the integrator part values didn't require a pencil and paper or a computer, either. I wanted to adjust the leading and trailing edge slopes of a 10-V output pulse from less than 100 μsec to 3 msec using a 10-turn front-panel potentiometer. Unlike the exponentially charging lowpass filter in Figure 9-2, the integrating amplifier maintains a constant charging current through the capacitor. In a time of 1 time constant, the output ramps up not 63% but 100% of the input voltage. Starting with -3.3 V input, it takes 3 time constants for the output to reach $+10$ V.

To produce a 3 msec/10 V ramp requires a time constant of 1 msec, made up primarily of the input potentiometer *R7*, a 100 kΩ 10-turn type, and feedback capacitor *C7*, 0.01 μF. The 100 μsec/10 V ramp requires a source resistance of 100 kΩ/30 $=$ 3300 Ω. Now here is an application for a Thevenin equivalent circuit to determine the portion of the input resistor supplied by *R5* and *R6*. Without a load, divider *R5* and *R6* attenuates the 10 V output pulse from U2 to 6.67 V p-p, offset to produce ±3.3 V. Its effective source resistance is *R5* and *R6* in parallel, 1330 Ω. Adding *R8*, 1000 Ω, increases the total integrator input resistance to 2330 Ω, which meets the requirement with some safety factor. Zener diodes *D1* and *D2* were added across the integrating capacitor *C7* to make the amplifier *A4* recover quickly from saturation at ±13 V.

The next blocks in the process are a pair of clipping circuits. The circuit involving *A5* and *A6* clips at 0 V and $+$ 10.5 V and uses feedback to attain sharp corners. A similar circuit, *A7* and *A8* with reversed diodes, clips at 0 V and -10.5 V. The output of either circuit is selected by switch *S1B* for the desired polarity.

In the positive clipping circuit, *A5* is a precision rectifier that clips at 0 V. Diode *D4* loaded by resistor *R11* conducts only positive signals, while diode *D3* in the feedback path prevents the output of *A5* from swinging more than 0.6 V negative and causing leakage through *D4*. The follower *A6B* buffers the output of the precision rectifier circuit and delivers feedback via *R10* to set the gain at -1.

The output from *A6B* ought to be a nice flat-top pulse produced by zener diodes *D1* and *D2* at *A4*. Zeners, however, are noisy. So another clipping circuit *A6A* using diode *D5* clamps the input to *A6B* at $+10.5$ V. *A6A* is a follower whose input is a 10.0 V reference. A 0.24 sec time constant lowpass filter *R13* and *C11* filters the reference, so the top of the pulse will have very low noise. During clamping *A5* goes open loop and saturates. Originally *D3* was chosen as a 13 V zener diode to speed recovery by preventing saturation. It leaked noise into the pulse and had to be changed to a conventional diode.

The negative clipping circuit operates the same way, but all the diodes are reversed. When *S1B* selects the output from the positive clipper at *A6B*, switch *S1A* selects logic pulses that are narrow at $+0$ V and wide at $+5$ V. When selecting negative output pulses from *A8B*, *S1A* selects narrow $+5$ V pulses.

All the reference voltages for *U1*, *U2*, *D5*, and *D8* are derived from a 10 V reference regulator *A1*. Several lowpass filters and followers are used to make separate low-noise reference voltages that will not interact with each other. All the followers use FET input op-amps, which can work with relatively high-impedance lowpass filters. Convenient values for the lowpass filters were 499 kΩ and 0.47 μF, producing a time constant of 0.24 sec and cutoff frequency of $160,000/240,000$ μsec $= 0.67$ Hz

At the arm of switch *S1B* we have high-quality pulses, either positive or negative. A 10-turn front-panel potentiometer *R19* adjusts the amplitude. The final block in the wave form generator is a follower amplifier that eliminates loading effects. You cannot simply connect a follower directly to a coaxial cable without the danger of oscillations. I use an isolation network at the output of the follower. The network consists of a 200 Ω resistor *R21* in series with the output terminal and a 220 pF bypass capacitor to ground *C19*. Now load capacitances from 0 to 1000 pF will cause only a 6-to-1 capacitive load variation instead of an infinite variation.

Resistor *R22* delivers DC feedback from the output terminal. High-frequency feedback above 300 kHz comes directly from the amplifier output via capacitor *C18*. The result is a follower with zero DC output resistance and a resistive-inductive impedance at high frequencies. Up to 1000 pF loading causes a small overshoot on a step.

The error amplifier was built on a separate circuit card and has its own local regulators *A101* and *A102*. The input circuit uses a low-offset op-amp *A105* connected in a 0.05% resistive bridge circuit to reject common mode voltages. Bypass capacitors *C105* and *C108* attenuate frequencies above 48 kHz by forming 3.3 μsec time constants with the bridge resistors.

The next amplifier *A106* attenuates high-frequency noise at 12 dB/octave above 48 kHz using a two-section RC lowpass filter. To achieve about 1.5 dB corner peaking, the first section has 1.5 times the desired 3.3 μsec time constant ($160,000/48$ kHz), while the second section has 2/3 of that time constant.

The task of the error amplifier after cleaning up the input signal is to subtract up to ± 10 V adjustable offset and then amplify the signal 10 times and clip it at ± 1 V. The $+10.0$ V reference, *A103*, feeds a unity gain inverter *A104A* to produce -10.0 V. Then

R115, a 10 K, 10-turn front-panel potentiometer spanning the two voltages produces the continuously variable ± 10 V offset. A lowpass filter *R117* and *C113* having a time constant of 10 msec corresponding to 16 Hz cutoff (160,000/10,000 μsec) eliminates high frequency noise from the reference. The follower *A104B* maintains the full ± 10 V offset, which is then added to the signal from *A106* using resistors *R107* and *R108*. A vernier pot *R116* feeding the lowpass filter *R118* and *C114* adds another $\pm 1\%$ of variable offset in via *R119*. The vernier pot increases the resolution from 0.1% to about 0.01%.

Finally, the error amplifier *A107* has a feedback resistor *R109* chosen to provide a gain of 10 relative to the signal from *A106*. The biased diode feedback network provides the ± 1 V clipping with near-instant overload recovery. As the output feeds an oscilloscope through a short cable, the only isolation network used is a resistor *R112*, 499 Ω.

The instrument works. Checking its own flat-top pulse shows that all contributors to a long-settling tail such as op-amp self-heating and reference tilt amount to less than 0.01%.

Now you can see that by using enough amplifiers and followers to isolate the functions of a circuit, the design became quite simple. The only place I needed to figure out a Thevenin equivalent circuit was between *U2* and *A4* in the wave form generator. Remembering $f_0 = 160,000/T$ was essential. It would have been convenient to have a table of lowpass filter values available for use in designing the error amplifier in the event I wanted to choose a precise amount of noise filter overshoot. The circuit could have been built with fewer op-amps, but there would have been interaction between portions of the circuit. It would have been much more complicated to figure out and perhaps would not have worked quite as well.

Final Notes

Part of your circuit design job should be writing a description of how it works before you build it. If you want to find the flaws in your design, there is nothing like trying to describe it to someone else.

Although computers, electronic parts, and surface-mount printed circuits have evolved since this chapter was written in 1991, the principles of circuit design are unchanged. The settling time testers built according to the preceding schematic are still in frequent use after sixteen years.

CHAPTER 10

Bandpass Filters

Steve Winder

You might not need a bandpass filter very often, but it's nice to know how to pull one out of your bag of tricks when you need it. I last used one a couple years ago, but I can cook one up as fast as I can apply solder! Here's a good cookbook from Steve Winder. Steve says he uses bandpass filters often in telecomm work, with a 300 Hz highpass to get rid of AC mains noise and a low-pass up to 3.4 kHz for antialiasing. /rap

In This Chapter

There are two categories of bandpass filters: wideband and narrowband. Filters are classified as wideband if their upper and lower passband cutoff frequencies are more than an octave apart. This is when the upper frequency is over twice that of the lower frequency. Wideband filters are ideally constructed from lowpass and highpass filters connected in series. The denormalization and scaling process for these has already been described in Chapters 6 and 7. This chapter describes how to design narrowband analog active or passive bandpass filters. Narrowband filters have upper and lower frequencies that are an octave or less apart.

Passive bandpass filter designs will be based on the tables of normalized lowpass component values found in Chapter 2 of the last publication cited in the References section. Formulae will be given for the denormalization and scaling of these component values to produce a bandpass design. The equations are more complex than for lowpass or highpass transformations, but this chapter gives examples of their application.

Active bandpass filter designs will be based on the normalized lowpass pole and zero locations given in Appendix B. Formulae will be given for denormalizing this pole and zero information, which will allow component values to be obtained. The equations are complex, but they are broken down into easier steps to simplify the process and reduce the chance of errors. This chapter gives examples of how to use the equations.

Lowpass-to-Bandpass Transformation

There is a close relationship between the bandwidth of a bandpass filter and the normalized lowpass filter from which it is derived. The bandwidth of a lowpass filter is from DC to the cutoff frequency, and the bandwidth of a bandpass filter is between the lower and upper cutoff frequencies. To obtain a particular bandwidth in a bandpass filter, first scale the normalized lowpass design to have this bandwidth and then transform this into a bandpass filter design. The resultant bandpass filter bandwidth will be the same as the lowpass filter from which it was derived. **Figure 10-1** illustrates this concept.

The relationship between the bandpass filter and its lowpass prototype not only applies to the $-3\,dB$ bandwidth. The width of the skirt in the bandpass filter response, at any given amount of attenuation, will be equal to the width of the skirt in the lowpass filter response frequency at which the same attenuation is achieved.

For example, suppose a bandpass filter with a center frequency of 10 kHz is desired. This filter must have a $-3\,dB$ bandwidth of 6.8 kHz and 40 dB attenuation at $Fc \pm 10\,kHz$, that is, the width of the skirt response at 40 dB attenuation is 20 kHz. The bandpass filter must be based on a lowpass filter design that produces the same response. That is, it must have 40 dB attenuation at a frequency of 20 kHz. The normalized stopband-to-passband frequency ratio of the lowpass filter is the same as that of the bandpass filter: 20 kHz divided by 6.8 kHz, which gives a ratio of 2.94. Thus, in a normalized lowpass prototype with a 1 rad/s passband frequency, 40 dB attenuation is required at a frequency of 2.94 rad/s.

Passive Filters

Passive bandpass filters are derived from the normalized lowpass model. The model is normalized for a passband that extends from DC to 1 rad/s and is terminated with a $1\,\Omega$ load resistance. The first process that you must carry out is to scale the lowpass model for the desired cutoff frequency, transform it into a bandpass filter, and finally scale for the correct load impedance.

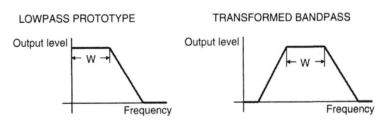

Figure 10-1: Lowpass-to-bandpass response transformation.

The design process starts with identifying the lowpass prototype. This may be Butterworth, Chebyshev, or another design. The filter order must also be determined. Starting with the specification given in the introduction, you need a filter with a 6.8 kHz, 3 dB bandwidth and with 40 dB attenuation at ±10 kHz. In addition, let the filter have a center frequency, F_0, of 198 kHz. Design a Butterworth bandpass filter that achieves this specification.

The stopband-to-passband ratio is 20/6.8 = 2.94, as explained in the previous example. Referring to the attenuation versus frequency curves for Butterworth filters, you can see that a fifth-order filter will provide the required performance. Start with a lowpass prototype, as shown in **Figure 10-2**.

The lowpass model must be frequency scaled to have a cutoff frequency of 6.8 kHz. This is done in the same way that lowpass filters are scaled, that is, the inductors and capacitors are divided by $2\pi Fc$, where Fc is the cutoff frequency. The divisor factor is therefore 42,725.66; results in the component values are shown in **Figure 10-3**.

To frequency translate the scaled lowpass prototype into a bandpass model, you must resonate each branch of the ladder at the center frequency, F_0. Series inductors become series LC circuits, and shunt capacitors become parallel tuned LC circuits. The capacitor and inductor values in the lowpass model are unchanged.

Remember that for a tuned circuit at resonance $F_0 = 1/2\pi\sqrt{LC}$, the inductor and capacitor values can be found by manipulating this equation. Hence the inductor required to tune the lowpass capacitor becomes $L_{BP} = 1/4\pi^2 Fo^2 C_{LP}$, and the capacitor required to tune the lowpass inductor becomes $C_{BP} = 1/4\pi^2 Fo^2 L_{LP}$.

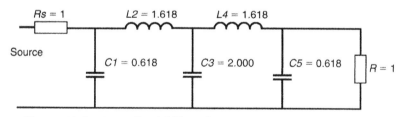

Figure 10-2: Normalized fifth-order Butterworth lowpass model.

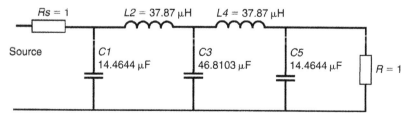

Figure 10-3: Scaled fifth-order Butterworth lowpass filter for 6.8 kHz.

Table 10-1: Bandpass component values.

Lowpass Component	Lowpass Value	Bandpass Component	Bandpass Value
C1	14.4644×10^{-6}	L1	44.669×10^{-9}
L2	37.87×10^{-6}	C2	17.0614×10^{-9}
C3	46.8103×10^{-6}	L3	13.8028×10^{-9}
L4	37.87×10^{-6}	C4	17.0614×10^{-9}
C5	14.4644×10^{-6}	L5	44.669×10^{-9}

For the bandpass filter tuned to 198 kHz, the frequency translating factor is $4\pi^2 Fo^2 = 1.547712 \times 10^{12}$. Using this information, the bandpass circuit component values are given in **Table 10-1**.

Putting these components into the circuit, you now have the bandpass filter shown in **Figure 10-4**.

The capacitor and inductor values given are for a normalized 1 Ω load. Denormalization of the bandpass model for higher load impedance requires component values to be scaled to have higher impedance. This is done in exactly the same way that lowpass or highpass filters are scaled. Inductor values increase in proportion to the load impedance, and capacitor values reduce in inverse proportion to the load. Capacitor values reduce because their impedance is inversely proportional to their capacitance values. As the load impedance increases, all the reactances in the circuit must increase in order to have the same response as the model.

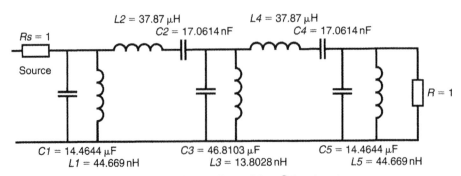

Figure 10-4: Bandpass filter with 1 Ω load resistance.

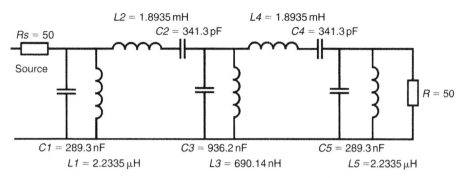

Figure 10-5: Bandpass filter, denormalized with 50 Ω load resistance.

The filter you have been designing is intended to provide a filter for a simple radio receiver, to pick up a carrier at 198 kHz. This requires a 50 Ω source and load impedance, to match the radio frequency components at its input and output (50 Ω is the standard impedance for *RF* circuits; 75 Ω is standard for television picture transmission). Impedance scaling is achieved by multiplying the inductor values by 50 and dividing the capacitor by 50. Finally, the filter circuit given in **Figure 10-5** is obtained.

This circuit is one of two possible configurations. This configuration was developed from the minimum inductor prototype and had two series resonant arms. Three parallel resonant shunt arms were connected to the common rail at one end and to either the source, the load, or the central node at their other end. This design gives low impedance outside the passband because the shunt arms have low impedance at DC and at frequencies above resonance.

If the design were, instead, developed from the minimum capacitor prototype, the end result would have used the same number of capacitors and inductors. The difference would have been that the filter would have had three series resonant arms between the source and load. Also, there would have been two parallel resonant shunt arms connected between the nodes of the series arms and the common rail. The alternative circuit is shown in **Figure 10-6**. This circuit was designed by FILTECH, which calculates the normalized element values and then scales them using double precision floating-point arithmetic. Although the transfer function of this filter is identical to the previous version, the input and output impedances of this version are high outside the filter's passband.

Having gone through this long-winded process, readers will be pleased to know that there are formulae that allow the whole process to be completed in one step. Of course there are slight complications: because of the different circuit topologies there are a number of formulae, and the difficulty is knowing which to use. I will give guidance on this subject, with examples in this chapter.

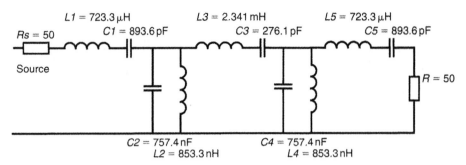

Figure 10-6: Bandpass filter, denormalized with 50 Ω load resistance.

Formula for Passive Bandpass Filter Denormalization

$$C_{Series} = \frac{F_U - F_L}{2\pi F_U F_L R X}$$

$$L_{Series} = \frac{RX}{2\pi \cdot (F_U - F_L)}$$

$$C_{Parallel} = \frac{X}{2\pi \cdot (F_U - F_L) \cdot R} \qquad [10\text{-}1]$$

$$L_{Parallel} = \frac{(F_U - F_L) \cdot R}{2\pi F_U F_L X}$$

The series and parallel subscripts indicate which circuit element is being considered. In the equations, the factor X is the normalized lowpass element value. The same value of X must be used for both components in a single branch. This is because each branch in the lowpass filter has one component, whereas branches in the bandpass have two components that are either series or parallel resonant. Both components in a single branch are related to a single component value in the lowpass prototype.

It might be helpful to redesign the fifth-order Butterworth filter to illustrate the use of these formulae. Since it is a symmetrical design, only the first three branches need to be calculated. As before, $R = 50$, $F_U = (198 + 3.4)\,\text{kHz} = 201.4\,\text{kHz}$, $F_L = (198 - 3.4)\,\text{kHz} = 194.6\,\text{kHz}$.

The first branch has a value $X = 0.618$ and could be a series arm or a shunt arm. Taking the shunt arm case first (parallel resonant) gives:

$$C_{Parallel} = \frac{X}{2\pi \cdot (F_U - F_L) \cdot R} = 0.618/(2.136283 \times 10^6) = 289.3\,\text{pF}$$

$$L_{Parallel} = \frac{(F_U - F_L) \cdot R}{2\pi F_U F_L X} = 340 \times 10^3/(15.218466 \times 10^{10}) = 2.23413\,\mu\text{H} \qquad [10\text{-}2]$$

The second branch has a value $X = 1.618$. Since the first arm was chosen to be a shunt arm, this arm must be connected in series. Calculating the values gives:

$$C_{Series} = \frac{F_U - F_L}{2\pi F_U F_L RX} = 6.8 \times 10^3/(1.992189 \times 10^{13}) = 341.3 \, pF$$

$$L_{Series} = \frac{RX}{2\pi \cdot (F_U - F_L)} = 80.9/42,725.66 = 1.8935 \, mH \qquad [10\text{-}3]$$

The third branch is a parallel shunt arm, the same as the first branch. This time the value of X is 2.0. Let's cheat by using the results of the first branch and multiplying them by a ratio of X_3 to X_1.

$$C_3 = 289.3 \times 2.0/0.618 = 936.2 \, pF$$
$$L_3 = 2.23413 \times 0.618/2.0 = 0.69035 \, \mu H, \text{ or } 690.35 \, nH \qquad [10\text{-}4]$$

The differences between these results and those obtained in Figure 10-5 are due to round-off errors in the tables of normalized values and during the calculations. The calculations were done by hand using a calculator. Floating-point arithmetic in a computer program such as FILTECH achieves more accurate results.

To obtain the circuit given in Figure 10-6, it is necessary to calculate the series arm first. This will use a value of $X = 0.618$.

$$C_{Series} = \frac{F_U - F_L}{2\pi F_U F_L RX} = 6.8 \times 10^3/(7.609233 \times 10^{12}) = 893.6 \, pF$$

$$L_{Series} = \frac{RX}{2\pi \cdot (F_U - F_L)} = 30.9/42,725.66 = 723.2 \, \mu H \qquad [10\text{-}5]$$

A shunt arm must be calculated next, using $X = 1.618$, followed by another series arm, using $X = 2.0$. Because of symmetry, the final two arms will have the same component values as previously calculated for the first two arms. The last arm will have the same component values as the first arm. The one-before-last arm will have the same component values as the second arm.

As in previous chapters on filters, further study of passive filters will be minimized in favor of active filters.

Active Bandpass Filters

Active filters can be designed using pole and zero locations, which are derived from the frequency response's transfer function. Op-amps are the "active" part of the circuit. These are used to buffer one stage from the next, which prevents interaction between stages. Each stage can therefore be designed to provide the frequency response of one pair of complex poles. Zeros are also required, above and below the passband. Active networks used in bandpass filter circuits also produce zeros. Because each filter stage is buffered from the next, the overall response is correct when all the stages are connected in series.

Bandpass Poles and Zeros

Normalized *lowpass* filter response's pole and zero locations are used as a starting point. Frequency translation is then required to convert these into normalized *bandpass* pole and zero locations. Frequency translation in both transfer functions and the S-plane are made by replacing s with s'' as given by the following equation:

$$s'' = \frac{\omega_0}{BW}\left[\frac{s}{\omega_0} + \frac{\omega_0}{s}\right] \qquad [10\text{-}6]$$

The passband center frequency is $\omega_0 = \sqrt{\omega_U \cdot \omega_L}$, and BW is the bandwidth, given by the difference between the upper and lower passband frequencies, $\omega_U - \omega_L$. This is not particularly easy to evaluate. However, Williams has published equations for finding the Q and resonant frequency, f_R, of each stage of a bandpass filter from a lowpass model. These are all that are needed to design active bandpass filters. I have manipulated Williams' equations slightly, to be consistent with those used to design bandstop filters. Bandstop filter equations will be given in the next chapter.

To start with you need to know the Q of bandpass filter Q_{BP} and the real and imaginary parts of the lowpass prototype pole location σ and ω. The pole positions can be found by using the formulae. The bandpass Q is the center frequency, f_0, divided by the bandwidth.

$$m = \frac{\sigma}{Q_{BP}}$$
$$J = \frac{\omega}{Q_{BP}} \qquad [10\text{-}7]$$
$$n = m^2 + J^2 + 4$$

The required $Q = \sqrt{\dfrac{n + \sqrt{n^2 - 16m^2}}{8m^2}}$

This gives the frequency scaling factor, $W = Qm + \sqrt{Q^2 m^2 - 1}$

And the frequencies are $f_{R1} = \dfrac{f_0}{W}$ and $f_{R2} = Wf_0$.

These are the pole transformation equations. Now the zero locations are needed, and, in an all-pole filter such as Chebyshev or Butterworth response, these are at the S-plane origin and at infinity. In Cauer and Inverse Chebyshev filters the zero locations have to be calculated, as follows:

$$k = \frac{\omega_\infty}{Q_{BP}}$$
$$h = \frac{k^2}{2} + 1$$

[10-8]

The zero scaling factor can now be found, $z = \sqrt{h + \sqrt{h^2 - 1}}$

The bandpass zero frequencies are then $f_{\infty,1} = \dfrac{f_0}{z}$ and $f_{\infty,2} = zf_0$.

What does the S-plane diagram look like now? An example of a fourth-order lowpass filter was given in Chapter 6, Figure 6-11. This had a Butterworth response, with poles on a unit circle at $-0.9239 \pm j0.3827$ and $-0.3827 \pm j0.9239$. Suppose the filter is required to have a passband between 9 rad/s and 11 rad/s ($BW = 2$; this is for illustration only and not intended to be a practical value). This gives $BW = 2$, $\omega_0 = 9.95$ rad/s, and $Q_{BP} = 4.975$. Notice that the geometric center frequency (9.95 rad/s) is not the same as the arithmetic center frequency (10 rad/s). Taking one pole from the first pair: $s = -0.9239 + j0.3827$, $\sigma = 0.9239$, and $\omega = 0.3827$.

The two bandpass poles produced from this are found from the following equations:

$$m = \frac{\sigma}{Q_{BP}} = 0.18571$$
$$J = \frac{\omega}{Q_{BP}} = 0.076925$$
$$n = m^2 + J^2 + 4 = 4.0404056$$

[10-9]

$$Q = \sqrt{\frac{n + \sqrt{n^2 - 16m^2}}{8m^2}} = 5.388756$$
$$W = Qm + \sqrt{Q^2m^2 - 1} = 1.039375$$

The frequencies are $f_{R1} = \dfrac{f_0}{W} = 9.57306$ and $f_{R2} = Wf_0 = 10.34178$.

The second pair of poles can be found in a similar way. Due to symmetry, $\sigma = 0.3827$ and $\omega = 0.9239$:

$$m = \frac{\sigma}{Q_{BP}} = 0.076925$$

$$J = \frac{\omega}{Q_{BP}} = 0.18571$$

$$n = m^2 + J^2 + 4 = 4.0404056 \qquad\qquad [10\text{-}10]$$

$$Q = \sqrt{\frac{n + \sqrt{n^2 - 16m^2}}{8m^2}} = 13.0556778$$

$$W = Qm + \sqrt{Q^2 m^2 - 1} = 1.09723$$

The frequencies are $f_{R1} = \dfrac{f_0}{W} = 9.068286$ and $f_{R2} = W f_0 = 10.917444$.

To help you visualize what has happened to the poles, I provide a pole-zero diagram in **Figure 10-7**. This diagram only shows the positive frequency poles; there are symmetrical negative frequency poles, but these have been omitted for clarity. Also, note that for a given Q the poles lie on a line that passes through the origin. The two poles just calculated both had a Q of about 5.4.

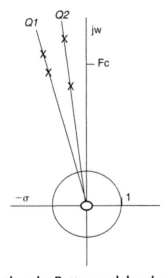

Figure 10-7: Fourth-order Butterworth bandpass pole locations.

The other poles had a Q of about 13, but are further from the bandpass filter's center frequency, Fc. Remember that the Q of a pole is given by the equation:

$$Q = \frac{\sqrt{\sigma^2 + \omega^2}}{2\sigma}$$ [10-11]

The Q of a bandpass pole is approximately $2\omega_0 Q_{LP}/BW$, where Q_{LP} is the normalized lowpass pole Q. **Figure 10-7** only shows the zeros at the origin; there are also zeros at infinity that cannot be shown (!).

The scene has been set. Let's now take a look at some basic bandpass active filter designs and show how the pole and zero locations are used to find component values. We'll return to the S-plane later when we discuss active Cauer and Inverse Chebyshev filters; these types both have zeros in the stopband.

Bandpass Filter Midband Gain

One of the main features of a bandpass filter is its center frequency, f_0. However, each stage of a bandpass filter has a resonant frequency, f_R, which could be above or below f_0. The gain of each stage is measured at these two important frequencies, f_0 and f_R, which gives gain G_0 and G_R, respectively. The gains of all stages are added together to give the overall filter gain at any particular frequency. Since the frequency response is symmetrical about the center frequency, there will be an equal number of stages resonant above and below the center frequency. In the example frequency response, illustrated by the graph in **Figure 10-8**, f_R is below f_0.

The gain of the filter at its center frequency can be found from the following equation, which also requires the stage's Q to be known. The terms f_R and Q can be found from the bandpass pole positions and using the relationship $G_R = 2Q^2$. The bandpass filter center frequency, f_0, is found from the filter's specification.

$$G_0 = \frac{G_R}{\sqrt{1 + Q^2 \left(\dfrac{f_0}{f_R} - \dfrac{f_R}{f_0} \right)}}, \text{ which simplifies to } G_0 = \frac{2Q^2}{\sqrt{1 + Q^2 \left(\dfrac{f_0}{f_R} - \dfrac{f_R}{f_0} \right)^2}}$$ [10-12]

This equation gives the midband gain of the stage being designed. Suppose that the bandpass filter design is required to have unity gain in the passband. The simplest way to do this is to have unity gain at the passband center frequency (f_0) in each stage, then $G_{RR} = G_R/G_0$. Suppose that $G_0 = 10$ and $G_R = 15$. Since I want a center frequency

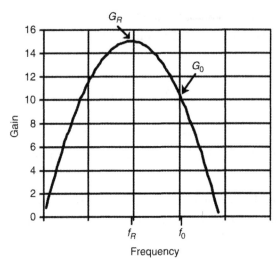

Figure 10-8: Gain versus frequency for a single stage.

gain of 1, not 10, the revised gain at resonance, G_{RR}, has to be scaled to be a tenth of G_R. In this case, $G_{RR} = G_R/G_0 = 15/10 = 1.5$. This means that the stage will need a potential divider, usually at its input, to reduce the "natural" gain of the stage from 15 to 1.5. If the desired midband gain is greater than unity, given by factor k, then G_{RR} must also be scaled by factor k: $G_{RR} = kG_R/G_0$. To achieve this scaling, the potential divider is modified to allow a greater proportion of the input signal into the filter stage.

If a number of stages are used, the overall midband gain will be the product of all the separate stage gains: $G_P = G1 * G2 * G3 *$ and so on. If each stage has a gain that is not unity at the filter center frequency, an inverting amplifier following the filter stages with a gain of $1/G_P$ could be used to restore the overall filter gain to unity.

Multiple Feedback Bandpass Filter

One of the simplest and most useful bandpass filters is the *multiple feedback bandpass* (MFBP) circuit. It is suitable for producing an all-pole response. This filter stage looks sort of like a lowpass and a highpass Sallen-Key filter combined into one, as shown in **Figure 10-9**. The advantages are impressive, though not obvious.

This filter is normally limited to a Q less than 20. This limitation is serious, but not disastrous, for many low-frequency applications. The performance depends on the bandwidth of the op-amp, which should be fairly high, compared to the resonant frequency desired. For good accuracy, the amplifier's GBW should be larger than $2Q^2 f_R$. Thus, an ordinary 0.5 MHz op-amp should not be used for a Q greater than 10 above

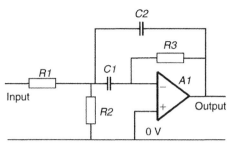

Figure 10-9: Multiple feedback bandpass (MFBP) filter.

1 kHz. Also, it is a good idea if the capacitors' tolerances are better than 5% and the capacitors are matched to 1%.

The advantages of this circuit tend to mitigate these limitations for low frequencies:

- The bandwidth depends only on $R3 \times C$. The bandwidth is defined by:

$$B = \frac{1}{(\pi R3 \times C)} \qquad [10\text{-}13]$$

  and is invariant of $R1$ or $R2$. Thus the bandwidth depends on selecting a good 1% resistor for $R3$ and getting good tolerance on the capacitors.

- The gain at resonance is simply equal to $R3/(2 \times R1)$. It does not depend on $R2$. Thus, its gain just depends on selecting two good 1% resistors (and it also depends on the matching of the two capacitors). For many applications, this gain does not need to be trimmed.

- The formula for the resonant frequency is a bit more complicated, but this is acceptable because $R2$ usually has to be trimmed with a pot.

The resonant frequency is:

$$f_R = [1/2\pi C](\sqrt{(R1 + R2)/(R1 \times R2 \times R3)}) \qquad [10\text{-}14]$$

or to put it in another light:

$$f_R = [1/2\pi C] \times (\sqrt{(R1 + R2)/R1})(\sqrt{(1/R2 \times R3)}) \qquad [10\text{-}15]$$

and the Q is:

$$Q = (1/2) \sqrt{(R3/R2)} \times \sqrt{(R1/(R1 + R2))} \qquad [10\text{-}16]$$

Fortunately, as $R2$ is adjusted to trim the resonant frequency, it does *not* change the bandwidth or the center-frequency gain. If this appeals to you, you will like this filter.

Note that trimming $R2$ does change the center frequency but not the bandwidth, so the Q is changed. However, if you know what gain and center frequency you want, trimming $R2$ is an easy way to get the desired response; the Q is trimmed to the right value when you just optimize the resonant frequency.

Another way to look at this is that when you know what bandwidth you want, you have chosen $R3$. Then the resonant frequency depends on the square root of the ratio of ($R1$ parallel to $R2$) versus $R3$. If you want a Q of 10 and $R3$ is 100 k, the value of $R2$ will be about 250 ohms, or 200 ohms plus a 100 ohm trimpot. It is thus not surprising that this does *not* provide a low-noise output. But modern low-noise op-amps will often provide acceptable performance. The factor of $\sqrt{((R1 + R2)/R1)}$ is very close to 1. Thus, a cookbook states that for $R3 = 100$ k, $R2 = 1$ k for $Q = 5$; 250 ohms for $Q = 10$; and 62 ohms for $Q = 20$. The $R2$ still must be trimmed to get good accuracy of f_R.

> **Note:** The outputs of a group of two, three, or more of these bandpass filters, with slightly spread center frequencies, can be connected in parallel to make a slightly wider bandwidth with two or three peaks since they are so easy to adjust. This can provide a steeper skirt. This signal can also be subtracted from the input to make a notch; see the next chapter.

Dual-Amplifier Bandpass Filter

The *dual-amplifier bandpass (DABP) filter* is more complicated than the MFBP structure, but it has the advantage that much higher Q factors can be achieved. Q factors of up to 150 are possible. The DABP is a bandpass filter stage with an all-pole response. The circuit diagram for a DABP filter is given in **Figure 10-10.**

The capacitance of $C1$ and $C2$ should be equal but may have any arbitrary value. In practice, the capacitors' values are chosen so that the resistors' values are all in a reasonable range, typically from $1\,k\Omega$ to $100\,k\Omega$. The limits are up to the designer, but remember that $R3$ and $R4$ load the output of op-amps $A1$ and $A2$. A high value of $R1$ will introduce noise and may degrade the signals because $R1$ is in series with the signal path.

Consider the equation: $R = 1/2\pi \cdot f_R C$, where $R = R3 = R4$, and $R1' = Q \cdot R$, assuming for the moment that $R2$ is open circuit (i.e., not there!). $R1$ is designated $R1'$ to show that this is for the case where $R2$ is not present. If the stage's Q is high, say 75, then $R1'$ will be 75 times $R3$ and $R4$. The value of C will have to be chosen so that R is close to the lower resistance limit. $R1'$ will then be close to the upper resistance limit.

Now, you will probably have realized that resistor $R2$ is used to adjust the gain. With $R2$ missing, the gain at resonance is 2. If lower gain at resonance is required, $R2$ must be in the

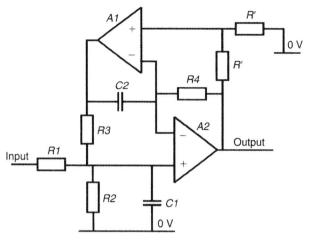

Figure 10-10: Dual-amplifier bandpass filter.

circuit. Resistors $R1$ and $R2$ form a potential divider, and their parallel resistance replaces $R1'$ in the equations given. The following equations use $R1'$ to determine $R1$ and $R2$:

$$R1 = \frac{2 \cdot R1'}{G_{RR}}$$

$$R2 = \frac{2 \cdot R1'}{2 - G_{RR}} \quad \text{(condition: } G_{RR} < 2\text{)} \qquad [10\text{-}17]$$

The revised gain at resonance G_{RR} can be found from the equation $G_{RR} = G_R/G_0$.

In the DABP case the resonant frequency gain is always equal to 2 by default due to internal feedback. Hence $G_R = 2$ and this can be used to find G_0, given the overall filter center frequency f_0 and the pole characteristics f_R and Q.

$$G_0 = \frac{2}{\sqrt{1 + Q^2 \left(\dfrac{f_0}{f_R} - \dfrac{f_R}{f_0} \right)^2}} \qquad [10\text{-}18]$$

Because the gain of each DABP stage at resonance is equal to 2, the gain at the filter center frequency may be less than unity. In this case, a separate amplifier stage may be needed if a unity-gain bandpass filter is required.

This circuit has independent adjustment of resonant frequency and Q. The parallel combination of $R1$ and $R2$ adjust the Q at resonance. Resistor $R3$ determines the resonant frequency.

Denormalizing DABP Active Filter Designs

As discussed earlier in this chapter, the resistor values used should all be in the range $1\,k\Omega$ to $100\,k\Omega$ where possible. This will prevent overloading of the op-amp's output and reduce noise pickup.

Consider a DABP filter stage design that uses the poles found earlier in this chapter for a bandpass filter with a passband from $9\,rad/s$ to $11\,rad/s$. The first pair of poles were found to be $\sigma = 0.9239$ and $\omega = 0.3827$.

From before, $Q = \sqrt{\dfrac{n + \sqrt{n^2 - 16m^2}}{8m^2}} = 5.388756$

$$W = Qm + \sqrt{Q^2m^2 - 1} = 1.039375 \qquad\qquad [10\text{-}19]$$

The frequencies are $f_{R1} = \dfrac{f_0}{W} = 9.57306$ and $f_{R2} = Wf_0 = 10.34178$.

Consider the pole with $f_{R2} = 10.34178$.

Let $C = 1\,\mu F$

$$R = R3 = R4 = \frac{1}{2\pi f_R C} = \frac{1}{2\pi \cdot 10.34178 \cdot 10^{-6}} = 15.389\,k\Omega \quad [10\text{-}20]$$

$$R1' = Q \cdot R = 82.928\,k\Omega.$$

Note that these values are less than the nominal $100\,k\Omega$ maximum and greater than the $1\,k\Omega$ minimum, so the value of C is suitable.

$$G_0 = \frac{2}{\sqrt{1 + Q^2 \left(\dfrac{f_0}{f_R} - \dfrac{f_R}{f_0} \right)^2}} = 1.8463811.$$

$$[10\text{-}21]$$

Hence, $G_{RR} = \dfrac{G_R}{G_0} = \dfrac{2}{1.8463811} = 1.0832$

An attenuator is needed to reduce the gain at resonance from 2 to 1.0832. The following equations use $R1'$ to determine $R1$ and $R2$:

$$R1 = \frac{2 \cdot R1'}{G_{RR}} = \frac{165,856}{1.0832} = 153.117\,k\Omega$$

$$R2 = \frac{2 \cdot R1'}{2 - G_{RR}} = \frac{165,856}{0.9168} = 180.907\,k\Omega$$

$$[10\text{-}22]$$

The parallel combination of $R1$ and $R2$ equal the value of $R1'$. Thus, although the individual resistance values of $R1$ and $R2$ exceed the recommended maximum of $100\,\text{k}\Omega$, the effective resistance into the op-amp will be $82.928\,\text{k}\Omega$.

State Variable Bandpass Filters

The state variable design can be used for all-pole responses. It has a lower sensitivity to the op-amp's gain-bandwidth product limitation, and stage Q factors of up to 200 are possible. It does, however, need three op-amps, as shown in **Figure 10-11**.

The equations for this filter allow the arbitrary choice of capacitor, C.

$$R1 = \frac{Q}{2\pi f_R C}$$
$$R2 = R3 = \frac{R1}{Q} \qquad\qquad \text{[10-23]}$$
$$R4 = \frac{R1}{G_{RR}}$$

The value of R' is arbitrary, but a typical value could be $10\,\text{k}\Omega$. G_{RR} is found from the equations given earlier in this chapter in the stage dealing with midband gain ("Bandpass Filter Midband Gain").

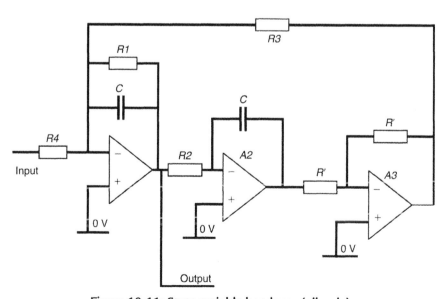

Figure 10-11: State variable bandpass (all-pole).

Denormalization of State Variable Design

The second pair of poles of the fourth-order design considered earlier in this chapter were $\sigma = 0.3827$ and $\omega = 0.9239$. The overall bandpass filter had a center frequency 9.95 rad/s and a Q_{BP} of 4.975.

$$m = \frac{\sigma}{Q_{BP}} = 0.076925$$

$$J = \frac{\omega}{Q_{BP}} = 0.18571$$

$$n = m^2 + J^2 + 4 = 4.0404056 \qquad [10\text{-}24]$$

$$Q = \sqrt{\frac{n + \sqrt{n^2 - 16m^2}}{8m^2}} = 13.0556778$$

$$W = Qm + \sqrt{Q^2 m^2 - 1} = 1.09723.$$

The frequencies are $f_{R1} = \dfrac{f_0}{W} = 9.068286$ and $f_{R2} = Wf_0 = 10.917444$.

I will use $f_{R1} = 9.068286$ to find the filter stage gain, G_0, given that the gain at resonance will be $G_R = 2Q^2 = 340.9014456$.

$$G_0 = \frac{340.9014456}{\sqrt{1 + Q^2 \left(\dfrac{f_0}{f_R} - \dfrac{f_R}{f_0} \right)^2}} = 129.9005655.$$

$$G_{RR} = \frac{G_R}{G_0} = \frac{340.9014456}{129.9005655} = 2.624326109$$

Let the capacitor be, $C = 1\,\mu F.$

$$R1 = \frac{Q}{2\pi f_R C} = 229.136\,k\Omega. \qquad [10\text{-}25]$$

This value is too high, so let $C = 2.2\,\mu F.$

$$R1 = 104.153\,k\Omega.$$

$$R2 = R3 = \frac{R1}{Q} = 7.977\,k\Omega.$$

$$R4 = \frac{R1}{G_{RR}} = 39.687\,k\Omega.$$

Let the value of $R' = 10\,k\Omega.$

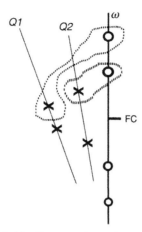

Figure 10-12: Cauer pole and zero pairing.

Cauer and Inverse Chebyshev Active Filters

Designing bandpass filters with a Cauer or an Inverse Chebyshev response is slightly more difficult because each filter stage must provide both poles and zeros close to the filter center frequency. Moreover, the pole and zero pairing must also be considered. A filter may have a number of poles and zeros and, in principle, any zero could be associated with any pole. In practice the pole-zero pairing affects performance. Pole and zero pairing is illustrated in **Figure 10-12**.

To design a Cauer or an Inverse Chebyshev filter, a different circuit topology is required. The Cauer response has zeros outside the passband, so a notch-generating circuit is required. Notches can be produced using a circuit that is an extension of the state variable filter and that is known as a *biquad*. This circuit is illustrated in **Figure 10-13**.

Note that in the bandpass biquad shown in Figure 10-13, $R5$ is connected to different nodes, dependent on whether the zero is above or below the resonant frequency. If the zero frequency, f_Z, is above the resonant frequency, f_R, connect nodes A and C. If the zero frequency, f_Z, is below the resonant frequency, f_R, connect nodes B and C.

The following equations give component values.

$$R1 = R4 = \frac{Q}{2\pi f_R C}$$

$$R2 = R3 = \frac{1}{2\pi f_R C}$$

$$R5 = \frac{f_R{}^2 R}{Q\left|f_R{}^2 - f_Z{}^2\right|}$$

[10-26]

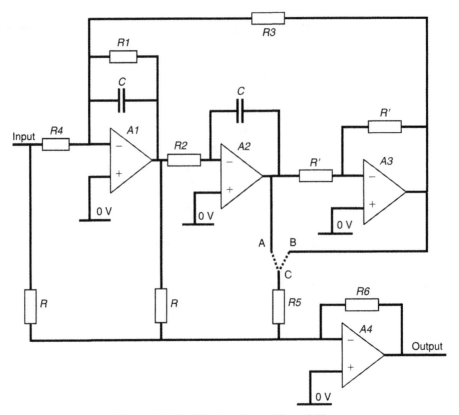

Figure 10-13: The bandpass biquad filter.

If the filter stage is the last of an odd-order filter (i.e., no zero is required), R5 is not in circuit and $R6 = R$.

If a zero is required, R5 is in circuit and the value of R6 is given by the following equation.

$$R6 = \cfrac{R}{\sqrt{\left| \cfrac{f_0^2 - f_Z^2}{(f_R^2 - f_0^2)^2 + \cfrac{f_R^2 f_0^2}{Q^2}} \right|^2}}$$

[10-27]

The resistors labeled R and R' can be any arbitrary value. A typical value may be in the range 1 kΩ to 100 kΩ, say 10 kΩ. The resistors labeled R have an effect on the input impedance of the filter stage.

Denormalizing Biquad Designs

The simplest approach with biquad filters is to scale the poles and zeros before using the design equations. Choose a convenient capacitor value and then use the equations to find the resistor values required by the design. If the resistor values are very small or very large, select a new capacitor value and try again. Again, aim to keep the resistor values between $1\,k\Omega$ and $100\,k\Omega$.

Consider the filter stage design needed to produce a pole at $f_R = 10.255\,rad/s$, with $Q = 21$. The filter center frequency $f_0 = 9.1\,rad/s$ and a zero at $14.2\,rad/s$ is required.

Let $C = 10\,\mu F$.

$$R1 = R4 = \frac{Q}{2\pi f_R C} = \frac{21}{2\pi \cdot 10.255 \cdot 10^{-5}} = 32.591\,k\Omega.$$

$$R2 = R3 = \frac{1}{2\pi f_R C} = 1.552\,k\Omega.$$

Let $R = 22\,k\Omega.$

$$R5 = \frac{f_R^2 R}{Q\left|f_R^2 - f_Z^2\right|} = 1.142\,k\Omega.$$

$$R6 = \frac{R}{\sqrt{\left[\dfrac{f_0^2 - f_Z^2}{(f_R^2 - f_0^2)^2 + \dfrac{f_R^2 f_0^2}{Q^2}}\right]}}$$

[10-28]

Substituting the values $f_0^2 = 82.81, f_R^2 = 105.165, f_Z^2 = 201.64$, and $Q^2 = 441$ gives $R6 = 22\,k\Omega/5.2135792 = 4.21975\,k\Omega.$

References

Williams, A., and Taylor, F.J., *Electronic Filter Design Handbook*, McGraw-Hill, New York, 1988.

Winder, S., *Analog and Digital Filter Design*, Elsevier-Newnes, Amsterdam, 2002.

CHAPTER 11

Bandstop (Notch) Filters

Steve Winder

Bandstop (notch) filters are tricky because they usually involve subtracting large signals. Thus any imperfections can cause a relatively large error. Trimming will probably be necessary to get a crisp notch with correct rolloff and a deep notch at the desired center frequency. Here Steve Winder shows us several good ways to make a notch. /rap

A Closer Look at Bandstop Filters

There are two categories of bandstop filters: wideband and narrowband. Filters are classified as wideband if their upper and lower passband cutoff frequencies are several octaves apart. This is when the upper frequency is many times that of the lower frequency.

Wideband filters are ideally constructed from odd-order lowpass and highpass filters connected in parallel. Odd-order filters are necessary because, outside their passband, these have both high input impedance and high output impedance. High impedance in the stopband prevents loading of the parallel-connected filter. Otherwise impedance mismatches could occur that would lead to an incorrect overall frequency response. The denormalization and scaling process for lowpass and highpass filters has already been described (in Chapters 6 and 7).

This chapter describes how to design narrowband active and passive bandstop filters to almost any specification. Narrowband filters have upper and lower frequencies that are less than about three octaves apart. The design of these filters uses the normalized lowpass filter pole and zero or component values as a starting point. In this chapter I use information from previous chapters and give examples where they help in understanding. I also provide formulae for passive designs in the denormalization and scaling of normalized component values and describe the method of denormalizing pole and zero

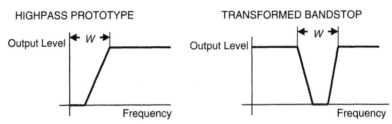

Figure 11-1: Lowpass-to-bandstop response transformation.

information for use with active filters. (See the information in Chapters 2 and 3 of the last publication cited in the References section.)

Bandstop filter design starts with normalized component values, which are converted into normalized highpass values. These highpass values are then scaled to give a new cutoff frequency, W. The new cutoff frequency must be made equal to the difference between upper and lower cutoff frequencies for the desired bandstop filter. In mathematical terms, $W = f_U - f_L$. **Figure 11-1** illustrates this concept.

The highpass filter's stopband frequency, to give a certain level of attenuation, is made equal to the bandstop filter's stopband width, N. An example will help to explain this idea.

Let's say that the stopband width is N Hertz to give 40 dB attenuation. The highpass filter is required to have 40 dB attenuation at a frequency of N Hertz. To find the filter order needed to achieve this response, the frequencies must be normalized before using graphs. The stopband where 40 dB attenuation occurs on the normalized frequency response curves is at W/N Hz. Using graphs for the normalized lowpass prototype, we can find the filter order needed for the bandpass design.

For example, suppose you want a bandstop filter where the difference between the upper and lower cutoff frequencies is 6.8 kHz and that gives 40 dB attenuation at $F_0^{\pm}$ 1 kHz, that is, the width of the skirt response at 40 dB attenuation is 2 kHz. Thus $W = 6.8$ kHz and $N = 2$ kHz. The normalized lowpass filter must give 40 dB attenuation at a normalized frequency ratio of 6.8 kHz divided by 2 kHz, which equals 3.4 rad/s. The normalized lowpass attenuation curves given in Chapter 2 of the last publication cited in the References section can be examined to find the filter order.

Passive Filters

Passive bandstop filters are derived from the normalized lowpass model. The model is normalized for a passband that extends from DC to 1 rad/s and is terminated with 1 Ω load resistance. The first process that you must carry out is to convert the lowpass model into a highpass prototype, scaled for the desired cutoff frequency. Then transform the highpass prototype into a bandstop filter with the correct center frequency. Finally, scale for the correct load impedance.

As in the case of all filters, the design process starts with identifying the lowpass prototype. This may be Butterworth, Chebyshev, or another design. The filter order must also be determined. Suppose you need a filter with a 2.4 kHz bandwidth between the 3 dB points and with 40 dB attenuation at ±250 Hz (a 500 Hz stopband width). In addition, the circuit is required to have a center frequency, F_0, of 320 kHz. Design a Butterworth bandstop filter that achieves this specification.

Passband to stopband ratio = 2.4/0.5 = 4.8. Referring to the normalized responses, a third-order filter will just about achieve the required 40 dB attenuation at 4.8 rad/s. Start with a lowpass prototype, as shown in **Figure 11-2**.

The lowpass model must be converted into a highpass model by replacing capacitors by inductors, and vice versa, using reciprocal values. In this case, the normalized highpass values of $L1$ and $L3$ remain equal to 1 Henry, but $C2$ becomes 0.5 farad. This normalized design is then frequency scaled to have a cutoff frequency of 2.4 kHz. This is done in the same way that lowpass filters are scaled. The inductors and capacitors are divided by $2\pi Fc$, where Fc is the cutoff frequency. The divisor factor is therefore 15,079.65 and results in the component values shown in **Figure 11-3**.

To frequency-translate into a bandstop model, resonate each branch of the ladder at the center frequency, F_0. Series capacitors become parallel tuned LC circuits. Shunt inductors

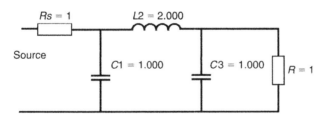

Figure 11-2: Normalized third-order Butterworth lowpass model.

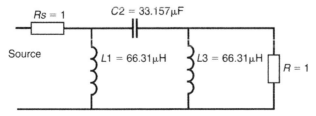

Figure 11-3: Scaled third-order Butterworth highpass prototype.

become series tuned *LC* circuits. The capacitor and inductor values in the highpass prototype are unchanged.

Remember that, at resonance, $F_0 = 1/2\pi\sqrt{LC}$, so the inductor required to tune the highpass capacitor becomes $L_{BS} = 1/4\pi^2 F_0^2 C_{HP}$, and the capacitor required to tune the highpass inductor becomes $C_{BS} = 1/4\pi^2 F_0^2 L_{HP}$.

For the bandstop filter tuned to 320 kHz, the frequency translating factor is $4\pi^2 F_0^2 = 4.04259 \times 10^{12}$. Using this information, the bandstop circuit component values are given in **Table 11-1**.

Putting these components into the circuit gives the bandstop filter shown in **Figure 11-4**.

Denormalization of the bandstop model for higher load impedance requires component values to be scaled to have higher impedance. This is done in exactly the same way that lowpass or highpass filters are scaled. Inductor values increase in proportion to the load impedance. Capacitor values reduce inversely proportional to the load. Capacitor values reduce because their impedance is inversely proportional to their capacitance value. As the load impedance increases, all the reactance values must increase their impedance in order to have the same response as the prototype model.

The filter design requires a 50 Ω source and load impedance to match the radio frequency components at its input and output. The normalized values of source and load impedance

Table 11-1: Bandstop component values.

Highpass Component	Highpass Value	Bandstop Component	Bandstop Value
L1	66.31×10^{-6}	C1	3.73045×10^{-9}
C2	33.157×10^{-6}	L2	7.46045×10^{-9}
L3	66.31×10^{-6}	C3	3.73045×10^{-9}

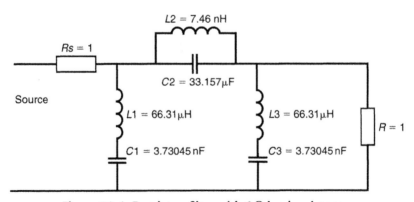

Figure 11-4: Bandstop filter with 1 Ω load resistance.

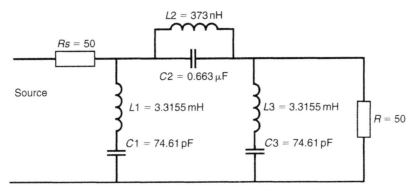

Figure 11-5: Bandstop filter, denormalized with 50 Ω load resistance.

are increased fifty-fold; therefore the impedance of the reactive components must also be increased fifty-fold. Multiplying the inductor values by 50 and dividing the capacitor values by 50 will result in the filter design shown in **Figure 11-5**.

This gives one of two possible configurations. This design was developed from the minimum inductor prototype and has one series arm that is parallel resonant. It also has two shunt arms that are series resonant. The series resonant shunt arms are connected across the input and the output terminals, so the input impedance will be low in the stopband.

If the design were, instead, developed from the minimum capacitor prototype, the end result would have used the same number of capacitors and inductors. The difference would have been that the filter would have had two parallel resonant arms wired in series between the source and load. Also, there would have been one shunt arm that was series resonant, connected between the common rail and the joining node of the two series arms.

The alternative circuit is shown in **Figure 11-6** and was designed by FILTECH (a filter design program that I helped to develop).[1] The FILTECH program calculates the normalized element values and then scales them using double precision floating-point arithmetic. The transfer function of this filter is identical to the previous version. However, the input and output impedance of this version are high in the filter's stopband.

Having gone through this laborious process, readers will be pleased to know that there are formulae that allow the whole process to be completed in one step. These formulae are similar to those used in the bandpass filter design process. Care must be taken to use the correct formulae for each stage of the design.

---

[1] FILTECH can be obtained from Adept Scientific in several countries: UK: http://directory.adeptscience. co.uk/productid/NOS-F/1/1194/NOS-F.html and in the USA: http://directory.adeptscience. com/productid/NOS-F/4/NOS-F.html

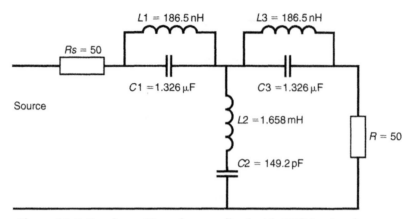

Figure 11-6: Bandstop filter, denormalized with 50 Ω load resistance.

Formula for Passive Bandstop Filter Denormalization

$$C_{Series} = \frac{1}{2\pi \cdot [F_U - F_L]RX}$$

$$L_{Series} = \frac{[F_U - F_L] \cdot RX}{2\pi F_U \cdot F_L}$$

$$C_{Shunt} = \frac{[F_U - F_L] \cdot X}{2\pi \cdot F_U F_L \cdot R} \qquad [11\text{-}1]$$

$$L_{Shunt} = \frac{R}{2\pi \cdot [F_U - F_L] \cdot X}$$

The series and shunt subscripts indicate which circuit element is being considered. A series subscript indicates the series arm (which is parallel resonant). A shunt subscript indicates the shunt arm (which is series resonant). In the equations, the factor X is the normalized lowpass element value. (Refer to the tables in Chapter 2 of the publication at the end of the References section.) The same value of X must be used for both components in a single branch. Remember that each branch in the all-pole lowpass filter has one component, whereas branches in the bandstop have two components that are either series or parallel resonant.

It might be helpful to redesign the third-order Butterworth filter to illustrate the use of these formulae. Since it is a symmetrical design, only the first three branches need to be calculated. As before, $R = 50$, $F_U = (320 + 1.2)$ kHz $= 321.2$ kHz, and $F_L = (320 - 1.2)$ kHz $= 318.8$ kHz.

The first branch has a value $X = 1.000$ and could be a series arm or a shunt arm. Taking the shunt arm case first (series resonant) gives:

$$C_{Shunt} = \frac{[F_U - F_L] \cdot X}{2\pi \cdot F_U F_L \cdot R} = 2.4 \times \frac{10^3}{3.2169} \times 10^{-13} = 74.6\,pF$$

$$L_{Shunt} = \frac{R}{2\pi \cdot [F_U - F_L] \cdot X} = \frac{50}{15,079.65} = 3.3157\,mH$$

[11-2]

The second branch has a value $X = 2.000$. Since the first arm was chosen to be a shunt arm, this arm must be series. Calculating the values gives:

$$C_{Series} = \frac{1}{2\pi \cdot [F_U - F_L]RX} = \frac{1}{1,507,964.5} = 0.663\,\mu F$$

$$L_{Series} = \frac{[F_U - F_L] \cdot RX}{2\pi F_U \cdot F_L} = 24 \times \frac{10^4}{6.43389 \times 10^{-11}} = 373\,nH$$

[11-3]

The third branch has the same prototype element values as the first branch. The filter is symmetrical, so the first and third branch component values will be the same. Symmetry is useful because if components have the same value, the cost of manufacturing is sometimes lower.

Differences between the results just obtained and those presented in Figure 11-5 are due to round-off errors, both in the tables of normalized values and during the calculations. The calculations were done by hand using a calculator. Floating-point arithmetic in a computer program achieves more accurate results.

To obtain the circuit given in Figure 11-6, it is necessary to calculate the series arm first. This will use a value of $X = 1.000$.

$$C_{Series} = \frac{1}{2\pi \cdot [F_U - F_L]RX} = \frac{1}{753,982.2} = 1.32629\,\mu F$$

$$L_{Series} = \frac{[F_U - F_L] \cdot RX}{2\pi F_U \cdot F_L} = 12 \times \frac{10^4}{6.43389 \times 10^{-11}} = 186.51\,nH$$

[11-4]

A shunt arm must be calculated next, using $X = 2.0$. Readers are invited to do the calculations themselves and compare their results with the values given in Figure 11-6. Because of symmetry, the final arm's component values are identical to those calculated above.

As mentioned in previous chapters, the specialized design of passive (*L-C*) filters is not a primary focus of this book on analog circuits because many kinds of filters can now be done with fast op-amps that could not be done in the past. Those that cannot be accomplished by op-amps may be properly part of RF design. For comprehensive information on passive filters using *L*s and *C*s, see the References section.

Active Bandstop Filters

Active bandstop filters can be designed using pole and zero locations from the frequency response's transfer function. Op-amps are the "active" part of the circuit. Op-amps have high input impedance and low output impedance. They also buffer one filter stage from the next, which prevents interaction. Each stage can therefore be designed to provide the frequency response of one pair of complex poles. Zeros may also be required in the stopband, and circuits that provide this function are usually more complex. Because stages are buffered from one another, when all the stages are connected in series the overall response should be that which is required.

Bandstop Poles and Zeros

Using the normalized lowpass response pole and zero locations as a starting point, frequency translation is required to find the normalized bandstop pole and zero locations. Frequency translation in transfer functions and the S-plane are found by replacing *s* with the following:

$$S'' = \frac{BW \cdot s}{s^2 + \omega_0^2} \qquad [11\text{-}5]$$

$\omega_0 = \sqrt{\omega_U \cdot \omega_L}$ and *BW* is the bandwidth, $\omega_U - \omega_L$.

This is not particularly easy to evaluate. However, as in the bandpass case, Williams and Taylor[2] have published equations for finding the *Q* and resonant frequency, f_R, of each section of bandstop filters from a lowpass model. These are all that are needed to design active bandstop filters. I have manipulated Williams and Taylor's equations slightly, consistent with the bandpass filter equations given in the previous chapter. To start with you must know the *Q* of bandstop filter, Q_{BS}, and σ and ω, which are the real and imaginary parts of the lowpass prototype pole location. The pole positions can be found

---

[2] Williams, A., and Taylor, F. J., *Electronic Filter Design Handbook*, McGraw-Hill, New York, 1988.

from formulae or tables. The bandstop Q is the center frequency, f_0, divided by the width of the stopband.

$$\omega_0^2 = \sigma^2 + \omega^2$$

$$A = \frac{\sigma}{\omega_0 \cdot Q_{BS}}$$

$$B = \frac{\omega}{\omega_0 \cdot Q_{BS}} \qquad\qquad [11\text{-}6]$$

$$f = B^2 - A^2 + 4$$

$$g = \sqrt{\frac{f + \sqrt{f^2 - 4A^2 B^2}}{2}}$$

(Does this remind you of a well-known quadratic solving equation? Try $a = 1$, $b = -f$, and $c = A^2 B^2$.)

$$h = \frac{AB}{g} \qquad\qquad [11\text{-}7]$$

This gives:

$$W = 0.5\sqrt{(A + h)^2 + (B + g)^2} \qquad\qquad [11\text{-}8]$$

and the frequencies are:

$$f_{R1} = \frac{f_0}{W} \quad \text{and} \quad f_{R2} = Wf_0. \qquad\qquad [11\text{-}9]$$

The pole's Q factor is given by $Q = \dfrac{W}{A + h}$.

Real poles have a Q factor of $Q = \sigma Q_{BS}$ and a resonant frequency at f_0.

Now to find the zero locations. In a prototype lowpass all-pole filter such as Chebyshev or Butterworth response, zeros are on the imaginary axis in the S-plane, at infinity. During transformation into a bandstop response they move to the center of the stopband. In a prototype lowpass Cauer and Inverse Chebyshev response, zeros are just outside the passband. When transformed into a bandstop response, the zero locations move into the stopband, placed symmetrically above and below the center frequency. Their locations have to be calculated, as follows: $J = 1/Q_{BS}Z$, where Z is the normalized lowpass zero frequency.

The zero frequencies are:

$$f_{\infty,1} = \frac{f_0}{2}\left[J - \sqrt{J^2 + 4}\right] \quad \text{and} \quad f_{\infty,2} = \frac{f_0}{2}\left[J + \sqrt{J^2 + 4}\right] \qquad [11\text{-}10]$$

So, what does the S-plane diagram look like now? In Chapter 6 an example of a fourth-order lowpass filter was given (Figure 6-11). This had a Butterworth response, with poles on a unit circle at $-0.9239 \pm j0.3827$ and $-0.3827 \pm j0.9239$.

Suppose the filter you want is required to have a stopband between 45 Hz and 55 Hz. This is for illustration only but could be used to remove power-line frequencies (50 Hz in Europe). This specification gives $BW = 10$, $f_0 = 50$ Hz, and $Q_{BP} = 5$. Taking one pole from the first pair: $s = -0.9239 + j0.3827$, $\sigma = 0.9239$, and $\omega = 0.3827$.

The two bandstop poles produced from this are found from the following equations:

$\omega_0^2 = \sigma^2 + \omega^2 = 1$ (since the poles are on a unit circle for the Butterworth response).

$$A = \frac{\sigma}{\omega_0 \cdot Q_{BS}} = 0.18478$$

$$B = \frac{\omega}{\omega_0 \cdot Q_{BS}} = 0.07654$$

$$f = B^2 - A^2 + 4 = 3.89176 \qquad [11\text{-}11]$$

$$g = \sqrt{\frac{f + \sqrt{f^2 - 4A^2B^2}}{2}} = 1.9727414$$

$$h = \frac{AB}{g} = 0.00716924$$

This gives $W = 0.5\sqrt{(A + h)^2 + (B + g)^2} = 1.029125693$.

The frequencies are $f_{R1} = \dfrac{f_0}{W} = 48.58493 \quad$ and $\quad f_{R2} = Wf_0 = 51.456284$.

The pole's Q factor is given by $Q = \dfrac{W}{A + h} = 5.361447$.

The second pair of poles can be found in a similar way. Due to symmetry, $\sigma = 0.3827$ and $\omega = 0.9239$:

$$A = \frac{\sigma}{\omega_0 \cdot Q_{BS}} = 0.07654$$

$$B = \frac{\omega}{\omega_0 \cdot Q_{BS}} = 0.18478$$

$$f = B^2 - A^2 + 4 = 4.028285277 \qquad \text{[11-12]}$$

$$g = \sqrt{\frac{f + \sqrt{f^2 - 4A^2B^2}}{2}} = 2.007046492$$

$$h = \frac{AB}{g} = 0.00704670333$$

This gives $W = 0.5\sqrt{(A+h)^2 + (B+g)^2} = 1.096709865.$

The frequencies are $f_{R1} = \dfrac{f_0}{W} = 44.59091$ and $f_{R2} = Wf_0 = 55.835493.$

The pole's Q factor is given by $Q = W/A + h = 13.1206264.$

To help you visualize what has happened to the poles, take a look at the S-plane diagram in **Figure 11-7**. This diagram only shows the positive frequency poles. There are symmetrical negative frequency poles, but these have been omitted for clarity. Also note that, for a given Q, the poles lie on a line that passes through the origin. The two poles

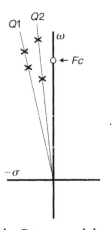

Figure 11-7: Fourth-order Butterworth bandstop pole locations.

just calculated both had a Q of about 5.4. The other poles had a Q of about 13.1 but are further from the bandstop filter's center frequency, f_0. Remember that the Q of a pole is given by the equation:

$$Q = \frac{\sqrt{\sigma^2 + \omega^2}}{2\sigma}$$

[11-13]

The Q of a bandstop pole is approximately $2f_0Q_{LP}/BW$, where Q_{LP} is the normalized lowpass pole Q. In the case of the normalized Butterworth filter poles given, $Q1_{LP} = 1/2\sigma = 0.54118$ and $Q2_{LP} = 1.3065$. The ratio f_0/BW is 5. The bandstop Q factors are then approximately: $Q1_{BS} = 10 \times 0.54118 = 5.41$. $Q2_{BS} = 10 \times 1.3065 = 13.1$.

The pole-zero diagram in Figure 11-7 is very much like the example given to describe bandpass filters. Bandstop filters do not have zeros at the S-plane origin (DC) or at infinity, they only have zeros at the stopband center frequency.

Taking a look at some bandstop active filter designs, I will show how the pole and zero locations are used to find component values. We will return to the S-plane later in this chapter when we discuss active Cauer and Inverse Chebyshev filters. Both these types have zeros in the stopband that are not at the center frequency.

The Twin Tee Bandstop Filter

The basic twin tee bandstop filter, with $R3$ and $2C$ shorted to ground rather than to $A2$, is one of the simplest bandstop filters, yet it is not often used. The reason for its lack of popularity is its poor Q factor; in fact, it has a Q of 0.25. One way to improve the Q factor is by adding amplifier $A2$, and applying positive feedback (with a gain less than 1.0) to the foot of $R3$ and $2C$, as shown. This means that changes in amplitude are amplified, which results in a sharper passband to stopband transition. The circuit diagram in **Figure 11-8** shows the amplified twin tee.

The component values can be calculated from:

$$R1 = R2 = \frac{1}{2\pi f_z C} \quad \text{where } C \text{ is any suitable value.}$$

$$R3 = \frac{R1}{2}.$$

[11-14]

The feedback factor, $k = 1 - 1/4Q$ for any desired Q factor. For example, suppose a Q of 5 is required, $k = 1 - 0.05 = 0.95$. If $R = 10\,\text{k}\Omega$, $kR = 9.5\,\text{k}\Omega$, and $(1 - k)R = 500\,\Omega$. The nearest preferred values in the E96 range are $9.53\,\text{k}\Omega$ and $499\,\Omega$, respectively. If

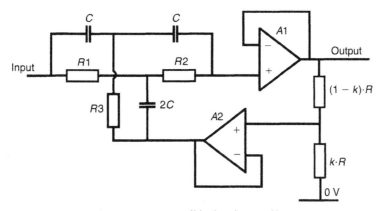

Figure 11-8: Amplified twin tee filter.

standard range resistors must be used, two components are required for each value, $9.1\,k\Omega + 390\,\Omega$ and $470\,\Omega + 27\,\Omega$, respectively.

The second op-amp can be omitted if the feedback resistors are much lower than the values of $R1$, $R2$, and $R3$. The node joining the feedback resistors can be connected directly to the junction of $R3$ and the shunt capacitor (value $= 2C$). This can lead to slight errors in the notch frequency due to an increase in impedance in the shunt path. Omitting the op-amp is probably not worth considering just to save space or to reduce costs, since dual op-amps are inexpensive and readily available.

Denormalization of Twin Tee Notch Filter

For this example, consider a 50 Hz notch filter with 10 Hz between the upper and lower passband edges. $Q = 50/10 = 5$. Using values $f_Z = 50\,Hz$ or $314.159\,rad/s$ and $C = 0.1\,\mu F$ (hence $2C = 0.2\,\mu F$), component values can be found by substitution into the following equations:

$$R1 = R2 = \frac{1}{2\pi f_Z C} = 31.83\,k$$

$$R3 = \frac{R1}{2} = 15.91\,k \qquad [11\text{-}15]$$

$$k = 1 - \frac{1}{4Q} = 0.95$$

Let $R = 20\,k\Omega$. Resistor element $kR = 19\,k\Omega$ and resistor element $(1 - k)R = 1\,k\Omega$.

Practical Implementation of Twin Tee Notch Filter

It might seem that using precision components such as 2% capacitors and 1% resistors might give reasonable performance, but this is not so. The accuracy of the center frequency and the Q will not be good, even with 1% capacitors, and for a higher Q, you won't get good depth from the notch. Go ahead and try it! For good results, it is necessary to trim this circuit with three trimpots.

For the example listed above, leave $R2$ at 31.6 k. Change $R1$ to 26.7 k and a 10 k pot. Change $R3$ to 13.0 k and a 5 k pot. Thus, $R1$ and $R3$ are given a +/− 18% trim range. Change $(1 − k) R$ to a 500 ohm pot, and set $k \cdot R$ at 4.99 k. Make sure the capacitors have better than a 5% tolerance, and for best results, match them to 1%. Put in a sine wave of the desired frequency.

If you simply try to adjust the pots, you will find that they all interact badly! To see how to trim the pots, put the filter's input and output into an *X-Y* (cross-plot) oscilloscope and start trimming, and watch which trims are affecting how much of gain and phase. You will soon find that you can trim for good depth at the desired frequency. The resistor $k \cdot R$ is primarily a trim on the Q, but it does interact with the other pots, whether you are using A2 or not. Set the input frequency to $f_Z (1 − 1/(2Q))$ and trim for the desired drop in the output amplitude. After a little practice, you can trim one of these circuits to excellent accuracy in less than a minute.

Bandstop Using Multiple Feedback Bandpass Section

One of the simplest bandstop filters suitable for all-pole responses is the multiple feedback bandpass (MFBP) circuit, described in Chapter 10, followed by a summing amplifier. The summing amplifier sums the output of the MFBP section (which is inverting) with the input signal. In frequency spectrum terms, the circuit is subtracting a passband from a wideband response to create a stopband. The circuit is illustrated in **Figure 11-9**; op-amp A2 and the three resistors labeled R form the summing amplifier.

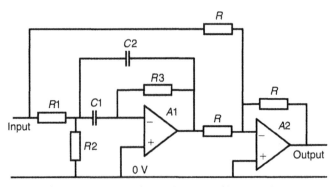

Figure 11-9: Bandstop (MFBP) filter section.

The MFBP circuit is typically limited to applications where the pole's Q value is less than 20. This limitation restricts its use considerably, but for simple applications it is easy to use. The performance of the MFBP circuit depends mainly on the op-amp employed. The gain-bandwidth product of the device should be well in excess of the resonant frequency multiplied by the resonant gain. In mathematical terms, $GBW \gg Q \times f_R$.

The gain at the bandpass circuit's resonant frequency is given by $R3/2 \cdot R1$ and is also modified by the ratio of the two capacitors. Therefore, it is a good idea to match the two capacitors within 1%. Clearly the gain at resonance must be unity. When the output from the bandpass section is summed with the input, both signals have the same amplitude and cancel each other to produce a notch. It might also be a good idea to add a trimpot as part of $R1$ if you want to get a good depth from the notch.

The center frequency is determined primarily by trimming $R2$. As stated in Chapter 10, the center frequency is:

$$f_R = \left(\frac{1}{2\pi C} \right) \sqrt{\left(\frac{R1 + R2}{R1} \right)} \sqrt{\left(\frac{1}{R3 \times R2} \right)} \qquad [11\text{-}16]$$

Because this depends so strongly on $R2$, $R2$ must be trimmed to get the correct resonant frequency of the notch. Fortunately, the interaction between the frequency trim and the gain trim is almost zero.

Denormalization of Bandstop Design Using MFBP Section

Consider one pole found earlier for the fourth-order Butterworth 50 Hz notch filter. For this pole, $f_R = 48.58493$ Hz or 305.26812 rad/s, having $Q = 5.361447$.

$$R3 = \frac{Q}{\pi f_R C} = 5590 \, \Omega$$

$$R1 = \frac{R3}{2} = 2795 \, \Omega \qquad [11\text{-}17]$$

$$R2 = \frac{R3}{4Q^2 - 2} = 49.47758 \, \Omega$$

In practice, obtaining resistor values close to those calculated could be difficult. Individual component selection and/or trimming might be needed to achieve a notch filter design with the desired amount of stopband loss.

Bandstop Using Dual-Amplifier Bandpass Section

A bandstop filter section can be made using a dual-amplifier bandpass (DABP) design. This is achieved by using a summing amplifier to subtract the bandpass response from the

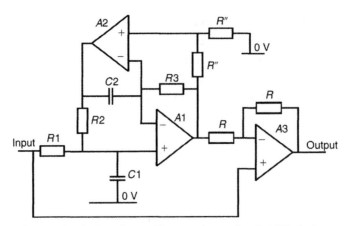

Figure 11-10: Bandstop filter section using DABP design.

input signal. The DABP topology is more complicated than using the MFBP structure, but it has the advantage that much higher Q factors can be achieved; Q factors of up to 150 are possible. The DABP is an all-pole response bandpass filter section, as described in Chapter 10. The bandpass response can be subtracted from the input signal by a summing circuit to create a bandstop response. The DABP filter has a noninverting output with a gain of two at the resonant frequency, so a slightly different summing circuit to the MFBP filter is required. In **Figure 11-10**, the input signal is applied to the noninverting input of the summing amplifier. The output from the bandpass section is applied to a resistor in series with the inverting input. The feedback resistor from the summing amplifier's output forms a potential divider to signals from the bandpass section. The bandpass section output will be at ground potential when no bandpass signals are present (outside the stopband). Therefore the summing amplifier forms a noninverting amplifier with a gain of two. The circuit diagram for a DABP filter is given in **Figure 11-10**.

The capacitance of $C1$ and $C2$ should be equal but may have any arbitrary value. In practice, the capacitors' values are chosen so that the resistors' values are all in a reasonable range, typically from $1\,k\Omega$ to $100\,k\Omega$. The limits are up to the designer, but remember that $R3$, R'', and R load the output of op-amps $A1$ and $A2$. High values of $R1$ will introduce noise and could degrade the signals because it is in series with the signal path.

Consider the equations:

$$R1 = \frac{Q}{2\pi \cdot f_R C}, \quad \text{and} \quad R2 = R3 = \frac{1}{2\pi \cdot f_R C}. \qquad [11\text{-}18]$$

In other words, $R1 = Q \cdot R2$.

If the section's Q is high, say 75, then $R1$ will be 75 times $R2$ and $R3$. The value of C should be chosen so that the value of $R1$ is close to the lower resistance limit. The value of $R2$ will then be close to the recommended upper resistance limit. This circuit has independent adjustment of resonant frequency and Q. The resistor $R1$ is used to adjust the Q at resonance. Resistor $R2$ and $R3$ determine the resonant frequency. At the resonant frequency, the gain is fixed at $G_R = 2$.

Denormalization of Bandstop Design Using DABP Section

Consider one pole found earlier for the fourth-order Butterworth 50 Hz notch filter.

For this pole, f_R = 51.456284 Hz or 323.3093676 rad/s, having Q = 5.361447.

Let C = 0.1 µF and let $R = R \le 10 \text{k}\Omega$.

$$R2 = R3 = \frac{1}{2\pi \cdot f_R C} = 4922\,\Omega \qquad [11\text{-}19]$$

$$R1 = Q \cdot R2 = 26,393\,\Omega.$$

State Variable Bandstop Filters

The state variable design can be used for all-pole responses or in any filter where a zero at the stopband center frequency is required. It has a lower sensitivity to the op-amp's gain-bandwidth product limitation, and section Q factors of up to 200 are possible. It does, however, need four op-amps, as shown in **Figure 11-11**.

The equations for this filter allow the arbitrary choice of capacitor, C.

$$R1 = R4 = \frac{Q}{2\pi f_R C}$$
$$R2 = R3 = \frac{R1}{Q} \qquad [11\text{-}20]$$

The value of R' is discretionary, but a typical value could be 10 kΩ. Resistors R also have an arbitrary value that could be set the same as R' if required. Note that the value of R has an effect on the filter's input impedance.

Denormalization of Bandstop State Variable Filter Section

Consider one pole found earlier for the fourth-order Butterworth 50 Hz notch filter. For this pole, f_R = 44.59091 Hz or 280.17295 rad/s, having Q = 13.1206264.

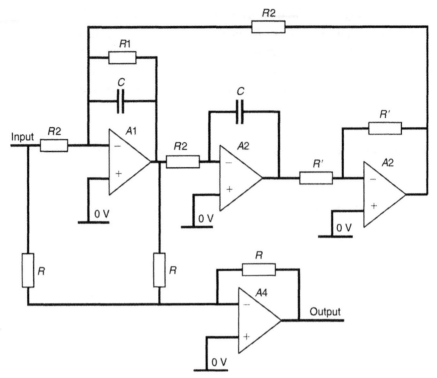

Figure 11-11: State variable bandstop (all-pole).

Let $C = 0.1\,\mu F$ and let $R = R' = 10\,k\Omega$.

$$R1 = R4 = \frac{Q}{2\pi f_R C} = 74{,}533\,\Omega$$

$$R2 = R3 = \frac{R1}{Q} = 5680\,\Omega$$

[11-21]

Cauer and Inverse Chebyshev Active Filters

Designing bandstop filters with a Cauer or an Inverse Chebyshev response is more difficult than for all-pole filters. This is because each filter section must provide both poles and zeros close to the filter's center frequency. Moreover, the pole and zero pairing must also be considered. A filter may have a number of poles and zeros and, in principle, any zero could be associated with any pole. In practice, the pole-zero pairing affects performance. The lowest-frequency pole should be paired with the lowest-frequency zero. In addition, the pole with the lowest Q should be used in the first stage; otherwise, signal

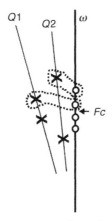

Figure 11-12: Cauer pole and zero pairing.

magnification by a large value of Q could cause overloading of subsequent stages. Pole and zero pairing is illustrated in **Figure 11-12**.

To design a Cauer or Inverse Chebyshev filter, a different circuit topology is required. The Cauer response has zeros in the stopband, so a tunable notch circuit is required. This can be achieved using a circuit that is an extension of the state variable filter and is known as a *biquad*. This circuit is illustrated in **Figure 11-13**. Note that, in the bandstop biquad, $R5$ is connected to a different node. This is dependent on whether the zero is above or below the resonant frequency. If the zero is above the resonant frequency, connect nodes A and C. If the zero is below the resonant frequency, connect nodes B and C.

The following equations give component values.

$$R1 = R4 = \frac{Q}{2\pi f_R C}$$

$$R2 = R3 = \frac{1}{2\pi f_R C} = \frac{R1}{Q}$$

$$R5 = \frac{f_R^2 R}{Q\left|f_R^2 - f_Z^2\right|} \qquad \text{[11-22]}$$

$$f_Z = \text{zero frequency.}$$

$$R6 = \frac{f_R^2 R}{f_Z^2} \quad \text{when } f_R > f_\infty \text{ or when } f_R < f_\infty.$$

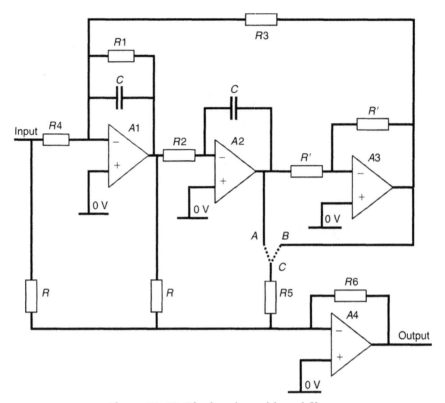

Figure 11-13: The bandstop biquad filter.

$R6 = R$ when an all-pole filter is required, since $f_R = f_\infty$ and $R5$ is not connected (see the "State Variable Bandstop Filters" section).

The resistors labeled R and R' can be any arbitrary value; a typical value may be in the range $1\,\text{k}\Omega$ to $100\,\text{k}\Omega$, say $10\,\text{k}\Omega$. The resistors labeled R have an effect on the input impedance of the filter section. The value of R should be several times higher than the input signal's source impedance.

Denormalization of Bandstop Biquad Filter Section

Consider a hypothetical design for a Cauer filter section that will produce a pole and a zero. This design has a pole $f_R = 280\,\text{rad/s}$, which has $Q = 15$. The zero for this design will be at $300\,\text{rad/s}$.

Let $C = 0.1\,\mu\text{F}$ and let $R = R' = 10\,\text{k}\Omega$.

$$R1 = R4 = \frac{Q}{2\pi f_R C} = 82,262 \ \Omega$$

$$R2 = R3 = \frac{1}{2\pi f_R C} = \frac{R1}{Q} = 5684 \ \Omega$$

$$R5 = \frac{f_R^2 R}{Q \left| f_R^2 - f_Z^2 \right|} = 4506 \ \Omega$$

$$R6 = \frac{f_R^2 R}{f_Z^2} = 8711 \ \Omega.$$

[11-23]

References

Williams, A., and Taylor, F.J., *Electronic Filter Design Handbook*, McGraw-Hill, New York, 1988.

Winder, S., *Analog and Digital Filter Design*, Elsevier-Newnes, Amsterdam, 2002.

CHAPTER 12

Current–Feedback Amplifiers

Sergio Franco

Again, Sergio Franco is another good explainer. What are the real differences between a conventional op-amp and a current-feedback Amplifier? Here's why and how they are not obvious! / rap

The Current-Feedback Concept

In their effort to approximate the ideal op-amp, manufacturers strive not only to maximize the open-loop gain and minimize input-referred errors such as offset voltage, bias current, and noise, but also to ensure adequate bandwidth and settling-time characteristics. Amplifier dynamics are particularly important in high-speed applications such as bipolar DAC buffers, subranging ADCs, S/H circuits, ATE pin drivers, and video and IF drivers. Being voltage-processing devices, conventional op-amps are subject to the speed limitations inherent to *voltage-mode* operation, stemming primarily from the stray capacitances of nodes and the cutoff frequencies of transistors. Particularly severe is the effect of the stray capacitance between the input and output nodes of high-gain inverting stages because of the Miller effect, which multiplies this capacitance by the voltage gain of the stage. By contrast, *current-mode* operation has long been recognized as inherently faster than voltage-mode operation. The effect of stray inductances in an integrated circuit is usually less severe than that of its stray capacitances, and BJTs switch currents more rapidly than voltages. These technological reasons are at the basis of *emitter coupled logic, bipolar DACs, current conveyors*, and the high-speed amplifier topology known as *current-feedback*.[1]

---

[1] Based on an article that appeared in *EDN Magazine*, Jan. 5, 1989, © Cahners Publishing Company 1990, A Division of Reed Publishing USA.

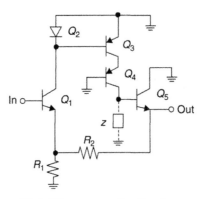

Figure 12-1: The current-feedback concept.

For true current-mode operation, all nodes in the circuit should ideally be kept at fixed potentials to avoid the slowdown effect by their stray capacitances. However, since the input and output of the amplifier must be voltages, some form of high-speed voltage-mode operation must be provided at some point. This is achieved by driving the nodes with push–pull emitter follower stages to rapidly charge or discharge their stray capacitances and by employing gain configurations inherently immune to the Miller effect, such as the cascode configuration.

The above concepts are illustrated using the simplified AC equivalent of **Figure 12-1** as a vehicle. The circuit consists of the emitter follower input stage Q_1, the current mirror Q_2 and Q_3, the cascode gain stage Q_3 and Q_4, and the emitter follower output stage Q_5. The feedback signal is the current fed from the emitter of Q_5 back to the emitter of Q_1 via R_2, indicating series-shunt feedback. A qualitative analysis reveals that the open-loop characteristics are set primarily by the equivalent impedance z between the collector of Q_4 and ground. The resistive component of z sets the open-loop DC gain, and the capacitive component controls the open-loop dynamics.

Variants of the basic topology of Figure 12-1 have long been used in high-speed applications such as active probes. Its adaptation to op-amp–like operation requires an input stage of the differential type. Moreover, to ensure symmetric rise and fall times, each stage must be capable of complementary push–pull action, and the npn and pnp transistors must have comparable characteristics in terms of the cutoff frequency f_t. Traditionally, monolithic pnp transistors have been plagued by much poorer performance characteristics than their npn counterparts. However, the development of truly complementary high-speed processes has made it possible to achieve monolithic speeds that were previously available only in hybrid form. The unique features and operation of the current-feedback (CF) amp are best appreciated by comparing them against those of its better-known counterpart, the conventional op-amp.

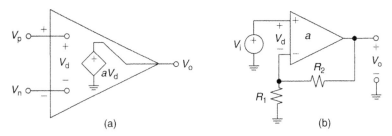

Figure 12-2: Circuit model of the conventional op-amp and
connection as a noninverting amplifier.

The Conventional Op-Amp

The conventional op-amp consists of a high input-impedance differential stage followed by additional gain stages, the last of which is a low output-impedance stage. As shown in the circuit model of **Figure 12-2a**, the op amp transfer characteristic is:

$$V_o = a(jf)V_d \qquad [12\text{-}1]$$

where V_o is the output voltage, $V_d = V_p - V_n$ is the differential input voltage, and $a(jf)$, a complex function of frequency f, is the open-loop gain.

Connecting an external network as in **Figure 12-2b** creates a feedback path along which a signal in the form of a *voltage* is derived from the output and applied to the noninverting input. By inspection:

$$V_d = V_i - \frac{R_1}{R_1 + R_2} V_o \qquad [12\text{-}2]$$

Substituting into Equation 12-1, collecting, and solving for the ratio V_o/V_i yields the *noninverting amplifier* transfer characteristic:

$$A(jf) = \frac{V_o}{V_i} = \left(1 + \frac{R_2}{R_1}\right)\frac{1}{1 + 1/T(jf)} \qquad [12\text{-}3]$$

$$T(jf) = \frac{a(jf)}{1 + R_2/R_1} \qquad [12\text{-}4]$$

where $A(jf)$ is the closed-loop gain and $T(jf)$ is the loop gain. The designation *loop gain* stems from the fact that if we break the loop as in **Figure 12-3a** and inject a test signal V_x with V_i suppressed, the circuit will first attenuate V_x to produce $V_n = V_x/(1 + R_2/R_1)$, and then amplify V_n to produce $V_o = -aV_n$. The gain experienced by a signal in going

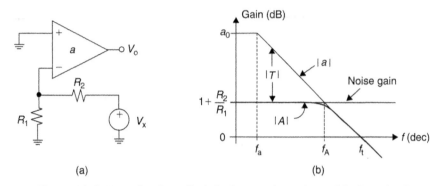

**Figure 12-3: Test circuit to find the loop gain and graphical method
to determine the closed-loop bandwidth f_A.**

around the loop is thus $V_o/V_x = -a/(1 + R_2/R_1)$. The *negative* of this ratio is the loop
gain, $T = -(V_o/V_x)$. Hence, Equation 12-4.

The loop gain gives a measure of how close A is to the ideal value $1 + R_2/R_1$, also
called the *noise gain* of the circuit. By Equation 12-3, the larger T, the better. To ensure
a substantial loop gain over a wide range of closed-loop gains, the manufacturer strives
to make a as large as possible. Consequently, since $V_d = V_o/a$, V_d will assume extremely
small values. In the limit $a \to \infty$ we obtain $V_d \to 0$, that is, $V_n \to V_p$. This forms the basis
of the familiar op-amp rule: *When operated with negative feedback, an op-amp will provide
whatever output is needed to force V_n to follow V_p.*

Gain-Bandwidth Trade-off

Large open-loop gains can physically be realized only over a limited frequency range.
Past this range, gain rolls off with frequency. Most op-amps are designed for a constant
rolloff of $-20\,dB/dec$ so that the open-loop response can be expressed as:

$$a(jf) = \frac{a_0}{1 + j(f/f_a)}$$ [12-5]

where a_0 represents the DC gain and f_a is the $-3\,dB$ frequency of the open-loop response.
For instance, the popular 741 op-amp has $a_0 \approx 2 \times 10^5$ and $f_a \approx 5\,Hz$.

Substituting Equation 12-5 into Equation 12-4 and then into Equation 12-3 and exploiting
the fact that $(1 + R_2/R_1)/a_0 \ll 1$, we obtain:

$$A(jf) = \frac{1 + R_2/R_1}{1 + j(f/f_A)}$$ [12-6]

$$f_A = \frac{f_t}{1 + R_2/R_1} \qquad [12\text{-}7]$$

where f_A is the closed-loop bandwidth and $f_t = a_0 f_a$ is the open-loop unity-gain frequency, that is, the frequency at which $|a| = 1$. For instance, the 741 op-amp has $f_t = 2 \times 10^5 \times 5 = 1\,\text{MHz}$.

Equation 12-7 reveals a gain-bandwidth trade-off. As we raise the R_2/R_1 ratio to increase the closed-loop gain, we also decrease its bandwidth in the process. Moreover, by Equation 12-4, the loop-gain is also decreased, leading to a greater closed-loop gain error.

These concepts can also be visualized graphically. By Equation 12-4 we have $|T|_{dB} = 20\log|T| = 20\log|a| - 20\log|1 + R_2/R_1|$, or:

$$|T|_{dB} = |a|_{dB} - |1 + R_2/R_1|_{dB} \qquad [12\text{-}8]$$

indicating that the loop gain can be found graphically as the *difference* between the open-loop gain and the noise gain. This is shown in Figure 12-3b. The frequency at which the two curves meet is called the *crossover frequency*. It is readily seen that at this frequency we have $T = 1\underline{|-90°} = -j$, so that Equation 12-3 yields $|A| = (1 + R_2/R_1) / |1 + j| = (1 + R_2/R_1)/\sqrt{2}$. Consequently, the crossover frequency is also the $-3\,\text{dB}$ frequency of the closed-loop response, that is, the closed-loop bandwidth f_A.

We now see that increasing the closed-loop gain shifts the noise-gain curve upward and causes the crosspoint to move up the $|a|$ curve, thus decreasing the closed-loop bandwidth as well as the loop gain. Clearly, the circuit with the widest bandwidth and the highest loop gain is also the one with the lowest closed-loop gain. This is the voltage follower, for which $R_2/R_1 = 0$ so that $A = 1/[1 + j(f/f_t)]$.

Slew-Rate Limiting

To fully characterize the dynamic behavior of an op-amp, we also need to know its *transient response*. If an op-amp with the response of Equation 12-5 is operated as a unity-gain voltage follower and is subjected to a suitably small voltage step, its dynamic behavior will be similar to that of an RC network. Applying an input step of magnitude ΔV_i as in **Figure 12-4a** will cause the output to undergo an exponential transition with magnitude $\Delta V_o = \Delta V_i$, and with the time constant $\tau = 1/(2\pi f_t)$.

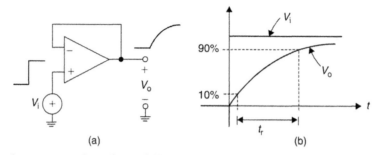

(a)   (b)

Figure 12-4: The voltage follower and its small-signal step response.

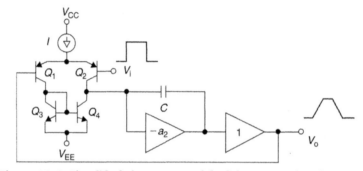

Figure 12-5: Simplified slew rate model of the conventional op-amp.

The *risetime* is defined as the amount of time t_r it takes for the output to swing from 10% to 90% of the step size. For an exponential transition, we have $t_r = \tau \times \ln(0.9/0.1) = 2.2\tau$. For the 741 op-amp we have $\tau = 1/(2\pi \times 10^6) \approx 160\,\text{nsec}$, and $t_r \approx 350\,\text{nsec}$.

The rate at which the output changes with time is highest at the beginning of the transition, when its value is $\Delta V_o/\tau$. Increasing the step magnitude increases this initial rate of change, until this rate saturates at a value called the *slew rate* (SR). This effect stems from the limited ability of the internal circuitry to charge or discharge capacitive loads, especially the internal frequency compensation capacitor.

To illustrate, refer to the circuit model of **Figure 12-5**, which is typical of many op-amps. The input stage is a transconductance amplifier consisting of the differential pair Q_1–Q_2 and the current mirror load Q_3–Q_4. The remaining stages are lumped together as an integrator block consisting of an inverting amplifier and the compensation capacitor C. Slew-rate limiting occurs when the transconductance stage is driven into saturation so that all the current available to charge or discharge C is the bias current I of this stage.

For example, the 741 op-amp has $I = 20\,\mu A$ and $C = 30$ pF so that $SR = I/C = 0.67$ V/μsec. The step magnitude corresponding to the onset of slew rate limiting is such that $\Delta V_i/\tau = SR$, or $\Delta V_i = SR \times \tau = (0.67\text{ V/μsec}) \times (160\text{ nsec}) = 106$ mV. As long as the input step is less than 106 mV, a 741 voltage follower will respond with an exponential transition governed by $\tau \approx 160$ nsec, whereas for greater input steps the output will slew at a constant rate of 0.67 V/μsec.

An important dynamic parameter in high-speed applications is the *settling time*, that is, the time it takes for the output to settle and remain within a specified band around its final value, usually for a full-scale output transition. Clearly, slew-rate limiting plays an important role in the settling-time characteristic of a conventional op-amp.

The Current–Feedback Amplifier

As shown in the circuit model of **Figure 12-6a**, the architecture of the CF amplifier differs from the conventional op-amp in two respects [1]:

1. The input stage is a *unity-gain voltage buffer* connected across the inputs. Its function is to force V_n to follow V_p, very much the way a conventional op-amp does via negative feedback. However, because of the low output impedance of this buffer, current can easily flow in or out of the inverting input, though we shall see that in the steady state (nonslewing) condition this current is designed to approach zero.

2. Amplification is provided by a transimpedance stage that senses the current delivered by the buffer to the external feedback network and produces an output voltage V_o such that:

$$V_o = z(jf)I_n \qquad [12\text{-}9]$$

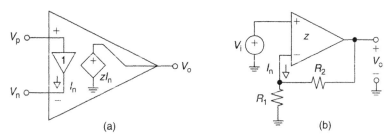

Figure 12-6: Circuit model of the current feedback amplifier and connection as a noninverting amplifier.

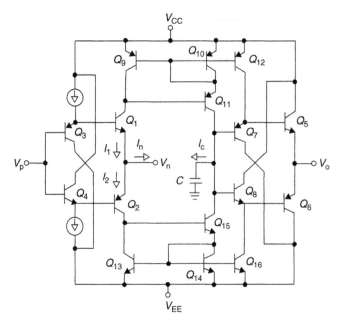

**Figure 12-7: Simplified circuit diagram of a current-feedback amplifier.
(Courtesy of Comlinear/NSC.)**

where $z(jf)$ is the transimpedance gain of the amplifier, in volts per amp or ohms, and I_n is the current out of the inverting input.

To appreciate the inner workings of the CF amp, it is instructive to examine the simplified circuit diagram of **Figure 12-7**. The input buffer consists of transistors Q_1 through Q_4. Whereas Q_1 and Q_2 form a low output–impedance push-pull stage, Q_3 and Q_4 provide V_{be} compensation for the push-pull pair, as well as a Darlington function to raise the input impedance.

Summing currents at the inverting node yields $I_1 - I_2 = I_n$, where I_1 and I_2 are the push–pull transistor currents. A pair of Wilson current mirrors, consisting of transistors $Q_9 - Q_{10} - Q_{11}$ and $Q_{13} - Q_{14} - Q_{15}$, reflect these currents and recombine them at a common node, whose equivalent capacitance to ground is denoted as C. By mirror action, the current through this capacitance is $I_c = I_1 - I_2$, or:

$$I_c = I_n \qquad\qquad [12\text{-}10]$$

The voltage developed by C in response to this current is then conveyed to the output via a second buffer, made up of Q_5 through Q_8. The block diagram of **Figure 12-8** summarizes the salient features of the CF amp.

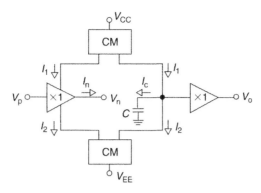

Figure 12-8: Current-feedback amplifier block diagram.

When the feedback loop is closed, as in Figure 12-6b, and whenever an external signal tries to imbalance the two inputs, the input buffer will begin sourcing (or sinking) an imbalance current I_n to the external resistances. This imbalance is then conveyed by the Wilson mirrors to the capacitor C, causing V_o to swing in the positive (or negative) direction until the original imbalance I_n is neutralized via the negative feedback loop. Clearly, I_n plays the role of the error signal in the system.

To obtain the closed-loop transfer characteristic, we exploit the fact that the input buffer keeps $V_n = V_p = V_i$. Applying the superposition principle, we have:

$$I_n = \frac{V_i}{R_1 \parallel R_2} - \frac{V_o}{R_2} \qquad [12\text{-}11]$$

This confirms that the feedback signal, V_o/R_2, is now in the form of a *current*. Substituting into Equation 12-9, collecting, and solving for the ratio V_o/V_i yields:

$$A(jf) = \frac{V_o}{V_i} = \left(1 + \frac{R_2}{R_1}\right) \frac{1}{1 + 1/T(jf)} \qquad [12\text{-}12]$$

$$T(jf) = \frac{z(jf)}{R_2} \qquad [12\text{-}13]$$

where $A(jf)$ is the closed-loop gain of the circuit and $T(jf)$ is the loop gain. This designation stems again from the fact that if we break the loop as in **Figure 12-9a** and inject a test voltage V_x with the input V_i suppressed, the circuit will first convert V_x to the current $I_n = -V_x/R_2$ and then convert I_n to the voltage $V_o = zI_n$ so that $T = -(V_o/V_x) = z/R_2$, as expected.

In an effort to ensure substantial loop gain and thus reduce the closed-loop gain error, the manufacturer strives to make z as large as possible relative to the expected range of values of R_2.

Consequently, since $I_n = V_0/z$, the inverting-input current will be very small, though this input is a low-impedance node because of the buffer. In the limit $z \to \infty$ we obtain $I_n \to 0$, indicating that a CF amp *will provide whatever output is needed to ideally drive* I_n *to zero*. Thus, the familiar op-amp conditions $V_n \to V_p$, $I_n \to 0$, and $I_p \to 0$ hold also for CF amps, though for different reasons.

No Gain-Bandwidth Tradeoff

The transimpedance gain of a practical CF amp rolls off with frequency according to:

$$Z(jf) = \frac{z_0}{1 + j(f/f_a)} \qquad [12\text{-}14]$$

where z_0 is the *DC* value of the transimpedance gain and f_a is the frequency at which rolloff begins. For instance, the CLC401 CF amp (Comlinear Co.) has $z_0 \approx 710 \text{ k}\Omega$ and $f_a \approx 350 \text{ kHZ}$. Moreover, since $f_a = 1/(2\pi z_0 C)$, it follows that $C = 1/(2\pi z_0 f_a) \approx 0.64 \text{ pF}$.

Substituting Equation 12-14 into Equation 12-13 and then into Equation 12-12 and exploiting the fact that $R_2/z_0 \ll 1$, we obtain:

$$A(jf) = \frac{1 + R_2/R_1}{1 + j(f/f_A)} \qquad [12\text{-}15]$$

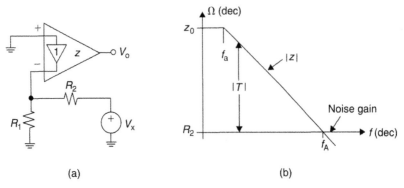

(a)   (b)

Figure 12-9: Test circuit to find the loop gain and graphical method to determine the closed-loop bandwidth f_A.

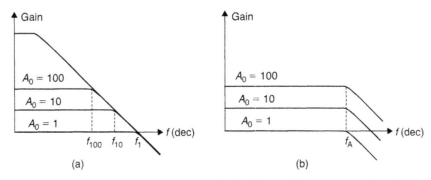

Figure 12-10: Comparing the gain-bandwidth relationship of conventional op-amps and current-feedback amplifiers.

$$f_A = \frac{z_0 f_a}{R_2}$$
[12-16]

where f_A represents the closed-loop bandwidth. With R_2 in the kilohm range, f_A is typically in the 100 MHz range. Retracing previous reasoning, we see that the noise-gain curve is now R_2, and that f_A can be found graphically as the frequency at which this curve meets the $|z|$ curve, as shown in **Figure 12-9b**.

Comparing with Equations 12-6 and 12-7, we note that the expressions for $A(jf)$ are formally identical; however, the bandwidth f_A now depends only on R_2, indicating that we can use R_2 to select the bandwidth and R_1 to select the gain. The ability to control gain independently of bandwidth constitutes a major advantage of CF amps over conventional op-amps, especially in automatic gain-control applications. This important difference is highlighted in **Figure 12-10**, where $A_0 = 1 + R_2/R_1$ denotes the DC value of the closed-loop gain.

Absence of Slew-Rate Limiting

The other major advantage of CF amps is the inherent absence of slew-rate limiting. This stems from the fact that the current available to charge the internal capacitance C at the onset of a step is proportional to the step *regardless* of its size. Indeed, applying a step of magnitude ΔV_i induces, by Equation 12-11, an initial current imbalance $\Delta I_n = \Delta V_i/(R_1\|R_2)$, which the Wilson mirrors then convey to the capacitor. The initial rate of charge is thus $\Delta I_c/C = \Delta I_n/C = \Delta V_i/[(R_1\|R_2)C] = [\Delta V_i(1 + R_2/R_1)]/(R_2C) = \Delta V_o/(R_2C)$, indicating an exponential output transition with time-constant $\tau = R_2C$. Like the frequency response, the transient response is governed by R_2 alone, regardless of the closed-loop gain. With R_2 in the kilohm range and C in the picofarad range, τ will be in the nanosecond range.

The time it takes for an exponential transient to settle within 0.1% of its final value is $t_s = \tau \ln 1000 \approx 7\tau$. For instance, in the case of a CLC401 CF amp with $R_2 = 1.5$ kΩ we have $\tau = R_2 C = 1.5 \times 10^3 \times 0.64 \times 10^{-12} \approx 1$ nsec, so that $t_r = 2.2\tau \approx 2.2$ nsec, and $t_s = 7\tau \approx 7$ nsec. These values are in reasonable agreement with the datasheet values $t_r = 2.5$ nsec and $t_s = 10$ nsec.

The absence of slew-rate limiting not only allows for faster settling times but also avoids slew-rate related nonlinearities such as intermodulation distortion. This makes CF amps attractive in high-quality audio amplifier applications.

Second-Order Effects

The above analysis indicates that once R_2 has been set, the dynamics of the amplifier are unaffected by the closed-loop gain setting. In practice it is found that bandwidth and risetime do vary with gain somewhat, though not as drastically as with conventional op-amps. The main cause is the nonzero *output impedance* of the input buffer, whose effect is to alter the loop gain and, hence, the closed-loop dynamics. Denoting this impedance as R_o, we shall refer to **Figure 12-11** to investigate the effect of R_o as well as the effect of external capacitances, either at the input or in the feedback path.

Consider first the case in which the external network is purely resistive so that $C_1 = C_2 = 0$. The circuit first converts V_x to the current through $R_2(I_x) = V_x/(R_2 + R_1\|R_o)$, then it divides I_x to produce $I_n = -I_x \times R_1/(R_1 + R_o)$, and finally it converts I_n to the voltage $V_o = zI_n$. Eliminating I_x and I_n and letting $T = -V_o/V_x$ yields:

$$T(jf) = \frac{z(jf)}{Z_2} \qquad [12\text{-}17]$$

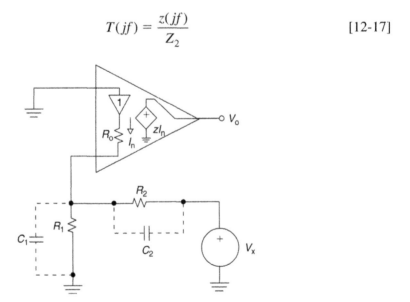

Figure 12-11: Test circuit to investigate the effect of R_o.

$$Z_2 = R_2\left(1 + \frac{R_o}{R_1 \,\|\, R_2}\right) \qquad [12\text{-}18]$$

Clearly, the effect of R_o is to *increase* the noise gain from R_2 to $Z_2 = R_2[1 + R_o/(R_1\|R_2)]$. This is shown in **Figure 12-12**, curve 1. Consequently, both the bandwidth and the rise time will deteriorate to a proportional degree. Replacing R_2 in Equation 12-16 with Z_2 as given in Equation 12-18, we obtain, after simple manipulation:

$$f_A = \frac{f_t}{1 + (R_o/R_2)A_o} \qquad [12\text{-}19]$$

where $f_t = z_0 f_a/R_2$ is the extrapolated value of f_A in the limit $R_o \to 0$, and $A_0 = 1 + R_2/R_1$ is the closed-loop DC gain. This equation indicates that bandwidth reduction due to R_o will be more pronounced at high closed-loop gains. This is shown in **Figure 12-13**.

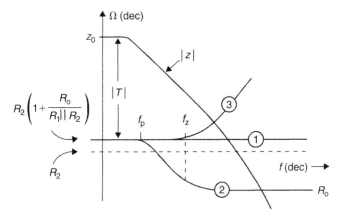

Figure 12-12: Noise-gain curves for the case of (1) purely resistive feedback, (2) a capacitance in parallel with R_2, and (3) a capacitance in parallel with R_1.

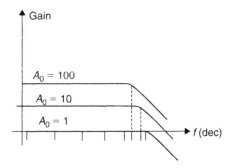

Figure 12-13: Effect of R_o on f_A as a function of A_0.

Example 1. A certain CF amp has $R_o = 50\,\Omega$, $R_2 = 1.5\,k\Omega$, and $f_t = 100\,MHz$. Find the bandwidths corresponding to $A_0 = 1$, 10, and 100.

Solution. By Equation 12-18 we have $f_A = 10^8/[1 + (50/1500)A_0] = 10^8/(1 + A_0/30)$. The bandwidths corresponding to $A_0 = 1$, 10, and 100 are, respectively, $f_1 = 96.8\,MHz$, $f_{10} = 75.0\,MHz$, and $f_{100} = 23.1\,MHz$. We observe that these values still compare favorably with a conventional op amp, whose bandwidth would be reduced, respectively, by 1, 10, and 100.

If desired, the external resistance values can be predistorted to compensate for the bandwidth reduction at high gains. Turning Equation 12-19 around yields the required value of R_2 for a given bandwidth f_A and DC gain A_o:

$$R_2 = \frac{z_0 f_a}{f_A} - R_o A_0 \qquad [12\text{-}20]$$

and the required value of R_1 for the given DC gain A_0:

$$R_1 = \frac{R_2}{A_0 - 1} \qquad [12\text{-}21]$$

Example 2. Redesign the amplifier of Example 1 so that $f_{10} = 100\,MHz$.

Solution. Since with $R_2 = 1.5\,k\Omega$ this device has $z_0\,f_a/R_2 = 100\,MHz$, it follows that $z_0 f_a = 10^8 \times 1500 = 1.5 \times 10^{11}\,\Omega Hz$. Then, for $A_0 = 10$ and $f_{10} = 100\,MHz$, we need $R_2 = 1.5 \times 10^{11}/10^8 - 50 \times 10 = 1\,k\Omega$, and $R_1 = 1000/(10 - 1) = 111\,\Omega$.

Besides a dominant pole at f_a, the open-loop response of a practical amplifier presents additional poles above the crossover frequency. As shown in Figure 12-12, the effect of these poles is to cause a steeper gain rolloff at this frequency, further reducing the closed-loop bandwidth. Moreover, the additional phase-shift due to these poles decreases the phase margin somewhat, and this could cause some peaking in the frequency response and ringing in the step response.

Finally, it must be said that the risetime of a practical CF amp does increase with the step size somewhat, due primarily to transistor current gain degradation at high current levels. For instance, the rise time of the CLC401 changes from 2.5 nsec to 5 nsec as the step size is changed from 2 V to 5 V. In spite of second-order limitations, CF amps still provide superior dynamics.

CF Application Considerations

Although the above treatment has focused on the noninverting configuration, the CF amp will work as well in most other resistive feedback configurations such as inverting amplifiers, summing and differencing amplifiers, I-V and V-I converters, and KRC active filters [2]. In fact, the derivation of the transfer characteristic of any of these circuits proceeds along the same lines as conventional op-amps. Special consideration, however, merits the cases in which the external network includes reactive elements, either intentional or parasitic.

Consider first the effect of the feedback capacitance C_2 in parallel with R_2 in the basic circuit of Figure 12-11. Replacing R_2 with $R_2\|(1/sC_2)$ in Equation 12-18 and expanding, we readily see that Z_2 now has a pole at $f_p = 1/(2\pi R_2 C_2)$ and a zero at $f_z = 1/[2\pi (R_0\|R_1\|R_2)C_2]$. The corresponding noise-gain curve is shown in Figure 12-12, curve 2, indicating that the crossover frequency is now pushed into the region of substantial phase shift due to the higher-order poles of z. If the overall shift reaches $-180°$ at this frequency, the loop gain will become $T = 1$ $|-180° = -1$, making A infinite, by Equation 12-12. When this condition is met, the circuit will oscillate. Even if the phase shift fails to reach $-180°$, the closed-loop response may still exhibit intolerable peaking and ringing. Hence, *capacitive feedback must be avoided with CF amps*. To minimize the effect of stray feedback capacitances, manufacturers often provide R_2 internally.

CF Amp Integrators

To synthesize the integrator function in CF form, which provides the basis for dual-integrator-loop filters and oscillators as well as other popular circuits, we must use configurations that avoid a direct capacitance between the output and the inverting input. One possibility is offered by the Deboo integrator, which belongs to the class of KRC filters and is therefore amenable to CF amp realization. Its drawback is the need for tightly calibrated resistances, if lossless integration is desired. The alternative shown in **Figure 12-14** not only meets the given constraint but also provides *active frequency compensation*, a highly desirable feature to cope with Q-enhancement problems in dual-integrator-loop filters [2]. Using standard op-amp analysis techniques, we can readily see that the unity-gain frequency of this integrator is $f_0 = (R_2/R_1)/(2\pi RC)$.

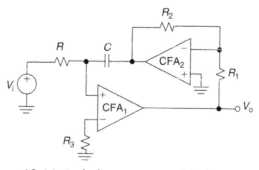

Figure 12-14: Actively compensated CF integrator.

This circuit can be realized in a cost-effective manner using a dual CF amp, such as the OP-260 (Precision Monolithics).

Stray Input-Capacitance Compensation

Next, let us investigate the effect of the input capacitance C_1 in parallel with R_1 in the basic circuit of Figure 12-11. Replacing R_1 with $R_1\|(1/sC_1)$ in Equation 12-18 and expanding, we readily see that Z_2 now has a zero at $f_z = 1/[2\pi(R_o\|R_1\|R_2)C_1]$. The corresponding noise-gain curve is shown in Figure 12-12, curve 3. If C_1 is sufficiently large, the phase of T at the crossover frequency will again approach $-180°$, bringing the circuit to the verge of instability.

As in the case of a conventional op-amp, the CF amp can be stabilized by using a feedback capacitance C_2 to introduce sufficient phase lead around the loop to compensate for the phase lag due to the input capacitance C_1. Though it was said earlier that capacitive feedback should be avoided with CF amps, this no longer holds when we want to combat the effect of an input capacitance.

The destabilizing effect of the stray input capacitance is of particular concern in current-mode DAC output amplification, where C_1 is the output capacitance of the DAC, typically in the range of a few tens to a few hundreds of picofarads. The situation is depicted in **Figure 12-15a**. The use of C_2 creates a noise-gain pole at $f_p = 1/(2\pi R_2 C_2)$. For a phase margin of 45°, C_2 is chosen to make this pole coincide with the crossover frequency f_c. Referring to Figure 12-15b, one can show that if f_z is sufficiently lower than f_c then $f_c \approx \sqrt{z_0 f_a f_z/(R_o + R_2)}$. Letting $f_c = 1/2\pi(R_o\|R_2)C_1$ and imposing $f_p = f_c$ yields:

$$C_2 = \left(\frac{R_o}{2\pi R_2 z_0 f_a}C_1\right)^{1/2}$$

[12-22]

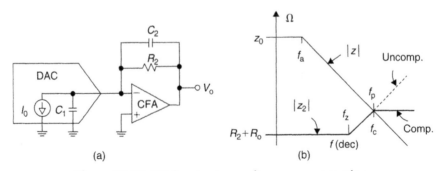

Figure 12-15: DAC output capacitance compensation.

Example 3. A DAC having $C_1 = 100$ pF feeds a CF amp having $R_2 = 1.5$ kΩ, $f = 150$ MHz, and $R_0 = 50\ \Omega$. Find C_2 for a phase margin of $45°$, and estimate the bandwidth of the amplifier.

Solution. Since $f_t = z_0 f_a/R_2$, it follows that $z_0 f_a = R_2 f_t = 1.5 \times 10^3 \times 150 \times 10^6 = 2.25 \times 10^{11}$ ΩHz. Then $C_2 = [50 \times 100 \times 10^{-12}/(2\pi \times 1.5 \times 10^3 \times 2.25 \times 10^{11})]^{1/2} = 1.54$ pF. The bandwidth is $f_A \approx 1/(2\pi R_2 C_2) = 1/(2\pi \times 1500 \times 1.54 \times 10^{-12}) \approx 69$ MHz. The value of C_2 may be increased for a greater phase margin, but this will also reduce the bandwidth of the amplifier.

Noise in CF Amp Circuits

Since CF amps are wideband amplifiers, they generally tend to be noisier than conventional op-amps. The noise characteristics are specified in terms of three input noise densities: the voltage density e_n, the inverting-input current density i_{nn}, and the noninverting-input current density i_{np}. Since the BJTs of CF amps are generally biased at much higher current levels than conventional op-amps, CF amps tend to exhibit lower voltage noise but higher current noise. Moreover, since the inputs are dissimilar because of the input voltage buffer, so are the current densities. Consequently, the datasheets report i_{nn} and i_{np} separately.

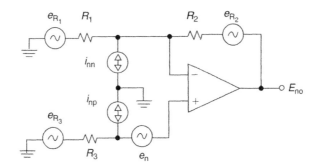

Figure 12-16: Noise model of a resistive CF amp circuit.

Figure 12-16 shows the noise model of a CF amp with resistive feedback. To find the overall input noise density e_{ni}, we use the superposition principle to find the contributions from the individual noise sources, and then we add up these contributions in rms fashion. The result is [2]:

$$e_{ni}^2 = e_n^2 + R_3^2 i_{np}^2 + (R_1 \| R_2)^2 i_{nn}^2 + 4kT\,[R_3 + (R_1 \| R_2)] \qquad [12\text{-}23]$$

where each component has been put in a form that lends itself to be amplified by the same noise gain $A(jf)$. The total rms output noise E_{no} above a given frequency f_L is

$$E_{no} = \left(\int_{f_L}^{\infty} |A(jf)|^2 \, e_{ni}^2 \, df \right)^{1/2}$$

[12-24]

where $A(jf)$ is given in Equation 12-15. Expressing the noise densities as:

$$e_n^2 = e_{nw}^2 \, (f_{ce}/f + 1), \quad i_{nn}^2 = i_{nnw}^2 \, (f_{cin}/f + 1), \quad \text{and} \quad i_{np}^2 = i_{np}^2 \times (f_{cip}/f + 1),$$

and substituting into Equation 12-23 and then into Equation 12-24 yields [2]:

$$E_{no} = \left(1 + \frac{R_2}{R_1}\right)$$

$$\times \left(\begin{array}{l} e_{nw}^2 \left(f_{ce} \ln \dfrac{f_A}{f_L} + 1.57 f_A - f_L \right) + R_3^2 i_{npw}^2 \left(f_{cip} \ln \dfrac{f_A}{f_L} + 1.57 f_A - f_L \right) \\[2mm] + (R_1 \| R_2)^2 i_{nnw}^2 \left(f_{cin} \ln \dfrac{f_A}{f_L} + 1.57 f_A - f_L \right) + 4kT \left[R_3 + (R_1 \| R_2) \right](1.57 f_A - f_L) \end{array} \right)^{1/2}$$

[12-25]

Example 4. Let the circuit of Figure 12-16 be a CLC401 CF amp configured for a 20 dB noninverting gain with $R_1 = 166.7 \, \Omega$ and $R_2 = 1.5 \, k\Omega$. Moreover, let $R_3 = 100 \, \Omega$. Find E_{no} for the case in which noise is observed over a 10 sec period.

Solution. We have $f_L = 1/10 = 0.1$ Hz. Using the datasheets values $z_0 = 710 \, k\Omega$ and $f_a = 350 \, kHz$, we obtain $f_t = z_0 f_a/R_2 = 165.7 \, MHz$. Substituting into Equation 12-19, along with the datasheet value $R_o = 50 \, \Omega$ yields $f_A = 124 \, MHz$. Substituting the datasheet values $e_{nw} \approx 2.4 \, nV/\sqrt{Hz}$, $f_{ce} \approx 30 \, kHz$, $i_{npw} \approx 2.6 \, pA/\sqrt{Hz}$, $f_{cip} \approx 30 \, kHz$, $i_{nnw} \approx 17 \, pA/\sqrt{Hz}$, and $f_{cin} \approx 40 \, kHz$ into Equation 12-25 yields $E_{no} \approx 0.57 \, mV$ rms, or $E_{no} = 0.57 \times 6 = 3.4 \, mV$ peak to peak.

Low Distortion for Fast Sinewaves Using CF Amps

Because CF amps do not have slew-rate limitations, some fast CF amps have excellent characteristics as fast low-distortion amplifiers for sinewaves. An LMH6504 can put out 2 volts p-p at 20 MHz with only −45 dB distortion. An LMH6723 draws only 1 mA of quiescent current yet can amplify flat out to 260 MHz (f_{-3dB}). Some are claimed to have excellent distortion for audio signals as high as 20 kHz and as large as 24 volts p-p. So the inherent freedom from slew rate permits some very good high-speed features, combined with low distortion.

Drawbacks of Current-Feedback Amplifiers vs. Conventional Op-Amps

Despite the many advantages of speed, slew rate, and low distortion, it is fair to note some of the disadvantages of CF amps:

- They cannot work rail-to-rail to either power supply rail, *not* at their inputs *nor* at their outputs. It is *inherent* that their emitter followers cannot do that, whereas many modern op-amps can. This can be important, especially for low-voltage operation.

- Current-feedback amplifiers have relatively low, poor voltage gain. Many modern op-amps have a gain of 100,000 to 1,000,000 or higher, but CF amps have a usable voltage gain of 1,200 to 4,000. This is not a disaster, but it's a definite weakness.

- Similarly, CF amps have mediocre PSRR and CMRR, 50 to 70 dB. This, too, is not a disaster, but you might have to trim for good circuit performance.

- CF amps can work at moderate impedance levels, but they slow down as the feedback impedance Zf goes up. They are generally unsuitable at high Z levels (even at 10 or 50 kilohms), whereas some op-amps work well with $Zf = 100$, 1,000, or 100,000 megohms.

- Because the small, fast transistors in a CF amp have such high power densities (several mA of transient current $\times$ several volts of V_{ce} for an emitter junction just a few microns square), the heating can cause significant heating tails, which gradually go away as the heat spreads out. This kind of error is shown on some datasheets but not all.

Notwithstanding these limitations, CF amps are finding favor in a strong niche market because they are very useful for fast tasks.

References

Current-Feedback Op-Amp Applications Circuit Guide, Comlinear Corporation Application Note OA–07, 1988.

Franco, S., *Design with Operational Amplifiers and Analog ICs*, 3ed., McGraw-Hill Book Company, 2002.

Nelson, D. and Evans, S., *A New Approach to Op-Amp Design*, Comlinear Corporation Application Note 300-1, March 1985.

CHAPTER 13

The Basics Behind Analog-to-Digital Converters

Bonnie Baker

In This Chapter

The analog-to-digital converter (ADC) is always in the back seat of the station wagon, looking at the analog signal through the rear window. In a way, I am soft on this device because this is the position I was at in my family's station wagon throughout my childhood, being one of six children.

The controller, in the front seat, can see the results of the converter's labor, but the question is, can those results be counted on? If the ADC reports the system data incorrectly, the controller is blind to errors that have been introduced by the converter and signal chain. This is true unless you are willing to allocate a lot of code to try to unscramble the mess (with no guarantee of success). But why not go to the source of the problem? Believe me when I say that the ADC can cause you a great deal of heartache if you don't understand the nuances. Your misunderstanding of how to use the ADC can leave the controller or processor struggling with erroneous or inaccurate data.

In this chapter, we are going to discuss the key specifications for ADCs and how they can impact the expected results from your converter. This list of specifications generally applies to all classes of converters. Then we will delve into the particulars of the Successive Approximation Register (SAR) ADC. This part of the discussion will start with an explanation about how the SAR converter works. The issues discussed will give you insight into how to use this type of converter effectively, the first time. There will be more performance specifications and characteristics discussed here with emphasis on how to design with or around some of the converter's shortcomings. This is followed with a user-friendly version of how a delta-sigma ($\Delta-\Sigma$) ADC works. The $\Delta-\Sigma$ topics will follow the same line of discussion as with the SAR converter. First, we will talk about the topology and in particular how it impacts your signal chain. Following this brief discussion, the performance specifications that are particular to the $\Delta-\Sigma$ converter will be discussed, with solutions on how to work with or work around the $\Delta-\Sigma$ converter limitations.

The primary ADC specifications are summarized in Appendix A, so if you forget about the particulars of a specification, this is a great place to look. Appendix A contains a glossary of common converter specifications.

There are numerous other converters that you can use for your application circuits, such as the Pipeline, FLASH, and Voltage-to-Frequency (V/F) converters, but these topologies are beyond the scope of this book.

The Key Specifications of Your ADC

Input Range of the ADC

The input range of the ADC can be a bit tricky. You will find variations of single-ended, differential, and pseudo-differential, while the input range is determined by the voltage reference (V_{REF}) of the converter.

An example of the configuration of an ADC with a single-ended input is shown in **Figure 13-1a**. This type of converter input is easy to use because there is no question of what to do with that pin. The input voltage range is equal to the full-scale range (FSR) of the converter. Additionally, the digital code at the output of this configuration is straight binary (see the "Straight Binary Code" section later in this chapter).

In **Figure 13-1b**, the input of the converter is configured as a pseudo-differential input. This simply means that the input to the converter is differential, but one of the input pins has a range that is limited to a few hundred millivolts above and below ground. This has the same output digital coding as the single-ended input device. The digital code at the output is straight binary. You might ask what this configuration would do for you? True to the spirit of differential input stages, this type of device will reject small common-mode noise. In simple terms, if a small signal, such as 50 Hz or 60 Hz, is an undesirable part of the signal you are trying to convert, this common-mode signal will be rejected or eliminated. This is a nice feature as long as you understand that you have to connect it properly to ground.

A third type of input for ADCs is the fully differential input stage. With this configuration both inputs can be brought from ground to the converter's FSR. This is nice because not only can you reject small common-mode signals, you also can convert a positive or negative analog signal to a digital output. You guessed it—the output code is in the format of two's complement (see the "Binary Two's-Complement Code" section later in this chapter).

So why is this important? There are some signal sources that are differential. One example is the signal from a Wheatstone bridge shown in **Figure 13-1c**. You will find that if one of the inputs goes positive, the other will go negative. This action will give you a gain of 2 V/V

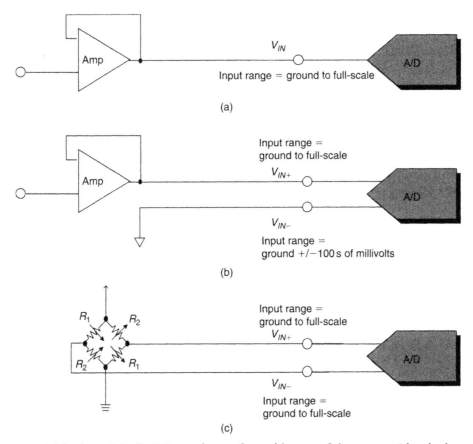

(a)

(b)

(c)

Figure 13-1: The input(s) of ADCs can be configured in one of three ways. The single-ended input (a) is configured for one input voltage referenced to ground. Another type of input stage has two inputs configured as a pseudo-differential stage (b) where the signal input is the noninverting input and the inverting input is used to reject small-signal system noise. The third type of input stage is the differential input (c) where the two inputs to the converter range from ground to the full-scale input voltage.

in the analog domain. It will also couple the noise in the environment on both lines. The ADC will then be able to filter that noise and convert the signal of interest from the bridge.

With this type of input, the FSR is double what you might expect. As you think about it, if the noninverting input goes to its full range and the inverting input is referenced to ground, the difference in this signal is $(V_{IN+} - V_{IN-}) = +V_{FS}$. Now if the input signal changes so that the noninverting input is referenced to ground, and the inverting input is taken to full scale, the difference in the input signal is $(V_{IN+} - V_{IN-}) = -V_{FS}$. So the actual FSR of this type of device is $+V_{FS} - (-V_{FS})$, or $2 \times V_{FS}$, where V_{FS} is the full-scale input voltage range.

Digital Coding of the Analog Signal

An analog-to-digital converter translates an analog input signal into a discrete digital code. This digital representation of the real-world signal can be manipulated in the digital domain for the purposes of information processing, computing, data transmission, or control system implementation. In any application where a converter is used, it is advantageous to have the code structure complement the microcontroller or processor's operands.

We are going to talk about the most common code schemes: straight binary and binary two's-complement code. For simplicity, all the following code examples are for a 4-bit converter. The median analog voltages in the tables are the equivalent analog voltages that are at the center of the digital code.

These codes are mathematically described using the full-scale (FS) input range of the converter. Usually the FS input range of an ADC is equal to or twice as much as the voltage reference applied to the device. In some instances, the voltage reference connection is tied internally in the device to the power supply. In all these configurations, you will need to refer to the product's datasheet for specifics.

The basic differences between these two types of code are:

- The least significant bit (LSB) size of the two's-complement code is twice as large as the LSB size of the straight binary code. This does not increase the number of codes that the converter can create.

- The analog FSR of the straight binary code is a positive voltage from ground to V_{REF}. The analog FSR of the two's-complement code is equal to the positive FSR plus the unsigned negative FSR.

- The digital output code of the two's-complement code is easier to use when running arithmetic calculations, such as subtraction.

Straight Binary Code

The straight binary code is more accurately called *unipolar straight binary*. This digital format for an analog-to-digital conversion is the simplest to understand. As the name implies, this coding scheme is used when only positive voltages are converted. This is a good output code for converters that are configured with a single input, as shown in Figures 13-1a and 13-1b. An example of this type of coding is shown in **Table 13-1**.

When this scheme is used to represent a positive analog signal range, the digital code for zero volts is equal to zero (0000 per Table 13-1). The definition of a positive voltage is the amplitude between the ADC ground or the inverting input and the noninverting input of the ADC. Given an ideal converter with no offset, gain, integral nonlinearity (INL),

Table 13-1: The unipolar straight binary code representation of zero volts is equal to a digital 0000. The analog full-scale minus one LSB digital representation is equal to 1111. With this code there is no digital representation for analog full-scale.

Median Analog Voltage (V)	Digital Code
0.9375 FS (15/16 FS)	1111
0.875 FS (14/16 FS)	1110
0.8125 FS (13/16 FS)	1101
0.75 FS (12/16 FS)	1100
0.6875 FS (11/16 FS)	1011
0.625 FS (10/16 FS)	1010
0.5625 FS (9/16 FS)	1001
0.5 FS (8/16 FS)	1000
0.4375 FS (7/16 FS)	0111
0.375 FS (6/16 FS)	0110
0.3125 FS (5/16 FS)	0101
0.25 FS (4/16 FS)	0100
0.1875 FS (3/16 FS)	0011
0.125 FS (2/16 FS)	0010
0.0625 FS (1/16 FS)	0001
0	0000

differential nonlinearity (DNL) error, or noise, the code transition from 0000 to 0001 occurs at the analog value of:

$$\text{First Code Transition} = (0 + 1 / 2 \text{ LSB})$$

The second code transition from 0001 to 0010 occurs at the analog value of:

$$\text{Second Code Transition} = (1 \text{ LSB} + 1 / 2 \text{ LSB})$$

where:

$$\text{LSB} = \frac{+\text{FS}}{2^n}$$

where LSB is equal to the least significant bit, n is equal to the number of converter bits, and +FS is equal to the analog FSR.

The digital output code versus analog input voltage is mapped in **Figure 13-2**. In this figure, the analog input voltage of $0\,\text{V}$ is converted to a digital code of 0000. If the analog input

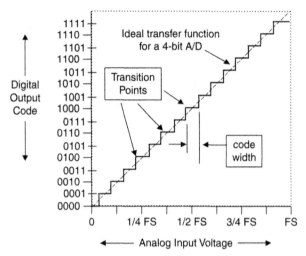

Figure 13-2: This is an ideal transfer function of a 4-bit ADC where the analog input is continuous and the digital output code is discrete.

voltage is changed at the input of the ADC to approximately 0.25 FS voltage, the ADC will produce a digital code of 0100. At approximately midpoint between the codes specified in Table 13-1, the converter will transition from a lower digital code to a higher digital code. A code width as defined in Figure 13-2 is equal to one least significant bit (LSB).

Binary Two's-Complement Code

In some applications, the unipolar ADC converts negative and positive values. This can only happen when the ADC has differential analog inputs, as shown in Figure 13-1c. These types of devices will output digital code in the binary two's-complement format. Binary two's-complement arithmetic is widely used in microcontrollers, calculators, and computers because simple subtractions and additions require less code. The addition of two binary numbers in two's complement is straightforward in that the two numbers are added together. The subtraction operation of numbers in two's complement is done by adding the two numbers together to get the subtracted solution.

For example:

Decimal	Two's Complement	Decimal	Two's Complement	Decimal	Two's Complement
+2	0010	+7	0111	+5	0101
+5	0101	−5	1011	−7	1001
+7	0111	+2	0010	−2	1100

Binary two's-complement coding is not as straightforward as straight binary. The codes are not continuous from one end to the other due to the discontinuity that occurs at the

Table 13-2: The binary two's-complement representation
of zero volts is also equal to a digital 0000. The analog
positive FS minus one LSB digital representation is equal to
0111, and the analog negative FS representation is 1000.

Median Voltage (V)	Code
0.875 FS (7/8 FS)	0111
0.75 FS (6/8 FS)	0110
0.625 FS (5/8 FS)	0101
0.5 FS (4/8 FS)	0100
0.375 FS (3/8 FS)	0011
0.25 FS (2/8 FS)	0010
0.125 FS (1/8 FS)	0001
0	0000
−0.125 FS (−1/8 FS)	1111
−0.25 FS (−2/8 FS)	1110
−0.375 FS (−3/8 FS)	1101
−0.5 FS (−4/8 FS)	1100
−0.625 FS (−5/8 FS)	1011
−0.75 FS (−6/8 FS)	1010
−0.875 FS (−7/8 FS)	1001
−1 FS	1000

analog bipolar zero. The two's-complement representation of a positive binary number is generated by logically complementing all the digits, which then converts it to the negative binary number counterpart, as shown in **Table 13-2** and **Figure 13-3**. With this code scheme, the most significant bit (MSB) is a sign indicator. A positive value is indicated with an MSB logic 0. An MSB value of logic 1 indicates that the output number is a negative value.

This system has an odd number of codes and only one zero state. Differential input ADCs (Figure 13-1c) are devices that can be operated in a single-ended, positive voltage input mode or a full-differential input mode. In the full-differential mode, the FSR of the device is equal to:

$$FSR = \{+IN_{MAX} - (-IN_{MIN})\} + \{-IN_{MAX} - (+IN_{MIN})\}$$

and the ADC's input voltage range is equal to:

$$A_{IN} = (+IN - (-IN))$$

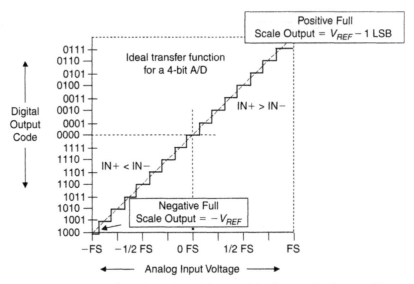

Figure 13-3: The transfer function of a 4-bit ADC bipolar analog input will produce a two's-complement code as a digital output.

The converter will produce digital code that represents the both negative and positive analog inputs as shown in Table 13-2.

Throughput Rate vs. Resolution and Accuracy

I once had a customer ask me on the company hotline for a 32-bit converter. I was taken off guard by his request. Why did he need 32 bits? I finally found out that he only needed 1 mV resolution out of 4.096 V. Well, that's easy. You can do this with a 12-bit converter that has a 4.096 V reference. With this converter, the FSR would be 4.096 V and the LSB size would be $V_{REF}/2^{12} = 4.096\,V/4096 = 1\,mV$. Why was the customer having problems finding a converter?

He told me that his bus was 32 lines wide. So my advice became, "Use a 12-bit converter and tie bus lines 0 through 19 to ground." He asked me if you could do that and I said, "Absolutely, and with this solution you will control your system noise as an added benefit!"

A throughput rate specification (also known as *data rate* for delta-sigma ADCs) defines the amount of time it takes for a converter to complete an entire conversion. The activities that are included in the throughput rate time are setup time, sample time, conversion time, and data transmission time. The two converter topologies on which we will center our attention will be the successive approximation register (SAR) and delta-sigma. When you think of the typical conversion times of these topologies, you can easily separate

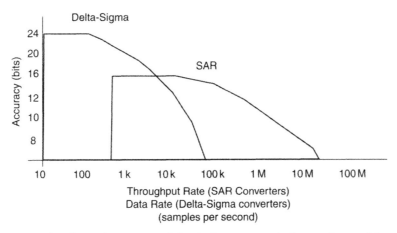

Figure 13-4: The throughput rate of the SAR converter is faster than a delta-sigma converter. In contrast, the delta-sigma converter is able to achieve higher accuracy as a trade-off for the slower speeds.

them into application classes (**Figure 13-4**). In general terms, the faster the throughput rate of the converter, the lower the resolution will be. But implied in this diagram is the same trend in accuracy. The resolution of a converter is simply the number of bits that the converter is capable of handling per conversion. The accuracy describes the number of bits that are repeatable from conversion to conversion. In all cases, the specified accuracy will be equal to or less than the resolution of the converter.

The data rate of the delta-sigma converter is generally slower than the throughput rate of the SAR converter. As we will see later in this discussion, the SAR converter only samples the input signal once and converts to a digital code, according to the sampled signal. The delta-sigma converter samples the input signal multiple times. It then implements various noise reduction algorithms to improve the number of bits in the converter, as well as the signal-to-noise ratio (SNR), but the trade-off with this type of sampling strategy is time.

Accuracy vs. Resolution

There are a few key specifications that you should become familiar with. Knowing these figures of merit will help you choose the right converter for your application and also identify the impostors. For instance, "2.7 V 16-Bit ADC with SPI® Serial Interface" is an example of what you might see at the top of a data sheet. Does this mean 16-bit accurate, noise-free for every conversion, accurate with respect to the input voltage, or does it mean 16-bit resolution where you are guaranteed that 16 bits will be transmitted out of the converter at the conclusion of a conversion? The latter is correct. The phrase "2.7 V

16-Bit ADC with SPI Serial Interface" as the title of the converter's datasheet only means that you will see 16 bits transmitted from the output of the converter.

You will find that those 16 bits can all be accurate or not, depending on the manufacturer. More than once in my career I have seen the last couple of codes or the LSBs of a 16-bit converter dither all over the place from conversion to conversion. So, I would define a converter like this as having 16-bit resolution, not as being 16-bit accurate. This is not a bad thing as long as you know what to expect.

So, resolution is defined as the number of bits that are transmitted out of the converter at the conclusion of the conversion. If you know this information about a converter, you can quickly calculate the theoretical LSB size with the following formula:

$$LSB = \frac{FSR}{2^n}$$

where n = number of bits. For a converter that has 16-bit resolution and an FSR of 5 V, the LSB size is 76.29 μV.

One of the more common questions about ADCs that I hear is, "How do I know that an ADC will give me a good, reliable code, and can I determine this from the converter's datasheet?" Of course, this depends on your definition of "reliable," but if you are looking for a repeatable output from conversion to conversion, you should refer to AC domain specifications. If you are looking for a converted code that represents the actual input voltage, DC specifications are more useful. But don't forget about the noise. DC specifications imply average accuracy (not repeatability). From conversion to conversion, these codes will vary, dependent on the internal noise of the converter.

AC Specifications Imply Repeatability

AC domain specifications, such as SNR, effective resolution (ER), signal-to-(noise + distortion) (SINAD), or effective number of bits (ENOB), provide information about ADC repeatability. Now, these specifications will tell you how repeatable your conversion is, but they will not tell you if the conversion is accurate. On the other hand, DC domain specifications, such as offset error, gain error, differential nonlinearity, and integral nonlinearity, provide information about how close, on average, the input signal is matched to an actual output code. These specifications do not imply repeatability, and noise could give you varying results from conversion to conversion.

Ideally, the SNR of a converter in decibels is equal to 6.02 n + 1.76 dB, where n is equal to the number of converter bits. This theoretical noise is a result of the quantization noise inherent in the converter. In practice, SNR is equal to 20 log (rms signal)/(rms noise), where rms means root-mean-square, equal to one standard deviation in a normal distribution. To determine the rms noise, the results of many conversions need to be collected.

As with the SNR, ER is measured by collecting a statistical sample of many conversions, but this time we don't have an AC input signal to the converter. The input signal is a clean, "noiseless" DC signal. If this DC signal has less noise than your converter (about 3×), you are good to go. The units of measure for ER are bits, which are referred to the output of the converter. The formula for ER is equal to:

$$ER = N - \log_2(\sigma)$$

where N is equal to the resolution of the converter and σ is equal to one standard deviation of the output data. With some converter literature, ER is sometimes called *ENOB*.

Whereas SNR or ER provides information about the device noise of the converter, SINAD and ENOB provide more information about ADC frequency distortions. SINAD is the ratio of the rms amplitude of the fundamental input frequency of the input signal to the rms sum of all other spectral components below one half of the sampling frequency (excluding DC). The theoretical minimum for SINAD is equal to the SNR, or 6.02 n + 1.76 dB. But in practice, an ADC will have some harmonic distortion of this input signal that is generated within the converter. The complementary specification to SINAD is ENOB. The unit of measure for SINAD is dB, and the unit of measure for ENOBs is bits. SINAD can be converted to ENOB with the following calculation:

$$ENOB = \frac{(SINAD - 1.76)}{6.02}$$

To this point in our discussion, the specification units are in terms of rms—statistically speaking, rms is one standard deviation of data that is shaped in a normal distribution curve. When the noise units are defined with rms units, the probability that the converter will give you a value of plus or minus one rms is ~68%. The relationship between the output noise of the converter, the normal distribution of a set of sampled outputs, and these statistical values are illustrated in **Figures 13-5a** and **13-5b**.

An rms specification is a statistical calculation from many samples or a population. The formula for one standard deviation is:

$$\sigma^2 = \frac{\Sigma(y - \eta)^2}{N}$$

where σ is the population standard deviation, y is a sample from the population, η is the population mean, and N is the set of population observations.

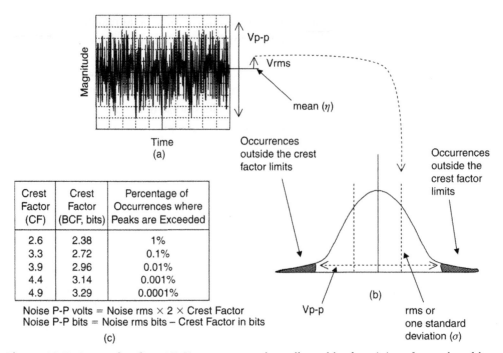

Crest Factor (CF)	Crest Factor (BCF, bits)	Percentage of Occurrences where Peaks are Exceeded
2.6	2.38	1%
3.3	2.72	0.1%
3.9	2.96	0.01%
4.4	3.14	0.001%
4.9	3.29	0.0001%

Noise P-P volts = Noise rms × 2 × Crest Factor
Noise P-P bits = Noise rms bits − Crest Factor in bits
(c)

Figure 13-5: A sample of an ADC's output can be collected in time (a) and translated into a histogram (b) where the mean and standard deviation of the samples can be calculated. With the standard deviation of these samples, a peak-to-peak value can be determined (c) with a multiple of 2 times the crest factor (CF) for output signal referred to input calculations or a subtracted bit crest factor (BCF).

A 68% probability of getting your expected output might not be the odds you want to work with. You might want to consider converting the specification limits to peak-to-peak (p-p) values. From the rms number, you can quickly calculate the p-p specification, which is very convenient if you are trying to get good repeatable results. This conversion is easily done with ER and ENOB specifications by multiplying your rms specification (in voltages) by two times the crest factor (CF, **Figure 13-5c**) or subtracting the bit crest factor (BCF, Figure 13-5c) from your rms specification (in bits). With this new calculation, your ADC has a better chance of producing your expected output. The industry standard crest factor for nonmilitary applications is 3.3. Also, be careful that the data you selected has the attributes of a normal distribution; otherwise, these calculations will not be as accurate as promised.

The calculation for the conversion of rms to p-p is:

$$V(\text{p-p}) = V(\text{rms}) \times 2 \times CF$$

or:

$$Bits(p\text{-}p) = Bits(rms) - BCF$$

For more details about these specifications, refer to Appendix A.

DC Specifications Imply Accuracy

If you are looking for a conversion from your converter that accurately represents the analog voltage, you should be looking at the DC specifications. The DC specifications that I am referring to are offset voltage, gain error (or FS error), differential nonlinearity, and integral nonlinearity. If your conversions are not repeatable as discussed in the "AC Specifications Imply Repeatability" section in this chapter, the accuracy of your converter is determined by the average of multiple samples. In this discussion, we assume that the converter is noise-free.

What do the LSB specifications mean when you are looking at ADCs? One day a fellow engineer told me that a 12-bit Converter X (manufacturer will remain unnamed) had just 7 usable bits. So essentially, the 12-bit converter was only a 7-bit converter. He based this conclusion on the device's offset and gain specifications. The maximum specifications were:

$$Offset\ error = \pm 3\ LSB$$
$$Gain\ error = \pm 5\ LSB$$

At first glance, I thought he was right. From the list above, the worst specification is gain error (± 5 LSB). Applying simple mathematics, 12-bits of resolution minus 5 is equal to 7 bits, right? Why would an ADC manufacturer introduce such a device? The gain-error specification motivates me to purchase a lower-cost, 8-bit converter. However, that didn't seem right. Well, as it turned out, it wasn't right.

Let's start by looking at the definition of LSB. Think of a serial 12-bit converter; it produces a string of twelve ones or zeros. Typically, the converter's first transmitted digital bit is the MSB (or LSB + 11). Some converters transmit the LSB first. We will assume that the MSB is first in this chapter. The second bit is MSB $-$ 1 (or LSB + 10); the third bit is MSB $-$ 2 (or LSB + 9), and so on. At the end of this string of bits, the converter finally transmits as MSB $-$ 11 (or LSB).

The terminology, LSB, is very specific. It describes the last position in the digital stream. It also represents a fraction of the full-scale input range. For a 12-bit converter, the LSB value is equivalent to the analog full-scale input range divided by 2^{12}, or 4096. If I put this in terms of real numbers, I have an LSB size of 1 mV with a 12-bit converter that has a full-scale input range of 4.096 V. However, the most instructive definition of LSB is that it can represent one code out of the 4096 codes possible.

Going back to the specifications and translating them into a 12-bit converter that has an input FSR of 4.096 V:

$$\text{Offset error} = \pm 3 \text{ LSB} = \pm 3\,\text{mV}$$
$$\text{Gain error} = \pm 5 \text{ LSB} = \pm 5\,\text{mV}$$

These specifications actually claim that the converter can have (worst case) an 8 mV (or 8 code) error introduced through the conversion process. This is not to say that the error occurs at the LSB, LSB -1, LSB -2, LSB -3, LSB -4, LSB -5, LSB -6, and LSB-7 positions in the output bit stream of the converter. The errors can be up to eight times one LSB, or 8 mV. Precisely stated, the transfer function of the converter could have up to eight codes missing out of 4096 codes. These codes will be missing at the lower or upper range of the codes. For example, a converter with an error of $+8$ LSB (($+3$ LSB offset error) + ($+5$ LSB gain error)) will produce possible output codes of 0 to 4088. The lost codes are from 4088 up to 4095. This is a small, incremental error of 0.2% at full scale. In contrast, a converter with an error of -3 LSB ((-3 LSB offset error) $-$ (-5 LSB gain error)) will produce codes from 3 up to 4095. The gain error in this situation produces an accuracy problem, not a loss of codes. The lost codes are 0, 1, and 2. Both of these examples illustrate the worst possible scenario.

The difference between the first measured transition point and the first ideal transition point is the offset voltage of the converter. If the offset error is known, it can easily be calibrated out of the conversion in hardware or software by subtracting the offset from every code. Gain error (full-scale error) is the difference between the ideal slope from zero to FS and the actual slope between the measured zero point and FS. Offset errors are zeroed out with this error calculation. Gain error is another ADC characteristic that can be calibrated out of the final digital code from the converter. Multiplying the final conversion by a constant does this. Although this calibration is possible, the software overhead may be too much. Typically, the offset error and gain errors do not track from converter to converter.

The real-life performance enhancements due to incremental improvements in an ADC's offset or gain specifications are negligible to nonexistent. To some designers, this seems like a bold assumption if precision is one of the design objectives. It is easy to implement a digital calibration algorithm with your firmware. However, more important, the front-end amplification/signal conditioning section of the circuit typically produces higher errors than the converter itself.

This discussion puts a new light on the conclusions reached at the beginning of this section. In fact, the 12-bit converter as specified above has an accuracy of approximately 11.997 bits. The good news is that a microprocessor or microcontroller can remove this offset and gain error with a simple calibration algorithm.

Differential nonlinearity (DNL) is the maximum deviation in code width from the ideal 1 LSB ($FS/2^n$) code width. The difference is calculated for each transition. This converter characteristic is very difficult to calibrate out. Even if you take the time to measure one converter for this error, the next converter from the same product family will have a slightly different DNL error from code to code. Integral nonlinearity (INL) is the maximum deviation of a transition point from the corresponding point of the ideal transfer curve with offset and gain errors zeroed. The INL performance of an ADC is actually derived from DNL tests.

Once again, the INL error is difficult to calibrate out of the final conversion, particularly from part-to-part of a product family.

For more details about these specifications, see Appendix A.

Successive Approximation Register (SAR) Converters

The SAR ADC arose out of industrial application requirements. This tried-but-true converter solution has spread across a variety of applications, including process control, medical, and earlier audio systems. In these applications, 8- to 16-bit conversion results were required.

The SAR ADC is nothing new to the data acquisition world. In the 1970s, the state-of-the-art SAR ADC was touted as a lower-power, more accurate, and less expensive device. These converters utilized R-2R resistive ladders in their design to achieve the differential linearity, integral linearity, offset, and gain specifications. They were able to achieve the promised performance because of careful IC layout practices and wafer-level resistor laser trimming. The core of this first-generation SAR ADC required an external sample-and-hold circuit but was exclusively built using a bipolar transistor process. This was a good marriage because the bipolar technology was best suited for low noise and high-speed performance. A good example of this type of converter is the industry standard, ADC700, manufactured by Texas Instruments.

In today's standards, this hybrid ADC would be considered too power-hungry. The current CMOS generation of SARs has succeeded in taking over the all-bipolar SAR. The architecture of this converter uses a capacitive redistribution input section, which inherently includes the sample/hold function. The capacitor arrays are more compact and much easier to match than the older nichrome R-2R ladder networks, which usually require an external sample-and-hold circuit on the analog front end. This new chip topology has lower-power operation, higher functionality, and is smaller in size.

All this is good news to the system designer who is looking for improved performance, higher integration, and an overall excellent cost/performance ratio. This generation of SAR converters not only includes the sample-hold function but also differential inputs and voltage-controlled gain capability through the voltage reference inputs. Since the integrated circuit design is implemented primarily with capacitors rather than resistors, the power dissipation and the chip size are lower than achieved in the past. The SAR converter also has taken a step toward increased functionality. In prior SAR ADC designs, the voltage reference circuit could be internal or external but in all cases was limited in voltage range. With this new topology, the device voltage reference is usually external and its range is much wider. This gives flexibility when selecting the desired LSB size. As mentioned before, the LSB size of a converter is:

$$LSB = \frac{FSR}{2^n}$$

where n = number of bits and FSR = the voltage reference voltage.

Under normal single-supply conditions, the voltage reference would be equal to 5 V. If this is the case, the LSB size of a 12-bit converter is equal to 1.22 mV (5 V ÷ 4096 codes). If the voltage reference for the converter is equal to 100 mV, the LSB size now becomes 0.0244 mV. This is a 50× reduction in LSB size (the converter noise does not necessarily track this reference change). If you have a very clean layout and voltage reference, this type of change could eliminate an analog gain stage.

A final advantage of the CMOS version of the SAR converter is that it is possible to integrate this circuit onto the microcontroller or processor chip. This is not feasible with the bipolar SAR converter unless you produced an expensive multichip, mixed-signal version.

The CMOS SAR Topology

The CMOS SAR ADC is a sampling system that takes one sample for every conversion. The analog input signal to a SAR converter first sees a switch and a capacitive array, as shown in **Figure 13-6**. The input node connects a capacitive array on one side and the noninverting input to a comparator on the other.

When the switch (S_1) is closed, the voltage input signal is sampled onto the internal capacitive array of the converter. After the sampling time is completed, S_1 is opened and the bottom side of the most significant bit (MSB) capacitor is connected to V_{REF} while the other capacitors are tied to V_{SS} (or the system ground). The charge from the MSB capacitor is redistributed among the other capacitors. The charge is distributed across the capacitor array, and the noninverting input of the comparator moves up or down according to the voltage presented at its input. The voltage at the noninverting input of the comparator, with respect to V_{SS}, is equal to $(1/2V_{DD} - V_{IN}) + 1/2V_{REF}$. If this voltage is greater than $1/2V_{DD}$, the MSB is equal to zero, which is transmitted out of the serial port, and the MSB capacitor is left tied to V_{REF}. The transmission of all bits to the serial port is synchronized with system clock (SCLK) through S_{DOUT} (serial data out). If the voltage

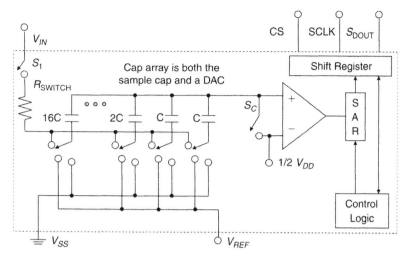

Figure 13-6: The modern SAR converter uses a capacitive array at the analog input. This capacitive array and the remainder of the device are easily manufactured in CMOS, making it easy to integrate with microcontrollers or microprocessors.

at the noninverting input of the comparator is less than $1/2V_{DD}$, the MSB capacitor is connected to V_{SS} and an MSB bit equal to one is transmitted out of the serial port.

As soon as the MSB value is determined, the converter starts to determine the MSB − 1 value. Connecting the MSB − 1 capacitor to V_{REF} while the other capacitors are tied to V_{SS} (except for the MSB capacitor) does this. Note that the MSB − 1 capacitor is not illustrated in Figure 13-6, but its value is 8C. With this change in the capacitive array connections, the value of the voltage at the noninverting input of the comparator is $[1/2V_{DD} − V_{IN}] + 1/2V_{REF}$ (MSB) + $1/4V_{REF}$. Now the voltage on the capacitive array is compared to the voltage at the inverting of the input comparator, $1/2V_{DD}$. In the analysis of this bit, if this voltage is greater than $1/2V_{DD}$, the MSB − 1 is equal to zero, which is transmitted out of the serial port. Additionally, the MSB − 1 capacitor is left tied to V_{REF}. If the voltage across the capacitive array is less than $1/2V_{DD}$, the MSB − 1 bit is equal to one. This bit value is transmitted out through the serial port. Once this is done, the MSB − 1 capacitor is connected to V_{SS}. This process is repeated until the capacitive array is fully utilized.

It is important to keep the power supply and PCB as quiet as possible during the conversion process of the SAR-ADC. The power supply can be a linear or switching supply, but in either case the noise to the power pin of the converter should be kept below the combination of the ADC power supply rejection capability and the power supply bypass capacitor frequency response. Switching activity on the PCB, particularly near the converter, should also be minimized during the conversion process. This critical time,

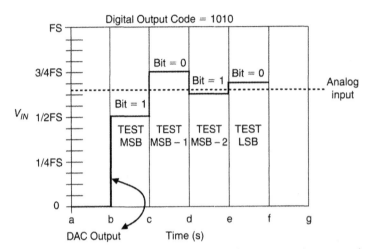

Figure 13-7: Another way of looking at the SAR conversion process is to examine the internal DAC output of the converter. The converter starts by converting the MSB of the analog input signal and then steps through each bit. Each bit conversion is timed with the system clock.

when these noise sources can have a negative impact, includes the signal acquisition time along with each bit decision.

Figure 13-7 shows another way of thinking about the SAR conversion by looking at the digital-to-analog converter (DAC) output. In this figure, the input is sampled between time (a) and (b). Starting at time (b), the analog voltage is tested against the DAC output voltage, which is now equal to 1/2 FS. If the analog input voltage is higher than 1/2 FS, a digital output code of 1 is sent out of the serial digital output. If the charge from the analog input voltage is lower than 1/2 FS, a digital output code of 0 is sent out of the serial digital output. In this case, the MSB value is 1. The capacitive array is switched to test MSB − 1 as discussed above.

Between time (c) and (d), the analog input charge is now compared to 3/4 FS. If the MSB was found to be a 0, the MSB − 1 bit would be compared to 3/4 FS. But as you can see in this case, with the MSB equal to 1, the MSB − 1 is determined to equal 0. This process continues until the final LSB code is determined.

Interfacing With the Input of the SAR Converter

Driving any A/D converter can be challenging if all issues and trade-offs are not well understood from the beginning. With the SAR converter, the sampling speed and source impedance should be taken into consideration if the device is to be fully utilized. Here we will discuss the issues that surround the SAR converter's input and conversion to ensure that the converter is handled properly from the beginning of the design phase. We are

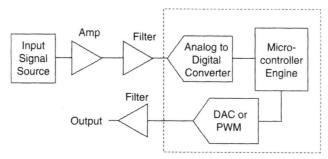

Figure 13-8: An input signal to the SAR converter should be buffered to reduce impedance-matching problems and filtered to reduce aliasing errors in the converter. The amplifier stage and filter stage in this diagram can be combined.

also going to review the specifications available in most A/D converter datasheets, and we'll identify the important specifications for driving your SAR. From this discussion, techniques will be explored which can be used to successfully drive the input of the SAR A/D converter. Since most SAR applications require an active driving device at the converter's input, the final subject is to explore the impact of an operational amplifier on the analog-to-digital conversion in terms of DC as well as AC responses. A typical system block diagram of the SAR converter application is shown in **Figure 13-8**. Some common SAR converter systems are data acquisition systems, transducer sensing circuits, battery monitoring applications, and data logging. In all these systems, DC specifications are important. Additionally, the required conversion rate is relatively fast (compared to delta-sigma converters), and having a lower number of bits that are reliably converted is acceptable.

For the input stage of the converter shown in **Figure 13-9**, the input signal could be AC, DC, or both. The operational amplifier is used for gain, impedance isolation, and its drive capability. A filter of some sort (passive or active) is needed to reduce noise and to prevent aliasing errors.

A model of the SAR ADC internal input sampling mechanism is shown in Figure 13-9. Critical values in this model are R_S, C_{SAMPLE}, and R_{SWITCH}. C_{SAMPLE} is equivalent to the summation of the capacitive array shown in Figure 13-6. Pin capacitance and leakage errors are minimal. The external source resistance and sample capacitor combines with internal switch resistance and internal sample capacitor to form an R/C pair. This distributed R/C pair requires approximately 9.5 time constants to fully charge to 12 bits over temperature. The ADS7829 (from Texas Instruments), 12-bit ADC requires 750 nsec to fully sample the input signal, assuming $R_S \ll R_{SWITCH}$.

Following these elements, the signal reaches the switch resistance, R_{SWITCH}, and the sample capacitor, C_{SAMPLE}. The sampling capacitor represents the bulked element that

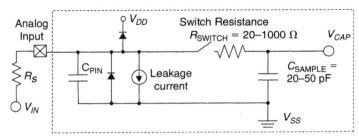

Figure 13-9: The combination of the external input resistance (R_S) and internal switch resistance (R_{SWITCH}) and the sample capacitor (C_{SAMPLE}) form a lowpass filter that has a risetime. If an accurate conversion is desired, the external input resistance should be minimized or the sampling time of the converter should be lengthened.

samples the input signal while the switch is closed. While the converter is sampling the input signal, the combination of the source resistance (R_S), the switch resistance (R_{SWITCH}), and the sampling capacitor (C_{SAMPLE}) forms a single-pole R/C network. The time constant of this network is:

$$t_{RC} = (R_S + R_{SWITCH}) \times C_{SAMPLE}$$

Assuming that the charge and voltage on the sample capacitor is zero at the time that the sample is acquired, the risetime of the voltage on that capacitor is equal to:

$$V_{CAP} = V_{IN}(1 - e^{-t/(R_S + R_{SWITCH})(C_{SAMPLE})})$$

The result of having a high external input resistance or fast conversion time can compromise the accuracy of the conversion. In either case, the converter will close its sampling switch before the sample capacitor has completely charged, as illustrated in **Figure 13-10**.

With the previous formula, we can determine the percentage of charge that will arrive at the sample capacitor with time.

If you apply this concept to a specific application where a 12-bit A/D converter from Figure 13-9 is used, you can calculate the number of bits that you have acquired from the input signal. This is shown in **Table 13-3**.

As calculated in the table, the accuracy of an analog-to-digital converter can be compromised if the device is not given enough time to sample. For instance, use the example of an ADS7829 (Texas Instruments) 12-bit A/D converter, which samples within 1.5 clock cycles using a clock rate of 2 MHz. The sampling time allotted by the converter is 750 nsec. This works very well with the numbers in Table 13-3. Now add a source resistance of 5 kΩ and you will find that the converter needs 1350 nsec to convert to 12 + 1 bits accurately.

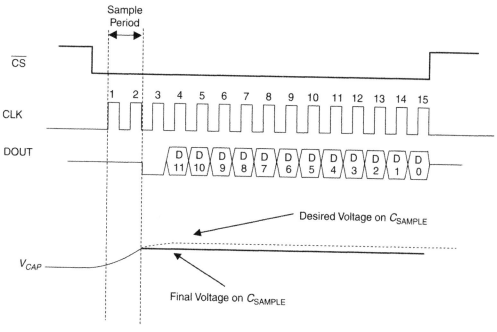

Figure 13-10: The converter is allowed to sample the input signal for a defined amount of time. The system clock to the converter determines this time. If the sampling time is too fast or the external input resistor is too high, the sample capacitor will not completely sample the input signal, resulting in an inaccurate conversion.

Table 13-3: The R/C pair of $R_{SWITCH} = 1\,k\Omega$ and $C_{SAMPLE} = 25\,pF$ requires approximately 9.5 time constants to fully charge to 12 bits over temperature.

# of Time Constants	1	5	8	9	10
$(R_S + R_{SWITCH}) \times C_{SAMPLE}$ in ns	25	125	200	225	250
% of Full-Scale Range on C_{SAMPLE}	63.2	99.3	99.966	99.9877	99.9955
% of Full-Scale Range on C_{SAMPLE} to go	36.8	0.67	0.034	0.0123	0.0045
ADC Bit Accuracy (bits)	1.4	7.2	11.5	13.0	14.43

The accuracy of a SAR ADC can be compromised if the device is not given enough time to sample. In the graph in **Figure 13-11**, the y-axis is the clock frequency in megahertz, and the x-axis is input (source) resistance in ohms. The sampling time of the converter for these clock frequencies is equal to 1.5 clocks. For example, a clock speed of 2 MHz would translate to a sample time of (1.5/2 MHz), or 750 nsec.

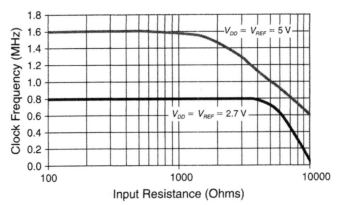

Figure 13-11: There is a level of source resistance that a SAR converter can tolerate with a given sampling frequency. But as the external source resistance increases, the converter reaches a point where the sampling capacitor needs more time to charge. These variables change from converter to converter. The product datasheet from your converter should be consulted.

There are two obvious solutions to the problem: one, reduce the source resistance, or two, increase the sampling time.

To keep the source resistance low, it is recommended that the converter be driven by an active element, such as an operational amplifier and an R/C filter. In this situation, the input signal could be AC, DC, or both. The operational amplifier can be used for gain filtering, impedance isolation, and its drive capability. When you drive the input of an ADC with an operational amplifier, whether it is a gain cell, a filter cell, or both, the amplifier can add offset, noise, gain errors, and distortion to the signal prior to the ADC.

Delta-Sigma ($\Delta-\Sigma$) Converters

I remember the days when I could count on enough good analog circuit questions to create quite an afternoon of adventures in the lab. I would use theory to predict the analog outcome in terms of stability, gain, or noise levels and try to match it with reality. The mixed-signal ADC and digital-to-analog converter (DAC) questions were riddled with digital timing problems to conquer but seldom the complex noise or stability calculations that pure analog circuits provided. Since the mixed-signal circuits still required an analog front end (gain and filtering), most calculations were still handled in the analog domain. The entrance of the 16-bit converter offered somewhat of a front-end noise reduction challenge, but it did not rival the battle being waged in the pure analog domain.

Then a new player joined the team. This player was literally thrown over the wall, at least for those who aren't IEEE fans. I couldn't imagine replacing these complex circuits with a digital-centric device, but I should have paid attention to the reports that were coming back from the conferences. These reports came from the IC designer to the poor, unsuspecting, IC user. It came in the form of a 1-bit digitizer that would output a 24-bit word. I once asked an audience of engineers in a technical seminar what 2^{24} was equal to. I expected one of the geniuses in the crowd to quickly shout out "16,777,216." The actual answer that I got was "4^{12}." And now, the person who called me and asked me to recommend a 32-bit converter can actually see the light at the end of the tunnel.

Having a converter that is capable of converting to a resolution of 16 million codes can be overwhelming at first, but let's back up and take the bird's eye view, then follow it with the details. With this approach, we can start with an intuitive level, which will get us a long way as we go into the forest to look for the trees.

Here Is How the Delta-Sigma ADC Works

The rudimentary delta-sigma ($\Delta-\Sigma$) converter is a 1-bit sampling system. A functional block diagram of an ADC $\Delta-\Sigma$ converter is shown in **Figure 13-12**. In this system, multiple bits are sent to the digital/decimator filter. In this filter there is a fair degree of mathematical manipulation performed.

An analog signal is applied to the input of the converter. This signal needs to be relatively slow because the $\Delta-\Sigma$ ADC samples the input signal multiple times; this technique is known as *oversampling*. The sampling rate (F_S) is tens or hundreds of times faster than the data rate (F_D) of the the digital results at the output ports. Each individual sample is accumulated over time with the previous samples. This collection is digitally averaged or filtered to achieve a statistical result of the sampled input signal.

The $\Delta-\Sigma$ ADC can be broken into four discrete segments (plus the serial interface). Keep in mind that you might have to modify this simple diagram for the individual converters that you are using. For instance, if your converter has been enhanced or simplified, the basic operation of these converters is the same from device to device. A good, working block diagram is shown in Figure 13-12.

This figure has four basic functions:

- Programmable gain amplifier (PGA); gain is achieved with capacitive double-sampling techniques (Figure 13-13)

- Multi-order charge-balancing ADC; charge is "balanced" across capacitors that surround an amplifier (Figure 13-14)

- Digital lowpass filter, which is often a multiple finite impulse response (FIR) filter (Figure 13-15)

- Decimation filter (Figure 13-16)

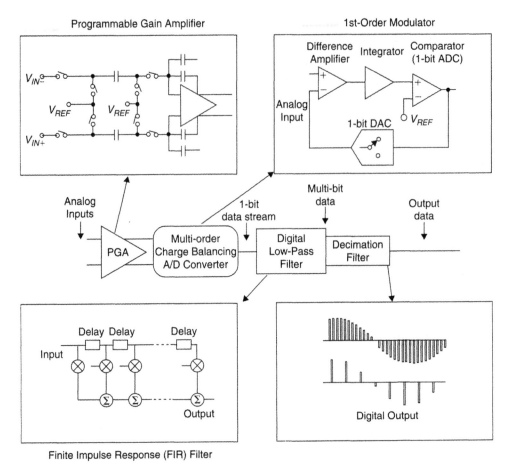

Figure 13-12: This block diagram of the delta-sigma converter has four segments: the Programmable Gain Amplifier (PGA), the multi-order charge-balancing ADC, a digital lowpass filter, and a decimation filter.

Programmable Gain Amplifier in the $\Delta-\Sigma$ Converter

The front end of some $\Delta-\Sigma$ converters have a PGA. This is not the classical analog PGA implementation where there is an amplifier surrounded by resistors which achieve an analog gain purely in the analog domain. This PGA takes the analog input signal and quickly converts it to a sampled signal. The gain settings for this stage are programmed through the digital interface. In this stage, we haven't completely left the analog domain, but we are halfway there. An example of this type of stage is shown in **Figure 13-13**.

The basic topology of the PGA stage is a differential switched capacitor amplifier. The switched capacitive topology uses a combination of oversampling and capacitor gain to achieve the possible gains of 1, 2, 4, 8, and 16. With this stage, the signal is clocked in on

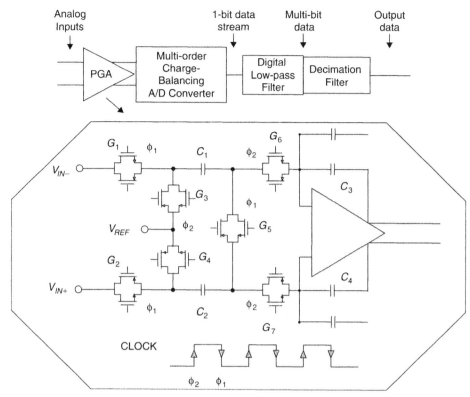

Figure 13-13: This switching network changes the analog input signal to a sampled voltage. This sampled voltage is taken in by the next stage, where the signal is finally digitized into a one or zero.

the rising edge of the sampling clock and transferred by the falling edge of the sampling clock to a second group of capacitors. For a PGA gain of one, the next rising clock edge sends the first signal forward to the modulator section of the A/D converter as well as sampling a second input signal. For a PGA gain greater than one, say a gain of two, the second rising clock edge does not send the signal forward to the digital filter section. The second stage of the PGA retains the original signal and adds it to the second sampled input. In this manner the charge is doubled. At the completion of the second sampling, the charge is finally transferred to the modulator section. This concept can easily be extended to gains of 4, 8, and 16. As the gain of the PGA stage increases, the number of cycles required to sample the signal also increases. One of two scenarios can be implemented for this change in gain. The first scenario would keep a constant sampling clock and increase the overall sample time by $16\times$ (for a gain increase of $16\times$). Another way that the PGA gain is implemented by increasing the number of samples taken by the input capacitor from 20 kHz for a gain of 1, to 320 kHz for a gain of 16. Adjusting the internal gain stage of the $\Delta-\Sigma$ converter is a technique that you can use to get an appropriate LSB voltage size for the transducer application.

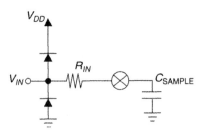

Figure 13-14: The input model of the Δ−Σ converter is similar to the input model of the SAR converter.

Initially, switches G_1, G_2, and G_5 are closed, and G_3, G_4, G_6, and G_7 are opened on the rising edge of the clock. During the time that the clock is high, charge is accumulated on C_1 and C_2. On the falling edge of the clock, switches G_1, G_2, and G_5 are opened and G_3, G_4, G_6, and G_7 are closed. The charge on C_1 and C_2 is transferred to C_3 and C_4. This cycle is then repeated. The most common way to implement a gain is to sample the signal multiple times and store the charge on C_3 and C_4. For example, if the input signal charge is stored on C_3 and C_4 twice, the signal will have a gain of two applied.

Assuming the external clock is a constant 10 MHz, PGA gain is implemented by increasing the number of samples taken by the input capacitor from 20 kHz for a gain of 1, to 320 kHz for a gain of 16. Adjusting the internal gain stage of the Δ−Σ converter is a technique that you can use to get an appropriate LSB voltage size for the transducer application.

In digitizing systems, the antialiasing analog filter has saved many designs from noise-ridden disasters. These filters serve the purpose of rejecting high-frequency (uninvited) noise in the analog system so that the digitizer doesn't alias unwanted signals into the bandwidth of interest. One would assume that the antialiasing filter would always be a permanent fixture, placed before the analog-to-digital converter. With the Δ−Σ converter, the internal digital network has nearly replaced this analog function.

Delta-sigma ADC manufacturers are promoting the requirement of a simple R/C lowpass filter at the input of the converter as the answer to all antialiasing problems. In fact, this filter does provide a small amount of high-frequency attenuation, but that is not the primary function of this simple lowpass filter. The most disruptive noise signals that are present at the input of the ADC are the switching currents coming in and out of the converter itself. The first stage of the unbuffered Δ−Σ converter is fundamentally a switched capacitor network. A model of the Δ−Σ input is shown in **Figure 13-14**.

Switching glitches can be easily measured with a few hundred ohms on the inputs and an oscilloscope. This glitch energy can disrupt the measurement of the small voltages at the inputs by upsetting the driving input circuitry. This input stage should be treated with the same care that the SAR converter receives. Any error that is introduced because of high source impedance affects the offset and gain of the conversion. A step toward solving this problem is to place this R/C filter on the inputs of the converter. Additionally, if the device has differential inputs, a 0.1 μF capacitor can be placed directly across the inputs. This is done to attenuate high-frequency noise that is present at the input pins of the device. Note that this technique is not recommended for analog operational amplifiers. The fact that this is called a PGA does not imply that input impedance is high. As a matter of fact, it is fairly low and dependent on the input capacitance and the over-sampling frequency.

Multi-order Charge-Balancing ADC

The charge-balancing ADC is the heart of this converter. It is responsible for digitizing the input signal and actually achieving the start of a low-noise conversion. In this stage, the architecture implements a noise-shaping function where low-frequency noise is pushed up to higher frequencies. This low-frequency noise appears outside the band of interest. This is one of the reasons that $\Delta - \Sigma$ converters are well suited for low-frequency, high-accuracy measurements.

The function of the multi-order charge-balancing A/D converter can be conceptualized with the first-order stage shown in **Figure 13-15**. The difference of the analog input voltage and the output of the 1-bit DAC provides an analog voltage at X_2. The voltage at X_2 is presented to the integrator. The output of the integrator progresses in a negative or positive direction. The slope and direction of the signal at X_3 are dependent on the sign and magnitude of X_2. At the time the voltage at X_3 equals the comparator reference voltage, the output of the comparator switches from negative to positive or positive to negative, dependent on its original state. The output value of the comparator (X_4) is clocked back into the 1-bit DAC as well as clocked out to the digital/decimator filter stage. At the time that the output of the comparator switches from a high to a low or vice versa, the 1-bit DAC responds on the next clock pulse by changing its analog output voltage to the difference amplifier. This creates a different output voltage at X_2, causing the integrator to progress in the opposite direction.

Many times the PGA stage of the multi-order charge-balancing stage is combined. But for discussion purposes, in Figure 13-15 the signal enters this stage from the PGA stage (Figure 13-13). Although this signal was sampled in the previous stage, the voltage magnitude can be anywhere between ground and the voltage reference. Depending on the 1-bit DAC's output voltage, the difference amplifier will produce a relatively high voltage or low voltage. This voltage is then integrated through the next stage (integrator). At the output of the integrator, a comparator will produce a one or zero. This is essentially

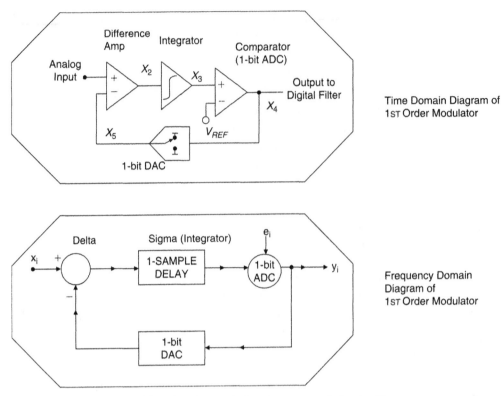

Time Domain Diagram of
1ST Order Modulator

Frequency Domain
Diagram of
1ST Order Modulator

Figure 13-15: The multi-order charge-balancing stage of the $\Delta-\Sigma$ converter samples the input signal from the PGA stage and converts that signal to a one or a zero. From a frequency domain perspective, the modulator passes the low frequency input signal and pushes the ADC noise into higher frequencies (see Figure 13-16).

the step where the signal is digitized, and in that step quantization noise is created. The comparator output is sampled by the DAC in sync with the sampling clock of the converter. This might or might not change the DAC output. The output of the comparator is also sampled in sync with the sampling clock of the digital filter.

The combination of the integrator and sampling strategy implements a noise-shaping filter on the digital output code. This noise shape is illustrated in **Figure 13-16**. In this diagram N is the number of modulator bits, N_{INC} is the increased resolution, k is the oversampling ratio, and M is the order of the modulator.

A digital filter is implemented in the next stage. The effects of the digital filter are illustrated in Figure 13-16.

Digital Lowpass Filter

The output of the charge-balancing modulator is a series of digital ones and zeros, which are sent to the digital filter function. The digital filter uses an oversampling and averaging

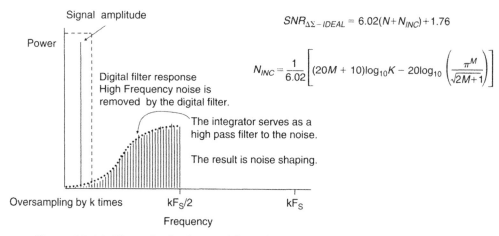

$$SNR_{\Delta\Sigma-IDEAL} = 6.02(N+N_{INC})+1.76$$

$$N_{INC} = \frac{1}{6.02}\left[(20M + 10)\log_{10}K - 20\log_{10}\left(\frac{\pi^M}{\sqrt{2M+1}}\right)\right]$$

Figure 13-16: The noise in the modulator is moved out into higher frequencies. This is called *noise shaping*. After the modulator, the digital filter is used to implement a lowpass filter where high-frequency noise is removed. The calculation for the SNR includes the effects of oversampling.

algorithm to further process the signal into the higher resolutions. The combination of the digital filter and the decimation filter stages directly affects the resolution and output data rate of the converter.

Typically, the digital filter is a finite impulse response (FIR) filter that essentially implements a weighted average on the digital output from the modulator. A first-order FIR filter is actually an averaging machine. A FIR digital filter is shown in **Figure 13-17**.

The first-order FIR filter in Figure 13-17 uses a moving-average process. This averaging process mathematically reduces the level of uncertainty in the output signal, but it does take time to acquire the samples. Theoretically, if you acquire four samples, you can change the digital output from an output with two possibilities (0 and 1) to an output that has four possibilities (00, 01, 10, 11). This is possible through the oversampling mechanism and averaging process.

For instance, if 4 bits are acquired, the possibilities after the averaging process are 0, 0.25, 0.5, and 0.75. Each oversample by a factor of four gives a 6 dB (or 1-bit) improvement in the converters SNR. So theoretically, given a 1-bit ADC, a 2-bit converter can be mathematically realized with 4^1 averaged samples. A 3-bit converter can be realized with 4^2 (or 16) averaged samples. A 4-bit converter can be realized with 4^3 (or 64) averaged samples. You can see that this technique only works so far—for instance, if a 24-bit converter is derived from a 1-bit ADC by averaging 4^{23} (or 70,368,744,178) samples. This would take a long time and long-term drift would be a problem. As a result, other techniques are used to obtain higher bit resolutions. These techniques include the modulator noise shaping (discussed previously) and multiple-level modulators.

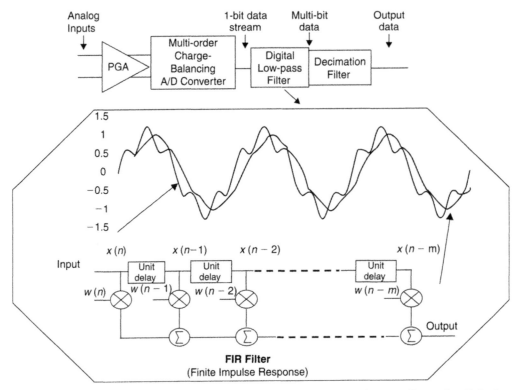

Figure 13-17: The filter illustrated in this diagram is a first-order FIR filter. The digital lowpass filter of the $\Delta-\Sigma$ converter further reduces the noise in the conversion process by averaging multiple digital codes in this rolling-averaging filter (FIR filter).

There are very few $\Delta-\Sigma$ converters on the market that have a first-order FIR filter. Most of these types of converters have a multi-order implementation of this function. A third-order filter is called a *sinc³ filter*. The transfer function of a *sinc³* filter with a cut-off frequency of 60 Hz is shown in **Figure 13-18**.

$\Delta-\Sigma$ converters have a variety of digital filters to choose from. In this discussion, we have talked about the FIR filter. Another type of discrete-time filter that is common is the infinite impulse response (IIR) filter (**Figure 13-19**).

Continuous-time filter approximations such as Butterworth, Bessel, Inverse Chebyshev, and so forth can be mapped into discrete-time IIR filters. But ultimately the design procedure for a discrete-time system begins with a set of discrete-time specifications. Continuous-time filter approximations are used as a convenient tool for meeting the discrete-time filter specifications. This type of discrete-time filter is transformed from the prototype continuous-time filter.

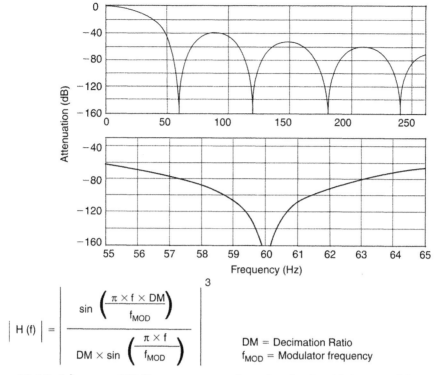

$$\left| H(f) \right| = \left| \frac{\sin\left(\dfrac{\pi \times f \times DM}{f_{MOD}}\right)}{DM \times \sin\left(\dfrac{\pi \times f}{f_{MOD}}\right)} \right|^{3}$$

DM = Decimation Ratio
f_{MOD} = Modulator frequency

Figure 13-18: A lowpass FIR filter attenuates the noise slowly with increased frequency. This is a third-order FIR filter known as a *sinc³* lowpass filter and is tuned for a 60 Hz notch frequency. This type of filter provides a comb response across the frequency spectrum.

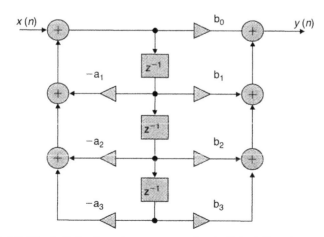

Figure 13-19: The infinite impulse response (IIR) digital filter has feedforward and feedback paths that assist in achieving sharper transitions regions. However, the trade-off for this advantage is possible instability.

The advantage of the FIR filter is its stability and linear phase response. It has a simple, straightforward design (Figure 13-17) and uses relatively lower power than the IIR filter. On the down side, the FIR filter has to be a higher-order filter to get the job done, so the latency (the amount of time required for a good conversion) is longer. With the IIR lowpass filter, you can use lower-order filters to accomplish the same cut-off frequencies as the FIR filter. As a result, there are few transistors in the implementation on silicon. On the down side, IIR filters have a nonlinear phase response and can be unstable.

Decimation Filter

The decimation filter is the function in the $\Delta-\Sigma$ converter, before the serial interface (**Figure 13-20**). The primary job of the decimation filter is to slow the data output rate and provide further signal bandlimiting. This seems like it would not be a good idea, but in fact, the converter user is not interested in the intermediary averaging steps that the converter implements. Instead, the user is only interested in the final results. So with this filter, the output data rate speed is reduced considerably with respect to the sampling frequency. The digital filter and decimation filter both reside in the same silicon space.

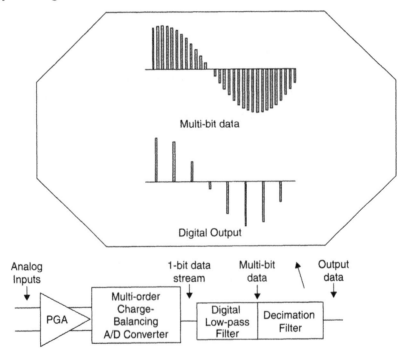

Figure 13-20: The purpose of a decimation filter is to slow the output data rate to match the output data rate to the input frequency bandwidth. The FIR lowpass filter inputs a multibit data stream to the input of the decimation filter. The simplest decimation filter filters out several conversion results while keeping enough to meet the output data rate of the converter.

The discrete, high-precision analog front end in the data acquisition circuit is not outdated but is being gently pushed further into its exclusive corner. This is not to say that the demand for precision data acquisition circuits has disappeared. The solutions to these problems are changing in orientation from the analog-dominated circuit to digital. Although the task of signal processing seems to be migrating to the digital domain, good analog engineering practices still apply. To my delight, it seems that the art of analog hardware design may be renamed but will never be obsolete.

What's the Big Deal About $\Delta-\Sigma$?

The $\Delta-\Sigma$ ADC is a new breed of device. Generally, this device samples analog signals at lower data rates than SAR converters, but it has the advantage of higher resolution. With this higher resolution, the classical analog front end can be eliminated. It is part of the initiative in which digital designs are encroaching into the analog-domain hardware. Initially, this was done with firmware-controlled internal timers, comparators, and I/O gates, in conjunction with external resistors and capacitors. Some of the basic analog functions such as D/A converters, ADCs, and integrators have been realized. An example of a $\Delta-\Sigma$ ADC design using a comparator, timer, external resistors, and capacitors is discussed in Chapter 15. The implementations of these functions are primitive, but they get the job done when the accuracy and fidelity of high-precision analog functions are not needed.

The integrated $\Delta-\Sigma$ ADC is quickly bridging this gap. Since the device is built using a complementary metal oxide semiconductor (CMOS), the available digital functions from controllers, processors, and memory devices can be exploited. To enhance the features of this digital capability, more analog circuits are migrating from bipolar processes into the CMOS world. Not only is this migration occurring, but the performance "quality" of these CMOS devices is also improving. Some of the features that you can find in a $\Delta-\Sigma$ ADC are shown in **Figure 13-21**.

The $\Delta-\Sigma$ ADC integrates a significant quantity of analog and digital functionality. These two functions are designed to interact extremely well. For example, an analog buffer can be digitally switched in or out of the input stage. A multiplexer can be found on some $\Delta-\Sigma$ ADCs, and the channels are programmed through the digital interface of the converter.

On the analog side of this device, voltage or current references are integrated into the chip, which allows for ratiometric operation. The advantage of a ratiometric system is that gain errors are eliminated because every element in the circuit uses the same reference. This applies to the excitation of the sensor all the way to the reference of the ADC. Sometimes, there is a PGA on the front end of the converter, and some converters have buffers that can be switched in or out of the circuit digitally.

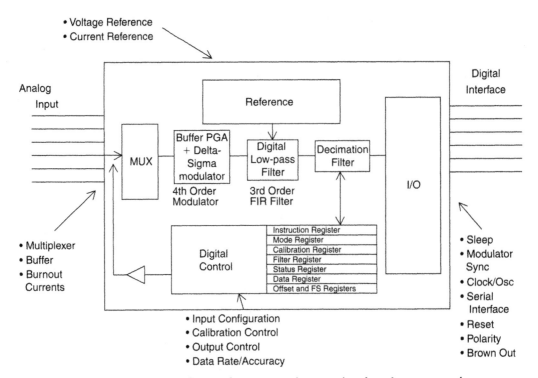

- Voltage Reference
- Current Reference

Analog
Input

Digital
Interface

Reference

Buffer PGA
+ Delta-
Sigma
modulator

Digital
Low-pass
Filter

Decimation
Filter

MUX

I/O

4th Order
Modulator

3rd Order
FIR Filter

Digital
Control

Instruction Register
Mode Register
Calibration Register
Filter Register
Status Register
Data Register
Offset and FS Registers

- Multiplexer
- Buffer
- Burnout
  Currents

- Sleep
- Modulator
  Sync
- Clock/Osc
- Serial
  Interface
- Reset
- Polarity
- Brown Out

- Input Configuration
- Calibration Control
- Output Control
- Data Rate/Accuracy

Figure 13-21: The $\Delta-\Sigma$ ADC has matured to a point that there are analog functions that cater to a variety of sensor requirements as well as digital functions that integrate controller digital functions and features. This combination is only possible in CMOS. Some modules that can be added to the $\Delta-\Sigma$ ADC are shown here. You won't find all of these features in any one $\Delta-\Sigma$ ADC. This would consume too much silicon and consequently be too expensive where all the features would not be used. Instead, you will find converters that use a select few of the features to target a converter to a specific application class.

In terms of digital features, all converters have a serial interface where the various modes can be programmed into the converter. For example, calibration algorithms can be implemented, the digital filter can be reset (or cleared), and status flags can inform the user about brownout events. Additionally, the digital filter corner frequency can be adjusted to match the application requirements.

Delta-Sigma ADC Specifications: Digital Filter Settling Time

In the system in **Figure 13-22**, the ADC must be a converter that has high resolution, low noise, multiple channels, and zero latency (zero-cycle latency) and low latency-time (settling time).

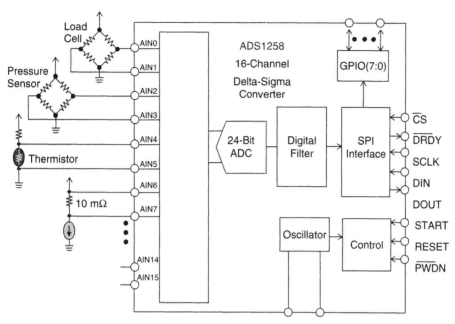

Figure 13-22: A 16-channel multiplexed application requires a $\Delta-\Sigma$ converter that has zero latency and fast latency time (ADS1258, Texas Instruments).

For ADCs, cycle latency is equal to the number of complete data cycles between the initiation of the input signal conversion and the initiation of the next signal conversion. The unit of measure for this definition of latency is (n)-cycle latency, where n is a whole number. Although fully settled data from the device shown in **Figure 13-23** (a one, input-channel, 24-bit $\Delta-\Sigma$ converter) is found with the third data set at the output of the converter, the latency is equal to a two-cycle latency.

This important distinction between two-cycle latency and three data output results is at the root of the definition of cycle latency.

The first data output occurs before the start of the next cycle, hence zero cycles. The second data output occurs before the end of the second full cycle. The fully settled data output is available before the end of the third full cycle. This device has a two-cycle latency.

Latency time is typically viewed as the time required for an ideal step input to converge, within an error margin, to a final digital output value. You would express this error band as a predefined percentage of the total output voltage step. The latency time of a conversion is between the time when the signal acquisition begins to the time that data is available to download from the converter. In contrast to the cycle latency specification, the latency time (or settling time) is never equal to zero.

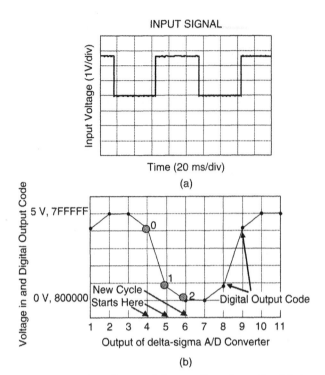

Figure 13-23: You can see the effects of the settling time of a third-order FIR filter if you synchronize an input square wave with the converter. The sampling starts for the first output data (b) on the falling edge of the input signal (a). The first output data code is approximately 80% too high. This result is a combination of one of the three filters containing the data from the actual input and the other two filters containing data taken with the previous input level of 5 V. The second output data code is now approximately 20% too high. This result is a combination of two of the three filters containing the data from the actual input and the other filter containing data taken when the input level was 5 V. The third output data code represents the input signal. All three stages of the digital filter contain data when the input signal is low.

Other Delta-Sigma ADC Differences from Vendor-to-Vendor

The $\Delta-\Sigma$ ADC has opened the door for a large variety of features because of the digital side of the converter. In essence, the real estate that is dedicated to analog is about 25% of the silicon, and digital occupies 75%. Having digital flexibility opens Pandora's box. With the $\Delta-\Sigma$ ADC, the feature set is limited due to silicon real-estate costs, not technology.

The fundamental features that most $\Delta-\Sigma$ ADCs have are sleep states, slave/master settings, internal/external clock options, variable sampling frequency, and voltage references. There are other features that some of the converters (but not all) have. For instance, some converters have an idle tone detect bit. Other converters have current references instead of the standard voltage reference; still other converters have self- and system-calibration capability, and the list goes on. Basically, the $\Delta-\Sigma$ ADC is the first of many devices that will be integrating analog and digital functions together under one roof.

In this chapter, we started by discussing the general key specifications for ADCs. We discussed how they impact the digital results from your converter. Then we looked at the successive approximation register (SAR) ADC and the delta-sigma ADC. We took a close look at the basic topologies of these converters and how they affect your conversion results. Now, going on to Chapter 14, we are going to apply these converters to "real-life" applications.

References

Analog Devices, "Using sigma-delta converters, part 1," AN388.

———, "Using sigma-delta converters, part 2," AN389.

Baker, B.C., "Anticipate the accuracy of your converter," *EDN Magazine*, March 18, 2004, reprinted with permission from *EDN Magazine*, Reed Business Information, copyright 2004.

———, "Giving $\Delta\Sigma$ converters a little gain boost with a front-end analog gain stage," *Application Bulletin*, AB-107, Burr-Brown Corporation, January 1997.

———, "How to get 23 bits of effective resolution from your 24-bit converter," *Application Bulletin*, AB-120, Burr-Brown Corporation, September 1997.

———, "Number of bits vs. LSB errors," *EDN Magazine*, July 8, 2004, reprinted with permission from *EDN Magazine*, Reed Business Information, copyright 2004.

———, "Synchronization of external analog multiplexers with the $\Delta\Sigma$ A/D converter," *Application Bulletin*, AB-116, Burr-Brown Corp., June 1997.

———, "Using operational amplifiers for analog gain in embedded system design," AN682, Microchip Technology.

———, "Voltage reference scaling techniques increase the accuracy of the converter as well as resolution," *Application Bulletin*, AB-110, Burr-Brown Corp., February 1997.

Bowling, S., "Understanding A/D converter performance specifications," AN693, Microchip Technology.

Burr-Brown, 1994 Application Seminar, Chapter 1.

Crystal, "Delta-sigma A-D converter conversion technique overview," AN10.

Jarmon, D., "A brief introduction of sigma-delta conversion," AN9054, Harris Semiconductor.

Johnson, J., "Switched-capacitor A-D converter input structures," AN30, Crystal.

Mitra, D'Sousa, C., "Using the analog-to-digital (A/D) converter," AN546, Microchip Technology.

Norsworthy, S.T., *Delta-Sigma Data Converters: Theory, Design, and Simulation*, IEEE Press, 1997.

Oppenheim, S., *Discrete-Time Signal Processing*, Prentice-Hall, 1989.

Sheingold, D.H., *Analog-Digital Conversion Handbook*, Prentice-Hall, 1986.

CHAPTER 14

The Right ADC for the Right Application

Bonnie Baker

Here, Ms. Baker explains what kind of sensors are naturally suitable for what kind of ADC. Will you need preamplifiers, buffers, or filters? Bonnie gives good examples! /rap

In This Chapter

Now that we have the ADC fundamentals out of the way, it's time to do some real work by looking at what these ADCs can do for us. In this chapter, we will spend some time looking at the places where signals come from and which ADC is the best for the application. Through this exercise we will get a better feel of how to prepare the signal for the microcontroller. Once this groundwork is established, we will move into looking at four specific applications. This chapter contains four real-world applications: temperature, pressures, light sensing, and motor control. In these applications, you will have to decide whether a SAR or Δ–Σ converter is appropriate. We will discuss the most common ADC problems and then quickly move to solutions. We will find that the SAR converter can service some of these applications. The Δ–Σ ADC will better service others. Some applications can utilize both converters, with a minimum number of trade-offs.

Classes of Input Signals

Before we dive into the details of applications, we need to define the origin of real-world analog signals. From there I will show you how to capture those signals with your circuits. **Figure 14-1** shows several possible signal sources, with respect to the frequency versus number of bits.

At the system level, an appropriate ADC can produce the proper number of noise-free bits, or you can use an analog front-end gain cell plus a lower-resolution ADC at a lower cost.

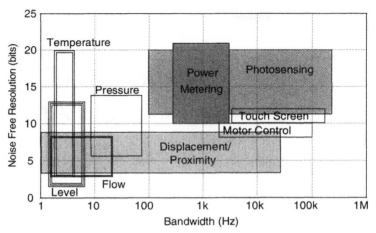

Figure 14-1: Input signals come to the ADC from various sources, but they mostly come from the physical world. The dynamics of these real-world signals define the sampling bandwidths. All these physical sources have their own set of sensors, and these sensors define the number of required noise-free bits.

There is a design architecture trade-off where cost, board space, and number of components come into play.

When it comes to sensing external analog events, at the very least you will want to measure temperature. Temperature testing is also the lowest common denominator in terms of frequency. This is because temperature doesn't change that quickly and the sensors reflect that characteristic. The physical environment of the sensor and the sensor package governs the speeds required for temperature-sensing events. Generally, temperature changes in the environment are slow ($>0.1\,\text{sec}/°\text{C}$). This is consistent with the package and temperature coefficient of the sensor. The sensor requires heat-up or cool-down periods if you want to make an accurate measurement.

The more common type of temperature sensors that you can use in your circuit is a thermocouple, resistance temperature devices (RTD), thermistors, or integrated silicon sensors. **Table 14-1** summarizes the general characteristics of these devices.

Some of you might wonder what these sensors look like. I find that a picture is worth a million words (see **Figure 14-2**). You will notice that size, and hence the thermal bulk, is quite big. Again, this is only because temperature typically changes very slowly.

The granularity of the temperature measurements in terms of bits per degree Celsius can be large or quite small, as shown in Table 14-1. Additionally, the number of bits required for your system can be low or relatively high. Of course, this depends on your requirements. Because there is a wide range of application conditions, you can use the

Table 14-1: Temperature-sensing applications use the thermocouple, RTD, thermistor, or integrated circuit (IC) sensing elements. The inexpensive thermocouple does not require excitation but requires a lookup table in the controller to linearize the results. The RTD sensor is the most accurate and operates over a wide temperature range, but it requires current excitation and it could be cost prohibitive for your application. The thermistor requires voltage excitation, with a $\pm\Delta 25°C$ linearity correction using series resistors. The silicon IC temperature sensor does not require external circuitry, but the accuracy is somewhat limited and the device is slow to respond to fast temperature changes.

	Thermocouple	RTD	Thermistor	Integrated Silicon
Temperature range	-270 to $1800°C$	-250 to $900°C$	-100 to $600°C$	-155 to $200°C$
Sensitivity	10s of µV/°C	$0.00385\ \Omega/\Omega/°C$ (Platinum)	Several $\Omega/\Omega/°C$	Based on a technology that is approximately -2 mV/°C sensitive.
Accuracy	$\pm0.5°C$	$\pm0.01°C$	$\pm0.1°C$	$\pm0.5°C$
Linearity	Requires at least a fourth-order polynomial or equivalent lookup table.	Requires at least a second-order polynomial or equivalent lookup table.	Requires at least third-order polynomial or equivalent lookup table. Can also be linearized to 10-bit accuracy over a 50°C temperature range.	At best within $\pm1°C$. No linearization required.
Ruggedness	The larger-gauge wires of the thermocouple make this sensor more rugged. Additionally, the insulation materials that are used enhance the thermocouple's sturdiness.	RTDs are susceptible to damage as a result of vibration, due to the fact that they typically have 26 to 30 AWG leads which are prone to breakage.	The thermistor element is housed in a variety of ways, but the most stable, hermetic thermistors, are enclosed in glass. Generally thermistors are more difficult to handle but are not affected by shock or vibration.	As rugged as any IC housed in a plastic package such as dual-in-line or surface outline ICs.
Responsiveness in stirred oil	<1 sec	1 to 10 secs	1 to 5 secs	1.5 to 30 secs
Excitation	None required	Current source	Voltage source	Typically supply voltage

(Continued)

Table 14-1: Continued

	Thermocouple	RTD	Thermistor	Integrated Silicon
Form of output	Voltage	Resistance	Resistance	Voltage, current or digital
Typical size	Bead diameter = 5 × wire diameter	0.25 × 0.25 in.	0.1 × 0.1 in.	From TO-18 to mini-DIP to SOT-23
Price	$1 to $50	$25 to $1000	$0.40 to $10	$0.40 to $10

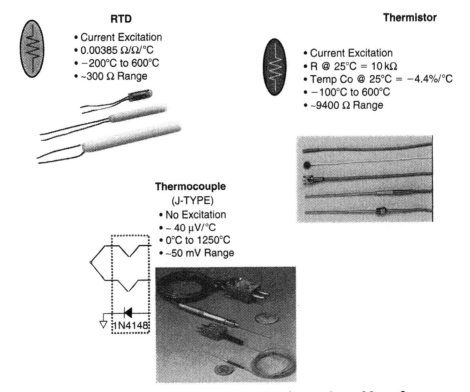

RTD
- Current Excitation
- 0.00385 Ω/Ω/°C
- −200°C to 600°C
- ~300 Ω Range

Thermistor
- Current Excitation
- R @ 25°C = 10 kΩ
- Temp Co @ 25°C = −4.4%/°C
- −100°C to 600°C
- ~9400 Ω Range

Thermocouple
(J-TYPE)
- No Excitation
- ~ 40 µV/°C
- 0°C to 1250°C
- ~50 mV Range

1N4148

Figure 14-2: Temperature sensors come in a variety of form-fit configurations. Each temperature sensor housing is designed to protect the sensor while still giving good thermal response.

SAR converter or the Δ–Σ converter for temperature measurements. There are many ways to implement the signal conditioning circuitry of a temperature sensor. Later in this chapter, we will examine SAR and Δ–Σ ADC options with an RTD sensor.

The other physical entities in Figure 14-1 are level, flow, and displacement or proximity. You usually don't measure flow directly with a sensor that specifically measures flow.

You can measure flow with temperature sensors or by measuring vibrations in a tube filled with a fluid. Displacement and proximity are other entities that don't have specific sensors dedicated for that purpose. You can measure these two phenomena optically with an LED and photodetector. You can also use an accelerometer.

The sensors for motor control, touch screen, and power metering are typically resistive. For example, you can place a very small resistor (tenths of ohms) in the "legs" of the motor control, MOSFET switching lines. The sensors of touch screens are generally resistive or capacitive. The resistive touch screen has gained attention because of the personal data assistant (PDA), and capacitive screens are used in "dirty" environments. In all cases, an ADC eventually digitizes the signal from the sensor. You can accomplish this with a SAR or Δ–Σ ADC.

The characterization (Figure 14-1) of different measurements in terms of sampling frequency versus number of bits is instructive to a point. One question that remains in this discussion is, "What are the output voltage ranges of these sensors?" **Figure 14-3** answers this question.

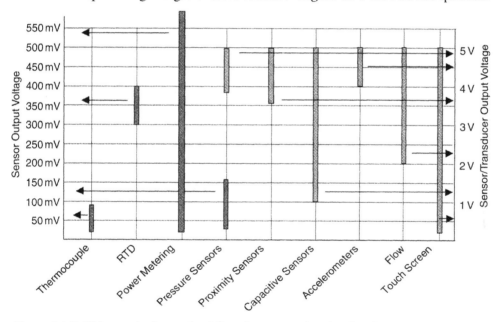

Figure 14-3: This graph shows the full-scale output signals of various sensors. It uses, where appropriate, a 1 mA or 5 V excitation. The output voltage level of various sensors, along with the sampling frequencies summarized in Figure 14-1, play a large part in determining which ADC you choose for the application circuit. There are two classes illustrated in this figure. The first class produces millivolt outputs across the sensor's full range. The left y-axis refers to this group of sensors. The second class produces voltage outputs across the sensor's full range. The right y-axis refers to the output ranges of this group. Most of the second group of devices are transducers.

There are two classes of output voltage ranges for sensors. The x-axis lists all the sensors. In the first class, the output voltages of the sensors range from approximately 50 mV to 600 mV. The scale for this range is on the left y-axis in Figure 14-3. The right-most y-axis sensors are the thermocouple, resistance temperature device (RTD), power metering, and pressure sensors.

The second class of sensor outputs ranges from about 0.25 V to 5 V. The y-axis for this output range is on the right side of the graph shown in Figure 14-3. Generally, the sensors that fit into this output range are pressure sensors, capacitive sensors, accelerometers, flow meters, and touch screens. You probably will note that pressure sensors fall into both classes of voltage output ranges. The pressure sensor in the first class is strictly a sensor, and the pressure sensor in the second class is actually a pressure transducer or a pressure sensor with an integrated signal conditioning system on-chip.

As a foundation, you should have a grasp of ADC specifications. You should also know which specifications are the most important for your application. This knowledge can save an enormous amount of time during your design phase by allowing you to simply pay attention to only the important details, not the whole list of performance characteristics and specifications.

The next section in this chapter gives you the ammunition to gain a basic understanding of how the SAR converter and Δ–Σ converter work in your sensor circuits. With this knowledge, you can quickly determine which ADC is right for your application. For example, if your ADC reports data at the wrong time, too fast, too slow, or too inaccurate, the controller or processor will struggle with too much data, not enough information, or erroneous data. The ADC can be the genius behind the system or its downfall.

Temperature Sensor Signal Chains

The SAR converter and the Δ–Σ converter are similar enough in performance in that they can both be applied to the same temperature circuits. Regardless of the converter you have chosen for your sensor, your circuit will always require some degree of analog circuitry. However, these two converters are different enough so that one or the other, not both, can only serve some sensor applications. Of the two devices, the SAR converter is the easiest to use and understand, but the Δ–Σ converter has more features and functions. The Δ–Σ converter has enough features to make you a hero or get you into serious trouble (as discussed in Chapter 13). **Figure 14-4** illustrates two system configurations where SAR converters and/or Δ–Σ converters are used.

In Figures 14-4a and 14-4b, the input sensor is a resistive Wheatstone bridge. I used the Wheatstone bridge out of convenience in this diagram. Other sensors can take the place of the Wheatstone bridge in these circuits. For example, an array of temperature sensors,

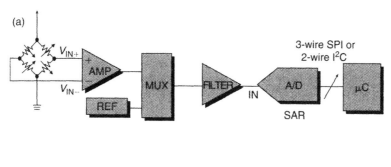

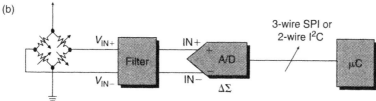

Figure 14-4: With many applications, you can use the SAR ADC or the Δ–Σ ADC to digitize the analog signal. Many times the SAR ADC requires front-end analog gain circuitry. The Δ–Σ ADC usually does not require this additional circuitry, but its slower speed will sometimes eliminate this type of device as an option.

optical sensors, and capacitive sensors (to name a few) would fit the bill. This resistive bridge in Figure 14-4 can model a variety of sensors, most commonly the pressure sensor or load cell.

In the top signal chain (Figure 14-4a), the AMP block subtracts and amplifies the two analog output signals from the bridge. This AMP block can just be operational amplifiers or an instrumentation amplifier, depending on the type of sensor in the circuit. In this system (Figure 14-4), the instrumentation amplifier will simultaneously sample the signals from the bridge and eliminate most of the common-mode noise. Many times, there is a voltage reference (V_{REF}) attached to the amplifiers or instrumentation amplifier. These voltage references come in handy in single-supply circuits. They can provide a reference point to the center of your supplies. They also are helpful when you want to achieve a ratiometric relationship between the signal and the ADC conversion.

Next in the signal line, in Figure 14-4a, is a MUX block, or multiplexer. It is not unreasonable in this type of system to have several signals digitized by the same ADC. You will frequently find a need to measure temperature, which will consume one channel of the multiplexer. Sensing other physical entities like pressure, force, light, and so on can use any one of the other channels.

You may or may not use a multiplexer in this signal path, but you will need an analog filter. You can implement the FILTER block one of three ways (see **Figure 14-5**).

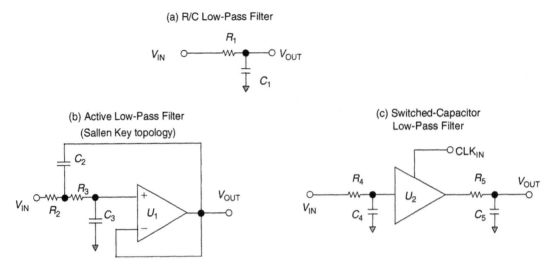

Figure 14-5: An easy lowpass filter design uses resistors and capacitors (a). The pole created by this filter is equal to $1/(2\pi RC)$. Multipole filters can use an active-filter topology (b). This second-order filter can be cascaded with more filter stages to build a higher-order filter. Switched-capacitor filters (c) are multipole, lowpass filters with very few external components. This type of lowpass filter can be a fifth (or higher) order filter.

There are more details about filters in Chapter 6, but I want to make a few sweeping statements at this point. The first and most fundamental way to design an analog filter is to use resistors and capacitors (Figure 14-5a). As discussed in more detail in Chapter 6, this filter can compromise the signal because of impedance-matching problems. As a second alternative, you can design an active lowpass filter using an operational amplifier, a few resistors, and capacitors (Figure 14-5b). A third and maybe more attractive solution for the digital designer is to use a switched-capacitor filter.

Although we cover these filters in detail in Chapter 6, I want to emphasize that a switched-capacitor filter is not your usual analog filter. They are sampling devices that keep the signal in the analog domain, but they also have the same aliasing problems as the ADCs. If you don't use special care with switched-capacitor filters, you could introduce as much or more noise than you are trying to eliminate. The primary function of the analog filter is to eliminate the higher-frequency signals that can be aliased by the ADC. You will usually need a fourth-, fifth-, or sixth-order lowpass filter in this signal path.

The analog FILTER precedes the ADC. The ADC digitizes the signal, which is free of higher frequencies because of the filter. Since there is a pre-gain stage in this signal path (AMP block), the ADC will most likely be a SAR converter. However, sometimes you might use a Δ–Σ converter because the input-signal voltage is so low. With the SAR converter, the ADC samples the signal once per conversion before digitization.

Following the ADC is a microcontroller (μC block). This block could also be a processor or field programmable gate array (FPGA). Many times the controller or processor will have an internal SAR converter. If you have a controller that has this function integrated, it can replace your stand-alone converter as long as you understand the possible errors introduced. An integrated ADC in the processor or controller runs the risk of being noisy because of the digital switching on the chip. Another issue with internal ADCs is that the IC manufacturers can have silicon limitations while they are trying to control costs. If this is the case, you can compromise the accuracy of various cells on the integrated chip.

The second signal path (Figure 14-4b) looks simpler than Figure 14-4a. In this signal path the sensor signal goes through a filter block into an ADC and then to the controller. Typically, the FILTER block is less complex than the FILTER block in Figure 14-4a. It is usually a signal-pole filter built with an R/C pair. Once the signal is through the filter, it goes to the ADC. Note that the input of the ADC is differential. The input differential stage of the ADC performs that same task as the analog instrumentation amplifier, which rejects common-mode noise.

Since this signal path seems to be simplified, you might think that it is your best option. As it turns out, in this signal path, the ADC is usually a Δ–Σ converter. Your circuit layout complexity increases if you add the Δ–Σ device. Usually multiple layer boards are needed with Δ–Σ ADCs where the SAR converter could easily reside on a two-layer board. The various blocks in Figure 14-4a have been absorbed into the Δ–Σ converter. For instance, many times Δ–Σ converters have a multichannel, multiplexer input stage.

Using an RTD for Temperature Sensing: SAR Converter or Δ–Σ Solution?

Before you pick the interface circuit for the temperature sensor, you need to know the sensor basics. The platinum RTD temperature-sensing element is the most accurate temperature sensor available. It is also more stable over time and temperature than the other types of temperature sensors. RTD element technologies are constantly improving, further enhancing the quality of this temperature measurement. Typically, a data acquisition system conditions the analog signal from the RTD sensor, making the analog translation of the temperature usable in the digital domain, as shown in **Figure 14-6**.

The acronym RTD means *resistance temperature detector*. The most stable, linear, and repeatable RTD is made of platinum metal. The temperature coefficient of the RTD element is positive: This is in contrast to the NTC Thermistor that has a negative temperature coefficient. An approximation of the platinum RTD (PRTD) resistance

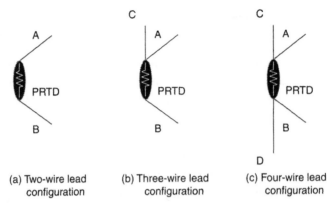

(a) Two-wire lead
configuration

(b) Three-wire lead
configuration

(c) Four-wire lead
configuration

Figure 14-6: This figure shows the three lead configurations for RTD elements. (a) Lead configuration provides one connection of each end of the sensor. This configuration is prone to absolute and temperature errors due to the lead resistance. The three-wire (b) configuration is the most commonly used configuration. This extra lead from the RTD element eliminates errors caused by currents through the leads. The most robust lead configuration is the four-wire lead (c). Wires A and B connect to the high impedance input of the front-end circuitry, and wires C and D conduct the excitation current.

changes over temperature can be calculated using the constant $0.00385 \, \Omega/\Omega/°C$. This constant is easily used to calculate the absolute resistance of the RTD at temperature with the formula below:

$$RTD(T) = RTD_0 + T \times RTD_0 \times 0.00385 \, \Omega/\Omega/°C \qquad (14\text{-}1)$$

where $RTD(T)$ is the resistance value of the RTD element at temperature in Celsius, RTD_0 is the specified resistance of the RTD element at $0°C$, and T is the temperature of the environment where the RTD is placed.

Typical specified $0°C$ RTD values are 10, 50, 100, 200, 500, 1000, or $2000 \, \Omega$. The new thick-film and thin-film platinum RTDs are less expensive than the wirewound PRTDs; however, they operate over a limited temperature range. Of these options, the $100 \, \Omega$ platinum RTD is the most stable over time and linear over temperature.

Since resistors are hard to measure directly, the RTD element requires excitation. A constant current source (as opposed to a voltage source) will provide a linear resistance to voltage conversion. When you excite the RTD element with a current source, the accuracy can be as good as $\pm 4.3°C$ over a $-200°C$ to $800°C$ temperature range. If the magnitude of the current source is too high, the element will self-heat, which will cause temperature measurement errors. Therefore, care should be taken to ensure that less than or equal to $1 \, mA$ of current is used to excite the element.

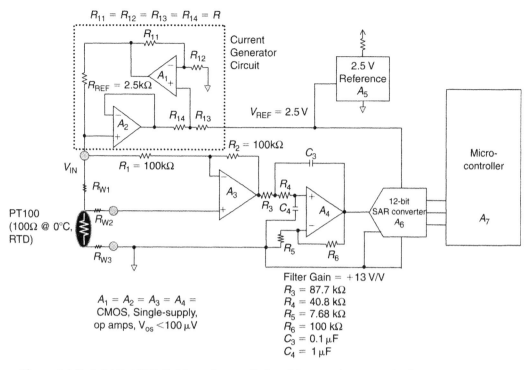

Figure 14-7: A SAR ADC digitizes the small signal from a three-terminal RTD temperature sensor as long as an amplifier circuit gains this small signal from the RTD. In this circuit, the operational amplifier, A_4, is used to implement a gain as well as a second-order lowpass filter.

The RTD Current Excitation Circuit for the SAR Circuit

For best linearity, the RTD sensing element requires a stable current reference. **Figure 14-7** illustrates one way to implement a stable current reference. In this circuit, a voltage reference, along with two operational amplifiers, generates a floating, 1 mA current source.

The evaluation of this circuit starts with the 2.5 V precision voltage reference (A_5, upper-right corner of Figure 14-7). A_5 connects directly to R_{13}. Since the noninverting input of the CMOS amplifier, A_1, is high impedance, the voltage drop across R_{13} and R_{14} is equal. The voltage at inverting input of A_1 is equal to the noninverting input voltage. That voltage is gained by $(1 + R_{11}/R_{12})$ to the output of the amplifier and the top of the reference resistor, R_{REF}. The voltage at the output of A_1 is equal to:

$$
\begin{aligned}
V_{OUTA1} &= (1 + R_{11}/R_{12}) \times (V_{REF} - V_R) \\
&= 2 \times (V_{REF} - V_R)
\end{aligned}
\tag{14-2}
$$

where V_{OUTA1} is the voltage at the output of A_1, and V_R is the voltage drop across R_{13}.

If $R_{11} = R_{12}$, the voltage at the output of A_2 is equal to $V_{REF} - 2V_R$. This same voltage appears at the inverting input of A_2 and across to the noninverting input of A_2.

Solving these equations, the voltage drop across the reference resistor, R_{REF} is equal to:

$$V_{RREF} = V_{OUTA1} - V_{OUTA2}$$
$$= 2 \times (V_{REF} - V_R) - (V_{REF} - 2V_R)$$
$$= V_{REF}$$

where V_{RREF} is the voltage across the reference resistor, R_{REF}, and V_R is the voltage drop across R_3 and R_4.

The current through R_{REF} is equal to:

$$I_{RTD} = V_{RREF}/R_{REF} \tag{14-3}$$

This circuit generates a DC, floating, current source that is ratiometric to the voltage reference and the ADC.

Absolute errors in the circuit will occur as a consequence of the initial offset voltages of the operational amplifiers, the output swing of A_1, mismatches between the resistors (R_{11}, R_{12}, R_{13}, and R_{14}), the absolute resistance value of R_{REF}, and the RTD element. Errors due to temperature changes in the circuit will occur as a consequence of the temperature drift of the same elements listed above. The primary error sources over temperature are the voltage reference (A_5), offset drift of the operational amplifiers (A_1 and A_2), and the RTD element.

RTD Signal Conditioning Path Using the SAR ADC

You can digitize changes in resistance of the RTD element over temperature with a SAR ADC. Figure 14-7 shows a current-excitation circuit for an RTD element. With this style of excitation, you can tune the magnitude of the current source to 1 mA or less by adjusting R_{REF}. With this circuit, a three-wire RTD element is selected (refer back to Figure 14-7). This configuration minimizes errors due to wire resistance and wire resistance drift over temperature. A_3 senses the voltage drop across the RTD element. A_4 then gains and filters the analog signal in preparation for the ADC input.

In the circuit using a SAR converter (Figure 14-7), the RTD element equals 100 Ω at 0°C. If the RTD senses temperature over from -200 to 600°C, the range of resistance from the RTD is nominally 23 Ω to 331 Ω. Since the RTD resistance range is relatively low, wire-resistance and wire-resistance change-over-temperature can skew the measurement of the RTD element. The three-wire RTD device reduces these errors.

The operational amplifier that contains A_3 subtracts wire-resistance error of R_{W1} and R_{W3}. In this configuration, R_1 and R_2 are equal and relatively high. You should select the value of R_1 to ensure that the leakage currents through the resistors do not introduce errors to the RTD element. The transfer function of this portion of the circuit is:

$$V_{OUT:A3} = (V_{IN} - V_{W1})\left(1 + \frac{R_2}{R_1}\right) - V_{IN}\left(\frac{R_2}{R_1}\right) \tag{14-4}$$

where $V_{IN} = V_{W1} + V_{RTD} + V_{W3}$, V_{Wx} is the voltage drop across the wire to and from the RTD, and $V_{OUT:A3}$ is the voltage at the output of A_3.

If $R_1 = R_2$ and $R_{W1} = R_{W3}$, the previous equation reduces to:

$$V_{OUT:A3} = V_{RTD} \tag{14-5}$$

A second-order, lowpass filter removes higher-frequency noise from the voltage signal at the output of A_3. This filter/gain stage is built using A_4, R_3, C_3, R_4, R_5, and C_4. The lowpass filter in this circuit should have a cut-off frequency below the mains frequency (<50 or $60\,Hz$), such as a cut-off of $10\,Hz$. This reduces amplifier and conducted noise as much as possible. You should choose the Chebyshev filter ($0.5\,dB$ ripple) because of its fast transition region in the frequency domain (see Chapter 6). The closest expected high-frequency noise in this circuit is $60\,Hz$. The attenuation of this filter at $60\,Hz$ is $39\,dB$ down from DC. The noise from the CMOS amplifiers could be as high as $29\,nV/\sqrt{Hz}$ (rms) @ $10\,kHz$. The noise due to the two amplifiers in the circuit signal path will be $48\,\mu V$ (rms) or $0.318\,mV$ (p-p) at the input to the 12-bit A/D converter. With a $10\,Hz$, second-order Chebyshev filter, only $\sim 2\,\mu V$ (p-p) remain in the signal. More critically, the noise that is injected by the mains frequency ($50\,Hz$ or $60\,Hz$) is reduced by $-24.5\,dB$ or $-27.9\,dB$ (inclusive). This is equal to an attenuation of $16.80\times$ or $23\times$ inclusive (see Chapter 8, "Noise: The Three Categories–Device, Conducted, and Emitted").

The LSB size of the 12-bit converter is equal to:

$$\begin{aligned} ADC_{LSB} &= \frac{V_{REF}}{2^{12}} \\ &= \frac{2.5\,V}{4096} \\ &= 0.610\,mV \end{aligned} \tag{14-6}$$

The sample speed of the SAR ADC can be as slow or as fast as need be. Your selection of this sampling speed will not affect the accuracy of the conversion.

Is the SAR ADC Right for This Temperature-Sensing Application?

The SAR ADC is a good fit for this type of application. You need to be willing to carefully gain the voltage from the resistive RTD, but the gain cells are easy to implement. The biggest challenge in this circuit is the current reference circuit. The SAR converter offers a low-power, low-cost solution that is easy to implement into this circuit.

RTD Signal Conditioning Path Using the Δ–Σ ADC

Delta-sigma A/D converters have an innate ability to resolve an analog input signal to a very small LSB voltage size. At first glance, high resolution doesn't seem to be an important specification for the RTD temperature sensor as long as you use an analog gain stage. To the contrary, close inspection of the interface circuit divulges a different story. The sensing element's output could be in the hundreds of millivolts. Worse yet, the output voltages that represent a change in temperature can be extremely low (sub-mV or μV). If the dynamic resolution of the A/D converter alone is relatively high, the total device count is lower by removing the front-end gain stage and reducing the complexity of the antialiasing filter (**Figure 14-8**).

For the given RTD of 100 Ω (@ 0°C) with an excitation current of 200 μA, the nominal full-scale output voltage range (−200°C to 600°C) would be 66.2 mV. The digitizing system that follows the RTD should reliably represent the temperature to 12-bit accuracy (given the error contributions of the RTD). The designer can choose to gain and actively filter the RTD voltage using analog techniques or put a digital engine to work: the Δ–Σ converter. This predominantly digital solution uses a passive antialiasing filter followed by the Δ–Σ device that gains and filters the signal with digital computational methods. The Δ–Σ converter makes this precision application possible because of its superior digital processing and system calibration capability.

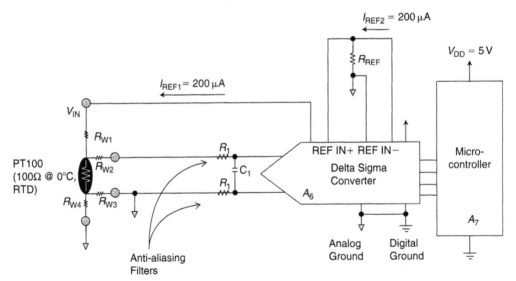

Figure 14-8: A Δ–Σ ADC directly digitizes the small signal from a four-terminal, RTD temperature sensor. In this circuit, a 200 μA current reference from the Δ–Σ ADC excites the temperature sensor. The 200 μA current source is ratiometric to the ADC reference voltage. The resistor/capacitive pairs (R_1/C_1) are used as antialiasing, lowpass filters.

The circuit in Figure 14-8 combines a high-precision, four-wire lead PRTD with a Δ–Σ converter. The reference current from the Δ–Σ converter excites the RTD, using two of the four pins. Using the remaining two wires of the thermal element, the voltage signal across the RTD is sensed using the differential inputs of the Δ–Σ ADC. This technique establishes a ratiometric relationship between the voltage reference to the ADC and the RTD temperature sensor.

In this application, I am using the AD7713 Δ–Σ ADC from Analog Devices, Inc. This device is perfectly suited for RTD sensing circuits because of the two current references. Additionally, this device will generate a 24-bit code for low-level voltage signals. This Δ–Σ ADC has self-calibration, system calibration, and background calibration options.

The internal Δ–Σ noise reduction techniques improve the performance of the circuit. The data rate of the converter is 10 Hz (as programmed) to reduce interference from the mains frequency as well as reduce aliasing of higher-frequency noise into the final signal output. This low frequency data rate is possible because of the slow responsiveness of the RTD element to temperature.

If the layout is good, the Δ–Σ ADC will have an effective resolution of 20 bits (rms) or 17.27 noise-free bits. If that is the case, you can select a subset of the possible bits. For instance, a 17-bit conversion has 32 different places where 12 bits can be extracted. This is how you eliminate the gain stage in Figure 14-2 of the SAR converter circuit.

Is the Δ–Σ ADC Right for This Temperature-Sensing Application?

The answer is yes. The sample rate of this type of converter matches the sensor and the physical event. Since this converter will give you a high number of noise-free bits, you can ignore the ones that don't fit into your range of interest and still get a 12-bit conversion out of the deal.

This is an example of how the Δ–Σ ADC is best suited for an RTD temperature measurement application. Although the Δ–Σ ADC is appropriate here, a different application with a different sensor could require a different Δ–Σ converter or a SAR ADC.

Measuring Pressure: SAR Converter or Δ–Σ Solution?

The second most common physical entity that we tend to want to measure is pressure. Pressure measurement devices can be classified into two groups: those for which physical pressure is the only source and those that require electrical excitation. The mechanical-style devices that are only excited by physical pressure, such as bellows, diaphragms, bourdons, tubes, or manometers, are usually suitable for purely mechanical systems. With these devices, a change in pressure will initiate a mechanical reaction—for example, a change in the position of mechanical arm or the level of liquid in a tube.

Electrically excited pressure sensors are most suited to a microcontroller or microprocessor environment. These kinds of sensors can be piezoresistive, linear variable differential transformers (LVDT), or capacitive sensors. Usually when you measure pressure you use the piezoresistive sensor.

The Piezoresistive Pressure Sensor

Pressure changes are usually slow, so lower-speed conversions are acceptable. Another characteristic of measuring pressure is that the actual sensor itself has a differential output device. This differential output is very helpful when you want to reject common-mode environmental noise. Therefore, differential analog input devices are very useful.

The piezoresistive is a solid-state, monolithic sensor that has silicon processing. *Piezo* means pressure; *resistance* means opposition to a DC current flow. There are 300 to 500 piezoresistive pressure sensors per wafer. Since these wafers generate a large number of sensors, they are less expensive than mechanical sensors.

A pressure reference consists of a cavity that is fabricated from two wafers sealed together. The topside of this fabricated sensor is the resistive material, and the bottom is the diaphragm. The piezoresistive bridges shown in **Figure 14-9** can have a voltage excitation or current excitation applied. Although the magnitude of excitation (whether it is voltage or current) affects the dynamic range of the output of the sensor, the maximum difference between V_{OUT+} and V_{OUT-} generally ranges from tens of millivolts to several hundred millivolts. The electronics that follow the pressure sensor change the differential-output, signal to a single-ended signal. The signal chain then gains and filters the signal in preparation for digitization.

The model for this sensor is a four-resistor element, the Wheatstone bridge (Figure 14-9). Four-element bridges have better sensitivity compared to a single element or two element sensors. If we apply a positive differential pressure to the four-element bridge, two of the

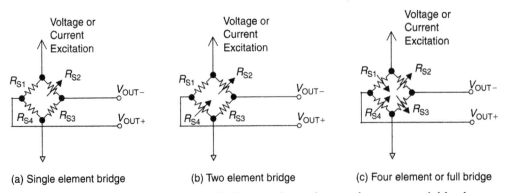

(a) Single element bridge   (b) Two element bridge   (c) Four element or full bridge

Figure 14-9: The resistive Wheatstone bridge configuration can have one variable element (a), two elements that vary with excitation (b), or four elements (c).

elements respond by compressing and the other two change to a tension state. When you apply a negative differential pressure to the sensor, the diaphragm moves in the opposite direction. Additionally, the resistors that were compressed go into a tension state, whereas the resistors that were in a tension state change into a compression state. Piezoresistive pressure sensors might or might not have an internal pressure reference.

The Pressure Sensor Signal Conditioning Path Using a SAR ADC

Figure 14-10 shows an example circuit of how to interface with this type of sensor with a SAR ADC.

There are several ways of capturing the small, differential, output signal of the sensor and transforming it into a usable digital code. One approach is to take the small differential output of the bridge, gain it, and convert it from differential to single-ended with an instrumentation amplifier (Figure 14-10). The signal then passes through a second-order, 10 Hz, lowpass filter. The lowpass filter eliminates out-of-band noise and unwanted frequencies in the system before the A/D conversion occurs. The stand-alone, 12-bit ADC follows the lowpass filter. The ADC transforms the analog signal into a usable digital code.

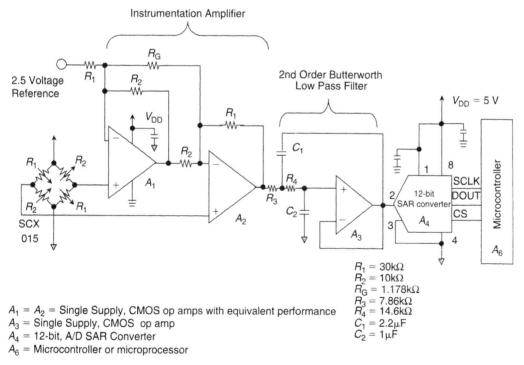

$A_1 = A_2 =$ Single Supply, CMOS op amps with equivalent performance
$A_3 =$ Single Supply, CMOS op amp
$A_4 =$ 12-bit, A/D SAR Converter
$A_6 =$ Microcontroller or microprocessor

$R_1 = 30k\Omega$
$R_2 = 10k\Omega$
$R_G = 1.178k\Omega$
$R_3 = 7.86k\Omega$
$R_4 = 14.6k\Omega$
$C_1 = 2.2\mu F$
$C_2 = 1\mu F$

Figure 14-10: The voltage at the output of the SCX015 pressure sensor is gained (Gain = 37.7 V/V) by the instrumentation amplifier (A_1 and A_2). Then a second-order 10 Hz, lowpass filter (A_3) filters the signal. Finally, a 12-bit ADC (A_4) digitizes the signal.

The microcontroller takes the converter code, further calibrates it, and translates if need be, for display purposes. In this signal path, one analog filter is required. This analog filter follows the instrumentation amplifier.

Is the SAR ADC Right for This Pressure-Sensing Application?

This circuit configuration requires a fair amount of analog circuitry, prior to the ADC (Figure 14-10). However, if the application requires a higher sample rate, this converter might give you the best solution.

Pressure Sensor Signal Conditioning Path Using a Δ–Σ ADC

A second way that the differential signal from the SCX015 pressure sensor is captured uses a Δ–Σ ADC as the core converter in the circuit. In this circuit (**Figure 14-11**), a single-pole, lowpass filter removes high-frequency noise for the differential output of the pressure sensor (SCX015). Immediately following these filters, the Δ–Σ ADC accepts the differential signal at its input. The resolution of the Δ–Σ ADC is considerably higher (usually 24 bits) than the resolution of the SAR converter in Figure 14-10. You can use this converter in the same manner as the converter in Figure 14-8. If you will remember, we did not use all the bits at the output of the Δ–Σ in Figure 14-8. We only selected the 12 bits that we needed.

Is the Δ–Σ ADC Right for this Pressure Sensing Application?

Bridge sensors are a perfect fit to the Δ–Σ ADC (Figure 14-11). The converter is able to accept a differential input while rejecting common-mode noise. This alone replaces the instrumentation amplifier illustrated in Figure 14-10. Additionally, the number of bits of

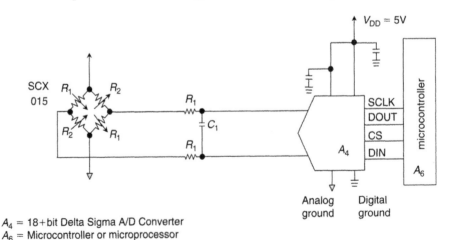

A_4 = 18+bit Delta Sigma A/D Converter
A_6 = Microcontroller or microprocessor

Figure 14-11: The voltage at the output of the SCX015 pressure sensor is filtered at both inputs to the Δ–Σ ADC by single-pole lowpass filters (R_1/C_1) and digitized by the Δ–Σ ADC (A_4).

this type of converter allows the designer to eliminate the analog gain stage, making this type of device a perfect fit.

Photodiode Applications

A close third on the popularity list of sensing circuits is the photosensing application. Photodiodes bridge the gap between light and electronics. Many times precision applications such as CT scanners, blood analyzers, smoke detectors, position sensors, IR pyrometers, and chromatographs utilize the basic transimpedance amplifier circuit that transforms light energy into a usable electrical voltage. These circuits use photodiodes to capture the light energy and transform it to a small current. This current is proportional to the level of illumination from the light source. A preamplifier then converts the current (in amperes) from the photodiode sensor into a usable voltage level.

Photosensing Signal Conditioning Path Using a SAR ADC

A practical way to design a precision photosensing circuit is to place the photodiode in a photovoltaic mode. A photodiode, in its photovoltaic mode, has 0 volts across it. This can be done by placing the device across the inputs of an amplifier and a resistor/capacitor pair in the feedback loop. **Figure 14-12** shows a single-supply circuit implementation of a photosensing circuit.

In this SAR photodetector circuit (Figure 14-12), the light source illuminates the photodiode, causing diode current to flow from cathode to anode. Since the input impedance of the inverting input of the CMOS amplifier is extremely high, the current generated by the photodiode flows through the feedback resistor, R_F. The current to voltage transfer function of this circuit is:

$$V_{OUT} = I_{SC} \times R_F + V_X$$

(14-7)

with a single-pole at $1/(2\pi R_F C_F)$, where V_{OUT} is the voltage at the output of the operational amplifier in volts, I_{SC} is the current produced by the photodiode with units in amperes, R_F is the feedback resistor with units in ohms, and C_F is the feedback capacitor with units in farads.

Once the signal becomes a voltage, it is easy to get a digital representation with an ADC. SAR converters are well suited for this circuit due to their higher sampling speed. Some current-input ADCs on the market also support this type of application.

In this circuit, the amplifiers A_1 and A_2 should have similar AC performance and noise specifications. It is best to make them the same amplifier chip. I am choosing the capacitor value, C_F, so that the amplifier circuit (A_2) is stable.

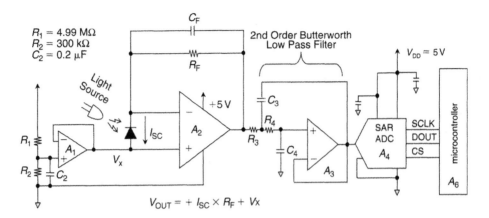

$$V_{OUT} = + I_{SC} \times R_F + V_X$$

$A_1 = A_2$ = single supply CMOS amplifier with similar frequency performance.
A_3 = single supply CMOS amplifier
A_4 = 12-bit SAR Analog-to-Digital Converter
A_6 = Microcontroller or microprocessor

Figure 14-12: The light source impinging on a photodiode generates a reverse current from the feedback resistor, R_F. The voltage diode's anode as well as the noninverting input of the amplifier are raised at a higher voltage than ground (~300 mV) to avoid amplifier output clamping near ground. A lowpass filter removes higher frequencies embedded in this output signal of the amplifier. The SAR ADC then receives this signal.

Is the SAR ADC Right for This Photodetection Application?

The SAR ADC is very appropriate for this application due to its conversion speed, compared to the Δ–Σ ADC. The conversion speeds of the SAR converter report the optical events more reliably, whereas the Δ–Σ ADC will generally report DC type signals.

Photosensing Signal Conditioning Path Using a Δ–Σ ADC

Figure 14-13 shows an example of this same application using a Δ–Σ converter. You can see that the Δ–Σ converter has a significant impact on the part count (between Figs. 14-12 and 14-13) in a positive way. The converter in Figure 14-13 manages the acquisition of the photodiode signal and converts it, using a switched-capacitor input. The device immediately converts the signal to a digital representation with the Δ–Σ modulator and digital filter.

The design of Δ–Σ ADC in Figure 14-13 specifically targets transimpedance amplifiers. From this example, you can see the specialized nature of some Δ–Σ converters. This particular device (DDC114 from Texas Instruments) only targets photosensing applications.

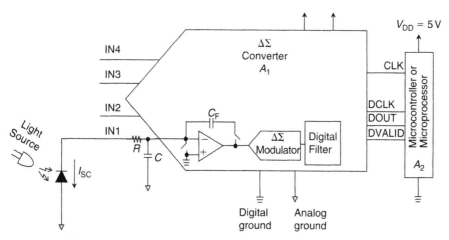

Figure 14-13: In this application circuit, the $\Delta-\Sigma$ converter replaces the active antialiasing filter in Figure 14-12 with a passive R/C filter. The $\Delta-\Sigma$ converter also replaces the transimpedance amplifier (A_2 and R_f). This new configuration has only two active devices plus the photosensor. In Figure 14-12, the part count is four active devices and eight passive parts.

Is the $\Delta-\Sigma$ ADC Right for This Photodetection Application?

The $\Delta-\Sigma$ ADC is well suited for this application. One advantage that this converter brings to the application is a reduced chip count.

Motor Control Solutions

The transition from mechanical, hydraulic, and DC drives to higher-efficiency AC motor drives is triggering an increase in design updating in factory equipment. The traditional AC motor-control system design used bulky and expensive current transformers for isolation and sophisticated digitizing systems to ensure that the equipment performed in a well-behaved manner. The A/D conversions in these AC induction motor systems were capable of digitizing a larger full-scale range than was needed for the application and in packages that required more board space than the design could afford. The current transformers (also called *Hall effect sensors*) required an additional assembly step for board installation, increasing the overall expense, which new designs don't easily tolerate.

New designs are approaching the isolation and digitization challenges presented by the motor control-application system from a different perspective in an attempt to lower costs and improve efficiency. The discrete, low-power, low-cost, differential input ADC maintains signal integrity in a potentially noisy environment. These sockets have 8-pin integrated devices that include the functions of common-mode rejection with the differential inputs, a gain stage, a sample/hold amplifier, and an ADC. All this functionality comes in package sizes as small as an MSOP. Digital optocouplers are replacing the bulky Hall effect sensors. These changes in the AC induction motor-sensing interface allow a more efficient, lower-cost, smaller real-estate solution to the age-old AC motor control design problem.

An AC induction motor control system block diagram is shown in **Figure 14-14**. In this motor-control servo-loop, a pulse width modulator (PWM) drive circuit switches the MOSFETs, which sends currents through the legs of the AC motor. A typical higher-speed PWM motor control system uses fast-switching, high-power bipolar transistors, IGBTs, or MOSFETs to drive the three alternating currents through the AC motor. A SAR ADC digitizes the AC motor speed, position, and all three motor-driving currents of the motor. The DSP chip receives the digitized information. The DSP engine evaluates the information received and initiates the next set of instructions to the PWM drive. The purpose of this local control system is to enhance the efficiency of the motor and to prevent catastrophic problems under all conditions, including temperature and load variations. The motor-control servo-loop implements this stability by adjusting the currents through the motor.

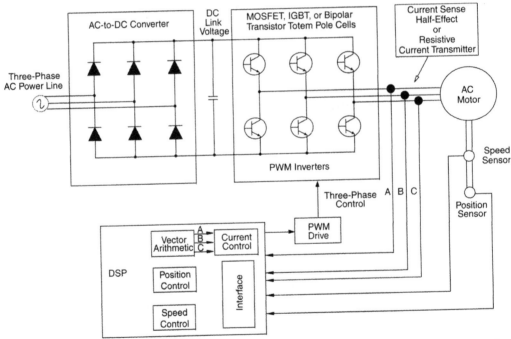

Figure 14-14: This is an AC motor control system configuration for three-phase control. The MOSFET, IGBT, or bipolar transistor switches drive current through the AC motor. The "current sensors" sense this current. The ADC changes the signal from the current sensors into a digital code. The DSP engine then receives the digital code from the ADC. The DSP engine calculates appropriate pulse width modulator (PWM) instructions, again driving the totem pole power transistors into action.

The PWM motor control system drives high-power bipolar transistors, IGBTs, or power MOSFETs that in turn performed the "push-pull" action through the three legs of the AC motor. Since there are three legs and 360 degrees available in the clock cycle, each leg is switched 120 degrees apart from the other two. The timing between the signals and the magnitude of the currents provide critical information to the control loop when motor efficiency is a priority. The power devices that drive the motor are capable of handling currents up to 20A and voltage drops across the device of several hundred volts to 600 V. An alternative design approach would be to simultaneously sense the current in two of the three legs to the motor. Sensing systems that monitor only two legs at the motor are less expensive to implement but are more susceptible to errors caused by grounding problems. A seemingly trivial circuit layout can be deceptive because of the higher than usual currents that are common in motor control circuits. Although the circuits that monitor three legs are slightly more expensive because of the addition sensor channel, they are the more accurate of the two.

Switching speeds for the totem pole power transistors can typically range from 2 kHz up to 200 kHz. This places high demands on the driver circuit's ability to make a full-scale change, particularly with motors that operate with higher voltages.

The motor-monitoring system is a critical link in the control system. This control system requires two different design approaches: the current-sensing portion and the position/speed sensors. The current-sensing circuitry simultaneously senses the three currents that are going to and from the motor. It consists of a sensing device, signal condition portion, and isolation device.

The ADC requirements vary depending on the application requirements. In an application in which higher sensitivity is required, a 12-bit converter is appropriate. In applications in which lower sensitivity is adequate, 8-bit to 10-bit converters are common. The controller receives the digital information from the ADC. The controller then assesses the condition of the motors. This processor(s) employ complex mathematics to perform vector transformations and PWM timing to finally regulate the motor to optimum efficiency. The processor(s) then feed back corrective action information to the PWM driver of the MOSFETs.

The motor controller's current sensor portion of the motor servo-loop presents three interesting design challenges. The first challenge is that the motor servo-loop requires that two to three signals are simultaneously sampled and converted to a digital representation within the limited cycle time of the motor drivers. High-end systems demand that the overall accuracy of these conversion systems is 12 bits. Simultaneous sampling can be an awkward problem to solve. Some designs use a front-end array of three sample/hold amplifiers and one high-speed 12-bit converter. The settling time of the sample/hold amplifiers and the throughput rate of the one converter limit the throughput rate of this configuration. To digitize a complete picture of the motor at a particular instant in time, the converter must digitize two or three separate voltages. Each A/D conversion added to this digitization process adds additional time to the throughput of the sensor stage.

A second issue that presents design challenges in the sensor stage circuit is the conversion of the sensor current to a good system-level signal. An example of the MOSFET switching section of the sensor stage circuit is shown in **Figure 14-15b**. The MOSFET devices are configured in a totem pole arrangement using one leg of the motor as the output load. When the totem pole switches from high to low (or low to high), the current output (IO) magnitude and direction change. The motor sensor is positioned at the output of the totem pole switch to capture the changes in IO. The sensor in these circuits has negligible effects on the operation of the motor. As an example, a small-value power resistor (R_{SENSE}), as shown in Figure 14-15b, can sense the pertinent information with very little voltage drop and power loss. In contrast, the voltage change at the output of the totem pole can be rather large. So, the trick is to sense the small voltage drop across R_{SENSE} while rejecting the large voltage excursions of the output of the MOSFET totem pole. Once the large common-mode interference from the motor is rejected, the signal is gained to a usable level for the ADC.

Third, if the signal is in the analog domain, electromagnetic interference (EMI) becomes an issue. The changes in the current magnitude and direction through the motor legs are an analog signal. Careful layout and design will save headaches later in the design. If the signal is in the analog domain, it should be a differential signal. The analog portion of the motor-sensing circuitry can most easily reject the noise present if the system has differential signals throughout. The best scenario would be to digitize the signal as soon as possible.

The previously mentioned transition from DC drives to higher-efficiency AC motors has been motivated by lowered costs in the AC motor control system coupled with the better power efficiency that has always been an AC motor characteristic. The target areas for cost reduction have been with the sensing circuitry, the ADC(s), and the microcontroller. Cost reduction of the sensing circuitry and the A/D conversion portion of the circuit have been largely driven by technology strides in terms of higher integration and reduced chip layout dimensions. The combination of these two phenomena has led to reduced chip sizes and a lower chip count for the application. In both cases, the end user enjoys these cost savings. **Figure 14-16** shows one possible design approach for the current sensing circuitry and ADC conversion portion.

In the sensing circuit shown in Figure 14-16, the ADCs are directly interfaced to the sensing resistors, R_{SENSE}. The input ranges of the converters are programmable by the voltage reference to the devices, ensuring that the full dynamic range of the ADCs is used. With the ADC in this circuit, the voltage reference input is 200 mV. This fully eliminates the need for an analog gain stage. Although the ADC is a 12-bit converter,

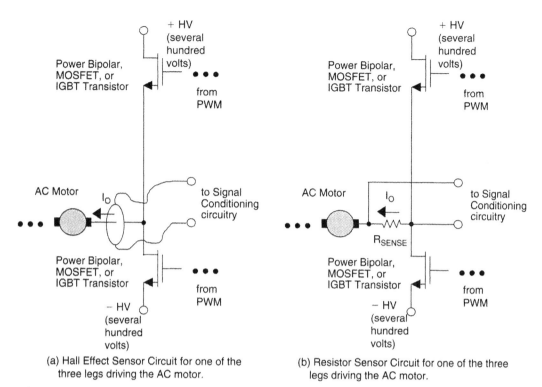

(a) Hall Effect Sensor Circuit for one of the
three legs driving the AC motor.

(b) Resistor Sensor Circuit for one of the three
legs driving the AC motor.

**Figure 14-15: The Hall effect sensor (a) magnetically senses the changes in current initiated
by the PWM inverter in one of the three driver legs to the motor. The transfer function of
a Hall effect sensor is current to current. The conditioning circuitry that follows is required
to convert the output current of the Hall effect sensor to a voltage. A sensing small-value
power resistor (b) can replace the Hall effect sensor. This resistor senses the changes in
current to and from the AC motor and immediately converts it to a voltage.**

the effective accuracy of the device with a 100 mV reference is 11 bits. The ADC
differentially senses the sense resistor, now, directly. The differential inputs of the ADCs
assist in maintaining good signal integrity through the completion of the digitizing
process.

The chosen method of isolation for this configuration is a digital optocoupler instead
of a Hall effect sensor. The digital section of the ADC is isolated from the processor
by way of three optocouplers. The digital output of each of the three converters is sent
through an array of three optocouplers, then to the microcontroller. All nine digital
optocouplers in this circuit must have excellent transient immunity. The throughput rate
of the optocouplers must also be fast enough for the clock input and the data output of the
ADC. In this application, all three converters are triggered at the same time. This meets
the design requirements of three simultaneous sampling channels, thereby reducing the
throughput rate of the sensor circuitry.

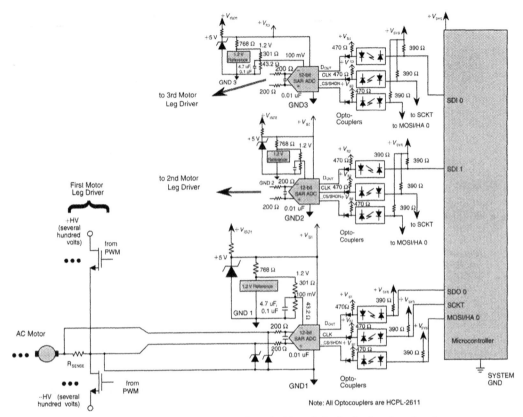

Figure 14-16: The ADC, on the isolated side of the isolation barrier, is part of the motor control sensing circuitry. The optocouplers transmit a digital signal across the barrier. Since an ADC digitizes the signal at the sensor, this circuit approach allows for better noise immunity.

A Few Final Words

So, here we are at the end of a chapter where we have discussed the ADCs for your circuits and how they fit with various analog signals. In particular, we concentrated on the SAR ADC and the $\Delta{-}\Sigma$ ADC.

It turns out that through this discussion, we were forced to look at the entire signal chain. You will find that this is a requirement when you design with an ADC in your circuit. It is one thing to find the "perfect" converter, and another thing to find the right converter for your signal chain.

As I mentioned in Chapter 13, the ADC is almost an afterthought. Your real task is to figure out what you need done in your signal chain and then pick the right converter, not the other way around. The punchline is, if the ADC reports incorrect data, the controller or processor will never know unless you write code that identifies errors in your analog system.

Generally, in terms of selecting the right product for the applications, SAR converters will require an antialiasing filter of second to fifth order, where the Δ–Σ ADC only requires an R/C pair to perform this kind of filtering. SAR converters have lower resolutions, but they are faster than the Δ–Σ ADC. If the signal is slow enough, the Δ–Σ ADC can easily fit the bill while eliminating a lot of the analog front end.

The converter should fit the application. Remember, with the ADC the best defense is a good, well-informed offense.

References

Baker, Bonnie C., "Anti-aliasing analog filters for data acquisition systems," AN699, Microchip Technology Inc.

————, "Number of bits vs. LSB errors," *EDN Magazine*, July 8, 2004 .

————, "Single supply temperature sensing with thermocouples," AN684, Microchip Technology, Inc.

Burr-Brown Corp., 1994 Application Seminar, Chapter 1.

Hyde, Darrell, "Evaluating thin-film RTD stability," *Sensors*, Oct. 1997, p. 79.

Li, Xum, "Producing higher accuracy from SPRTs (standard platinum resistance thermometers)," *Measurement & Control*, June 1996, p. 118.

Madden, J.R., "Refresher on resistance temperature devices," *Sensors*, Sept. 1997, p. 66.

Omega Engineering Inc., *OMEGA Temperature Measurement Handbook & Encyclopedia*, Stamford, CT., 1996.

Omega Engineering Inc., "Practical Temperature Measurements," *Omega Catalog*, p. Z-11.

Oppenheim, Schafer, *Discrete-Time Signal Processing*. Prentice Hall, Inc., 1989.

Steele, Jerry, "Get maximum accuracy from temperature sensors," *Electronic Design*, Aug. 19, 1996, p. 99.

Sulciner, James, "Understanding and using PRTD technology, part I: history, principles, and designs," *Sensors*, Aug. 1996, p. 10.

Sulciner, James, "Understanding and using PRTD technology, part II: selection," *Sensors*, Sept. 1996, p. 43.

Tandeske, Duane, *Pressure Sensors*, Marcel Dekker, Inc., 1991.

CHAPTER 15

Working the Analog Problem From the Digital Domain

Bonnie Baker

Here Bonnie shows how a comparator and a few Rs and Cs can help a microcontroller make a surprisingly good Digital-to-Analog Converter (DAC)—or a slow but simple ADC. /rap

In This Chapter

When you move your analog problem solving from hardware to firmware, a few useful processor or controller peripherals will take you a long way. These peripherals include the pulse width modulator (PWM), the comparator, a timer or two, and the I/O gates. Although all these peripherals sound like they are digital, we are going to use them to an analog advantage in our circuits.

The PWM might or might not be a part of your controller or processor arsenal. If the PWM function is an internal function from your digital chip manufacturer, you can use it to produce fairly accurate voltage references. If you don't have a PWM on board, you can generate the signal in firmware. The voltage source generated by this tool is as accurate as your on-board timer and power supply voltage. The only thing this voltage reference requires is a PWM generator and an analog filter.

You will find that there is probably an integrated comparator in your controller or processor. If not, an I/O gate can take over this function for some applications in which you are looking for a trigger or level indicator of an analog signal. I prefer working with a comparator because the threshold is usually more predictable than your run-of-the-mill digital I/O pins. You can design functions such as a window comparator or just a standard comparator with this peripheral.

An internal timer (or two) is absolutely necessary if you plan to implement the functions that we are going to talk about in this chapter. The timer of your controller or processor intimately connects to the device's clock. The clock of the device can be as accurate as you would like. Some controller or processor clocks are an internal R/C pair. These types of clocks are accurate only to a point. Other clocks inside controllers or processors are much more accurate than their R/C cousins. The accuracy of these clocks can be as good as 1% to 2% over temperature. If you are interested in finessing the accuracy of the clock, you will have to resort to using a resonator or crystal oscillator. Accurate clocks come in handy if you are implementing digital filters in the controller or processor and you want to reject the noise that is riding on a particular frequency. A popular choice for a rejection frequency of a digital filter is 50 Hz or 60 Hz. But clock accuracy is not a critical specification if you are designing a D/A converter. With this type of circuit, clock jitter affects the accuracy.

Finally, to close this topic of converting analog to digital, this chapter will use passive components. In particular, we use resistors and capacitors. The techniques and concepts that I show you are not new. As a matter of fact, you learned about them in your first year of college. I always smile when I am able to use something from those days. When I am able to do that, it seems that simplicity is in charge.

Pulse Width Modulator (PWM) Used as a Digital-to-Analog Converter

A PWM comes in handy in a variety of functions and applications, but using it to generate an analog reference voltage is the most useful thing that I can think of (being the analog engineer that I am). In this section of the chapter, I show you how to build a "poor man's" digital-to-analog converter (DAC). The controller/DAC uses very few external parts, which makes the cost of this DAC very low.

Looking at This Reference in the Time Domain

It is possible to use the PWM to create an analog voltage. You will need not only the PWM function but an analog filter to accomplish this task. **Figure 15-1** shows the timing diagram of a PWM module.

With a functioning PWM, the clock sets the fundamental frequency to the controller. After that, you can adjust the duty cycle by changing the ratio of $T_{ON}:T_{OFF}$. Figure 15-1 shows the output of a PWM as it would appear at the output of an I/O gate. When the signal is ON, the output voltage is full scale. The actual value of this full-scale voltage is dependent on the microcontroller or microprocessor and power supply voltage. If the power supply (V_{DD}) is equal to 5 V, the ON-voltage magnitude is ideally 5 V but in reality a few hundred millivolts below 5 V. When the signal is OFF, the output voltage is ideally 0 V but in reality a few hundreds of millivolts above ground. The actual value of these

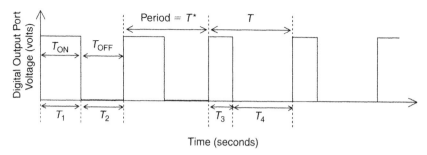

*The fundamental frequency of PWM signal = 1/T

Figure 15-1: A typical PWM waveform has a consistent period (T) with a varying ratio of on and off times. You can apply an analog filter to the output of the PWM signal to generate an analog voltage. Once you filter the PWM signal, longer T_{ON} times will generate higher voltages.

voltages is dependent on the specific controller or processor that you have chosen to use. The actual values of these voltages will affect the accuracy of the output analog voltage that we are going to create with an analog filter.

The number of divisions (K) that your clock can produce during the PWM period determines one part of the accuracy and granularity of your reference. For example, if your clock can only divide the PWM period (T) by 64, the highest granularity that you will get out of your adjustable voltage reference is $1/K = 1/64$ of your full-scale range. For instance, if you are using a 5 V power supply, the smallest (ideal) change in the reference voltage is:

$$
\begin{aligned}
\Delta V_{REF-MIN} &= V_{DD}/K \\
\Delta V_{REF-MIN} &= 5\,V/64 \\
\Delta V_{REF-MIN} &= 78\,mV
\end{aligned}
\qquad [15\text{-}1]
$$

This assumes that your PWM cycles between 0 V and 5 V. In this particular example, this is a 6-bit analog adjustable reference (or DAC), where $64 = 2^6$.

Based on the number of time divisions in the period, T, the ideal number of bits (or the resolution) of this DAC is:

$$
\begin{aligned}
\text{DAC resolution} &= \log\,(K)/\log\,(2) \quad \text{(in bits)} \\
\text{DAC accuracy} &= 6.02\,N + 1.76 \quad \text{(in dB, where } N \text{ is the number of bits)}
\end{aligned}
\qquad [15\text{-}2]
$$

The only problem with this voltage reference at this point is that the signal is still in the digital domain.

Changing This Digital Signal to Analog

An analog filter after the PWM pulse generates a DC voltage. The value of this voltage depends on the ratio of $T_{ON}:T_{OFF}$ and the power supply voltage. If the signal is ON more than it is OFF, after the filter, the output voltage will be above midscale (midscale=$V_{DD}/2$). Alternatively, if the signal is ON less than it is OFF, the output voltage will be below midscale. **Figure 15-2** shows this relationship graphically.

If you properly filter the PWM signal on the output port of the controller, it is theoretically possible to produce any analog voltage between ground and V_{DD} at the output of the filter. Wouldn't it be wonderful if the ideal theories stuck? But this system has errors. They are the quantization error from the controller clock, the output swing of the I/O gate, the ripple rejection of the lowpass filter, and any offset errors from the lowpass, filtering amplifier. If we are in a single-supply environment, the output swing of the amplifier will never reach the rails, so you will lose a few hundred millivolts near ground and the power supply. But let's see how close we can get to reality.

Figure 15-3 shows the suggested circuit diagram for the PWM voltage reference.

The most challenging part of this design is to decide on the kind of filter to use. In this discussion I assume that you have read through Chapter 6, so I won't engage in another discussion on the analog filter terminology.

There are two filter specifications at the top of the list to address: corner frequency and filter order. You can reduce the higher-frequency ripple by making the corner frequency lower than the fundamental frequency of the PWM. The magnitude of that reduction depends on your application requirements. If you can tolerate a ripple of 5% (which is essentially a little better than a 3-bit system), your filter requirements will be relaxed. In contrast, if you need a reference that complements a 9-bit system, a ripple that is a little less than 0.1% is the way to go.

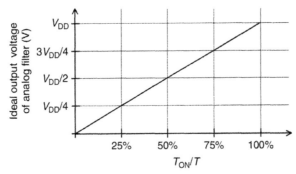

Figure 15-2: If a PWM signal is sent to an output port of the microcontroller and then through an analog lowpass filter, the output voltage of the filter with respect to V_{DD} is ratio-matched to the ON time (T_{ON}/T) of the PWM signal.

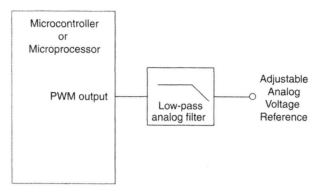

Figure 15-3: This is the hardware implementation of a PWM voltage reference using the controller or processor to generate the PWM signal. The analog lowpass filter changes the PWM signal to a DC voltage.

Defining Your Analog Lowpass Filter for Your PWM-DAC

Figure 15-1 shows the output signal of the PWM generator. Although the ratio of $T_{ON} : T_{OFF}$ can change, the fundamental frequency never does. The fundamental frequency of the signal in Figure 15-1 is equal to $1/T$. If you want to look at this signal in the frequency domain, an FFT graph is a great tool to use.

Figure 15-4 shows a simplified FFT plot of the signal in Figure 15-1.

In Figure 15-4, you can see that the FFT plot breaks the response of the PWM into its equivalent frequencies. Also in Figure 15-4 there is the frequency response of a lowpass filter. The corner frequency of this filter is lower than the PWM fundamental frequency. Your system requirements determine the analog filter, corner frequency, and order. Let's look at an example with some numbers to help us understand this a little better.

Using the following parameters:

PWM fundamental frequency, f_{PWM} = 20 kHz
Required minimum ripple, A_{STOP} = −56 dB (8-bit accuracy to $\pm\frac{1}{4}$ LSB) [15-3]

The calculation of a single-pole analog filter for this circuit is straightforward. This formula will give you your corner frequency:

$$f_{C\text{-1st order filter}} = \frac{f_{PWM}}{\sqrt{((10^{-ASTOP/20})^2 - 1))}}$$

$$f_{C\text{-1st order filter}} = \frac{20\,kHz}{\sqrt{((10^{56/20})^2 - 1))}}$$

$$f_{C\text{-1st order filter}} = 31\,Hz$$
[15-4]

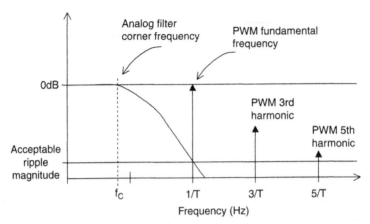

Figure 15-4: The trick to reducing the ripple at the output of the PWM generator is to apply an analog filter that averages this switching frequency to a DC result. The primary frequency generated at the output of the PWM generator is equal to $1/T$. The primary frequency also has the highest magnitude of all the harmonics. So when we're designing the analog lowpass filter, this fundamental frequency response dominates the calculations and results.

Since this is a single-pole filter, the circuit should have an R/C pair followed by a buffer amplifier. To achieve a 31 Hz corner frequency, I choose an arbitrary capacitor value of 1 µF. With this value the required resistor would be:

$$R_{31\,Hz\;filter} = 1/(2\pi\,f_C\,C_{31\,Hz\;filter})$$
$$R_{31\,Hz\;filter} = 1/(2\pi\,31\;Hz \times 1\,\mu F) \qquad\qquad [15\text{-}5]$$
$$R_{31\,Hz\;filter} = 5.184\;k\Omega \text{ (a 5 k}\Omega \text{ resistor will do the trick)}$$

If you need to back calculate the number of bits that you are actually getting out of your filter, you can use this formula:

$$\text{Resolution of system in bits} = (A_{STOP} - 1.76)/6.02 \qquad\qquad [15\text{-}6]$$

Figure 15-5 shows the circuit diagram for this lowpass filter.

This circuit implementation uses an amplifier with a single-pole implemented in the feedback loop (R_2 and C_2). There are two reasons to select this circuit configuration:

An R/C filter that is referenced to ground has considerable clipping. Erroneous signals are generated to the output.

Most voltage references provide a low resistance output. You can only achieve this if you use an amplifier in the circuit.

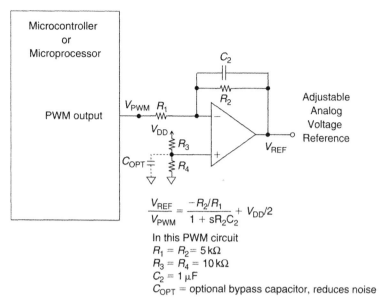

$$\frac{V_{REF}}{V_{PWM}} = \frac{-R_2/R_1}{1 + sR_2C_2} + V_{DD}/2$$

In this PWM circuit
$R_1 = R_2 = 5\,k\Omega$
$R_3 = R_4 = 10\,k\Omega$
$C_2 = 1\,\mu F$
C_{OPT} = optional bypass capacitor, reduces noise

Figure 15-5: The first-order filter uses an inverting amplifier configuration with an R/C pair in its feedback loop. This is an inverting amplifier circuit in a single-supply system. In this environment, you should reference the signal to halfway between V_{DD} and ground by using R_3 and R_4 in a voltage divider configuration.

The amplifier introduces DC errors to your D/A converter. The amplifier's offset voltage goes directly to the amplifier output. This offset error will appear on every D/A converter output setting. In other words, if you expect a D/A converter output of 1.0 V and the offset voltage of your amplifier is 5 mV, the actual output of your D/A converter will be 0.990 V (assuming you are using the inverting configuration in Figure 15-5).

Besides the offset error of the amplifier, the output swing of the amplifier is limited near the positive and negative rails (in a single-supply environment the negative rail is usually ground). There are two approaches to defining output voltage swing. The output current determines how close the output can go to the rail (V_{OH} and V_{OL}); this is the most common specification. These output swing values are usually tens of millivolts from the rails. The definition of the amplifier's open-loop gain (A_{OL}) specification is over a smaller output swing. This smaller range is within the linear output voltage range of the amplifier. These output swing values are usually hundreds of millivolts from the rail. The output voltages of the DAC in these regions will manifest these amplifier errors.

If you need your voltage reference to remain stable under transient conditions, you might want to increase the filter corner-frequency or order. If this is the case, a higher-order filter is a good alternative because you already have an amplifier in the circuit. Designing these filters is easy if you use the free lowpass filter software from various operational amplifier manufacturers. Some of the tools you might want to consider are:

Active Filter Synthesis Program	www.circuitsim.com
FilterPro Program	www.ti.com
FilterLab Program	www.microchip.com
FilterCAD Program	www.linear-tech.com
FilterWizard Program	www.analog.com
Webench Filter Program	www.national.com/appinfo/amps/webench_filters.html

For more information about analog filters, definitions, and design, refer to Chapters 6, 7, 10, and 11.

Pulling the Time Domain and Frequency Domain Together

If you'll recall the time domain discussion, we found that the number of clock divisions that were possible throughout the period, T, would affect the accuracy of your reference. We were able to define the accuracy in terms of bits and decibels. The discussion was in terms of time and voltage.

In contrast, when we talked about the accuracy of this reference in the analog domain we quantified the accuracy of the system in terms of decibels (dB). We did this during the examination of the response of the PWM and filter in the frequency domain. We then were able to connect the desired decibels to bits.

Assuming an ideal lowpass filter:

$$\Delta V_{REF-MIN} = V_{DD}/K \qquad (K \text{ is number of time division in } T)$$
$$\text{DAC resolution} = \log (K)/\log (2) \qquad (\text{in bits}) \qquad\qquad [15\text{-}7]$$
$$\text{DAC accuracy} = 6.02 N + 1.76 \qquad (\text{in dB, where } N \text{ is the number of bits})$$

Using a lowpass filter, assuming K is infinite:

$$\text{Resolution of systems} = (A_{STOP} - 1.76)/6.02 \qquad (\text{in bits})$$

With these design equations, you can design a D/A converter that has a relatively slow output. The frequency-limiting factors in this design are the clock speed of your controller's fundamental PWM signal and the cut-off frequency of the analog lowpass filter. If you are interested in improving the frequency response of this system, you can use a faster clock without compromising the resolution, or you can use fewer clock divisions in the PWM period. This will reduce the resolution. On the analog side of this discussion, you can use a higher-order filter. If you use this option in your design, the Bessel approximation type will have the best settling time, given the input is a pseudo-square wave.

Using the Comparator for Analog Conversions

A comparator is the most common A/D converter that you will find in your processor or controller. A comparator takes an analog voltage, analyzes its value, and determines whether or not it is above or below a reference. This function is the building block of many high-level A/D converters. Converters that use comparators include the SAR, delta-sigma ($\Delta-\Sigma$), voltage-to-frequency, and dual slope A/D converters, just to name a few. Later in this chapter (under the "Using the Timer and Comparator to Build a $\Delta-\Sigma$ A/D Converter" section), I will use a comparator inside a modulator feedback loop to implement a first-order $\Delta-\Sigma$ A/D converter function.

Some of the comparator specifications of interest are the input range, input offset voltage, and input hysteresis. The diagram that I refer to during this comparator specification discussion is shown in **Figure 15-6**.

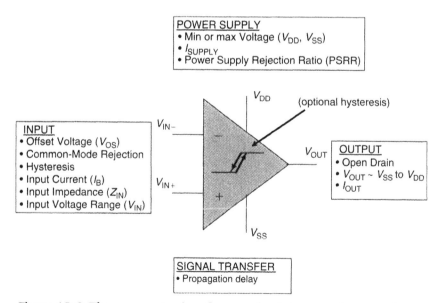

Figure 15-6: The comparator has the same basic input and output pins as the operational amplifier. There are several differences between the comparator and operational amplifier. The differences of the comparator include a hysteresis in the input stage, propagation delay, and a digital output stage.

Input Range of a Comparator (V_{IN+} and V_{IN-})

In a single-supply circuit, the input range of a comparator can range from one rail (ground) to the other (V_{DD}) and be able to have a reference through that entire range. But you should be aware that this specification is not always rail to rail. Like the operational amplifier, some comparators have one internal differential pair when others have two differential pairs in parallel. This dual differential pair will have rail-to-rail input operation. If you want more details about these types of input stages, refer to Chapter 4.

Input Hysteresis

All comparators have a hysteresis across the input stage, but in noisy environments it may be necessary to enlarge the hysteresis region. A larger hysteresis will prevent the "chattering" effect on the output of the comparator when there is noise in the input signal. If you have this kind of noise in your system, **Figure 15-7** shows a circuit that will extend the hysteresis quantity.

You can program the trip points of the hysteresis. With V_{DD}, V_{OH} and V_{OL} known, you can calculate the resistor values with:

$$V_{AVG} = V_{TLH} + (V_{THL} - V_{TLH})/2$$
$$V_{AVG} = V_{DD} \times R_2/(R_1 + R_2)$$
$$R_{EQ} = R_1 \times R_2/(R_1 + R_2) \qquad [15\text{-}8]$$
$$DR = (V_{THL} - V_{TLH})/V_{DD}$$
$$R_3 = R_{EQ} [(1/DR - 1)]$$

Where

V_{AVG} is the average between the two threshold voltages, V_{TLH} and V_{THL}.

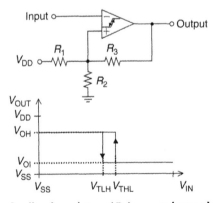

Figure 15-7: Using a feedback resistor, (R_3) can enlarge the input hysteresis of a comparator from the noninverting to the output terminals. You use R_1 and R_2 to set a voltage reference for the comparator.

V_{TLH} is the threshold voltage when the input signal travels from low to high. This will cause an output from high to low on the output of the comparator.

V_{THL} is the threshold voltage when the input signal travels from high to low. This will cause a transition low to high on the output of the comparator.

R_{EQ} is the parallel equivalent of resistors R_1 and R_2. This relationship sets up the reference voltage for the comparator circuit.

DR is the feedback divider ratio around the comparator.

Window Comparator

A window comparator comes in handy when you are monitoring an analog voltage and you would like to know when that voltage extends outside of a predetermined range. This type of circuit can provide an interrupt for your controller as the signal moves above or below a safety limit. You will design this circuit to identify instances where the input signal violates your limits while ignoring minor fluctuations inside your safety range.

The window comparator in **Figure 15-8** requires two comparators and three resistors. The three resistors form a voltage divider for the references to the comparator across the

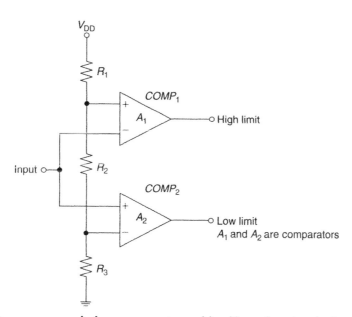

Figure 15-8: You can use a window comparator to identify analog signals that have exceeded a high or a low limit. You can accomplish this with two comparators and by setting the threshold voltage with a voltage divider across the power supply. In this circuitry you will probably not require rail-to-rail input comparators.

supply voltage. If you want a more accurate system, you should use a precision voltage reference at the top of the resistor ladder instead of the power supply.

The outputs of $COMP_1$ and $COMP_2$ respond to the analog signal on the input of this circuit. In the event that the analog input signal goes higher than the voltage at the noninverting input of $COMP_1$, the output of $COMP_1$ will go low. This is because you are putting the analog input signal into the inverting input of the comparator. During the time when $COMP_1$ is responding to this high input voltage, $COMP_2$ remains at a logic high. Conversely, when the analog input signal goes lower than the voltage at the inverting input of $COMP_2$, the output of $COMP_2$ will go low. The analog "safe" zone, where both comparator outputs are high, is set with the resistor R_2.

R_1, R_2, and R_3 form a voltage divider that sets the high and low threshold voltages. The design equations for this circuit are:

$$V_{TH} = V_{DD} \times (R_2 + R_3)/(R_1 + R_2 + R_3)$$
$$V_{TL} = V_{DD} \times R_3/(R_1 + R_2 + R_3)$$

[15-9]

Where

V_{TH} is the high limit of the window comparator and
V_{TL} is the low limit of the window comparator.

Combining the Comparator With a Timer

Comparators alone can convert an analog signal to a digital output. This might seem to be a primitive, limited function, but when you start to combine the comparator with other peripherals in the controller or processor, its capabilities multiply. For instance, you can combine a comparator with a timer (and a few external components). With a little effort, this combination suddenly becomes a fairly good A/D converter.

Figure 15-9 illustrates a good example of using a comparator and timer to implement an A/D converter function.

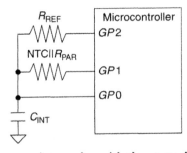

Figure 15-9: You can measure a resistor value with the controller's internal timer as long as you know the value of the reference resistor, R_{REF}.

You can perform a conversion in the following fashion:

1. Set *GP1* and *GP2* to high impedance inputs.

2. Set *GP0* to a low impedance, low output to discharge C_{INT}.

3. Set *GP0* to a high impedance input and *GP1* to a low impedance high output, and start your counter. Since *GP1* is high, the capacitor will start to charge.

4. When *GP0* changes to 1, log the number of counts into your t_{NTC} register (t_1, Figure 15-10).

5. Set *GP1* and *GP2* to inputs again.

6. Set *GP0* to a low output to discharge C_{INT} again.

7. Set *GP0* to an input and *GP2* to a high output, and start your counter.

8. When *GP0* changes to 1, log the number of counts into your t_{REF} register (t_2, Figure 15-10).

With this configuration, you are measuring the rise time of two R/C combinations. The first R/C combination is R_{NTC} and C_{INT}. R_{NTC} can be any type of resistor, but if it is a negative temperature coefficient (NTC) thermistor, you can measure temperature inexpensively. R_{PAR} is a resistor that you select to put in parallel with R_{NTC}. This resistor linearizes the response of the NTC thermistor, making it easier to measure the temperature range of interest.

You use a second R/C combination as your control or reference circuit. You will know the value of R_{REF} with this resistor combination. The most stable type of resistor to choose for this element is a standard wirewound, 1%, resistor. Finally, you should use a film polypropylene capacitor or any type of capacitor with low dielectric absorption for C_{INT}.

Figure 15-10 shows the time response of the algorithm, 1 through 8. The transfer function of this system is:

$$R_{NTC} = R_{REF} \times t_{NTC} / t_{REF} \qquad [15\text{-}10]$$

With this circuit, if V_{TH} is ratiometric with V_{DD}, the conversion time is independent of the power supply voltage, V_{DD}. V_{TH} is set by the input of an I/O port or by a comparator. Either type of input will respond reliably. You will note that the transfer function of the calculation is independent of the capacitor and the clock frequency of your controller. In this system, the resolution is dependent of the R/C time constants and the granularity of your timer. You determine the accuracy with R_{REF}.

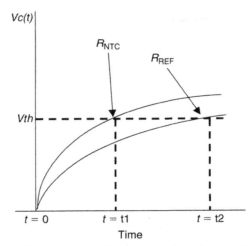

Figure 15-10: When you pull the gate that is connected to R_{NTC} or R_{REF} high after it has been in a high impedance setting, the capacitor, C_{INT}, starts to charge. The controller monitors this risetime until the voltage on top of C_{INT} equals the threshold voltage, V_{TH}. You will log the amount of time that it takes to charge the capacitor as t1.

Using the Timer and Comparator to Build a $\Delta-\Sigma$ A/D Converter

This portion of this chapter describes how to implement a $\Delta-\Sigma$ A/D converter function using a microcontroller. Although many microcontrollers do not have a built-in A/D converter, you can use the comparator function, internal voltage reference, and timers to digitize an analog signal.

Some standard controllers have a comparator module consisting of two comparators. You can use a controller's internal voltage reference source with the comparators to establish thresholds. If this internal voltage reference is too noisy, drifts too much with temperature, or is not available, you can use an external reference. By combining these elements, you can design a first-order modulator and a first-order filter. This combination emulates the function of an analog-to-digital $\Delta-\Sigma$ conversion.

You can quickly implement this method of conversion in firmware, with very few additional external components. Therefore, the cost of hardware implementation is minimal, particularly for such a high-resolution converter solution. The input range is very flexible and adjusted with external resistors. Although this method is not particularly strong in terms of DC accuracy, it is well suited for ratiometric applications.

$\Delta-\Sigma$ Theory

The function of the classical $\Delta-\Sigma$ analog-to-digital converter uses two circuit segments: a modulator and a digital filter. The modulator section acquires an input signal as shown

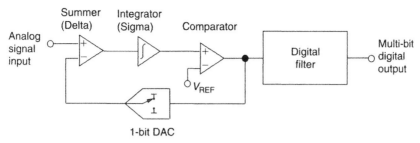

Figure 15-11: This is the modulator for a Δ−Σ A/D converter. In the stand-alone Δ−Σ A/D converter, low-frequency noise moves out into higher frequencies. This first-order modulator can achieve accurate conversions up to about 10 bits without additional digital filtering.

in **Figure 15-11**. This is the same type of modulator that was discussed in Chapter 13 (Figure 13-18).

In the discussion in Chapter 13, I mentioned that this block diagram was simply a concept for discussion. Stand-alone Δ−Σ A/D converters have more than one modulator segment and they also accept differential input signals. The modulator block in Figure 15-11 is a first-order modulator that only accepts single-ended inputs.

In this block diagram, the circuit subtracts the input signal from a digital-to-analog (D/A) converter signal in the negative feedback loop. Then the differentiated signal passes through an integrator and finally to one of the two inputs of a comparator. The comparator acts like a one-bit quantitizer. The comparator sends its signal back to the differentiator by way of a one-bit D/A converter. Additionally, the output of the comparator passes through a digital filter. The complexity of this digital filter is up to you, the controller programmer. With time, the output of the digital filter provides a multibit conversion result. As mentioned before, this fundamental circuit concept generates a large variety of the converters that provide high resolution relatively inexpensively.

The next logical step for this type of A/D converter is to move it into the controller. A basic controller is not able to execute this type of function, but a few additional peripherals make it possible. **Figure 15-12** shows the circuit diagram for this type of microcontroller implementation. The circuit in Figure 15-12 transforms the theoretical concept in Figure 15-11 to reality.

Figure 15-12 shows a circuit that has the integrator function in the Δ−Σ block diagram of Figure 15-11. The external capacitor, C_{INT}, implements this integrator function. The absolute accuracy of this external capacitor is not critical, only its stability from integration to integration, which occurs in a relatively short period of time. When *RA3* of the microcontroller is set high, the voltage at *RA0* increases in magnitude. This occurs

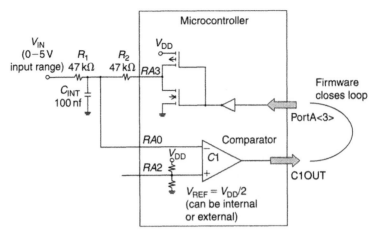

Figure 15-12: If a microcontroller has an internal comparator and timer, you can build a $\Delta-\Sigma$ converter with two additional external resistors and one capacitor. In this configuration, a lowpass filter is a byproduct of the external resistors and capacitor input network.

until the output of the comparator ($C1OUT$) is triggered low. At this point, the driver to the $RA3$ output is switched from high to low. Once this has occurred, the voltage at the input to the comparator ($RA0$) decreases. This occurs until the comparator is tripped high. At this point, $RA3$ is set high and the cycle repeats. While the modulator section of this circuit is cycling, two counters keep track of the time and of the number of ones versus zeros that occur at the output of the comparator.

The comparator is part of the controller as well as its voltage reference. You can implement the one-bit D/A converter in firmware by driving $RA3$ in accordance with the output of the comparator (CMCON<6>, PIC16C623 from Microchip Technology). The firmware drives the D/A converter output at $RA3$. Two counters implement a first-order digital filter (also known as an *averaging filter*).

The Controller Implementation

With the circuit in Figure 15-12, it is possible to conceptualize the $\Delta-\Sigma$ function. **Figure 15-13** summarizes the controller implementation of this circuit in a flowchart.

Normally the output of the comparator is connected to $RA3$. This keeps the voltage at $RA0$ equal to the reference voltage of the comparator, in preparation for the next conversion. At the start of this flowchart, the result and counter variables are cleared.

You should check the comparator at the beginning of each loop. If the voltage on the capacitor is less than the input voltage, $RA3$ is set high, which will put charge into the capacitor, raising the voltage. If the voltage on the capacitor is greater than the input

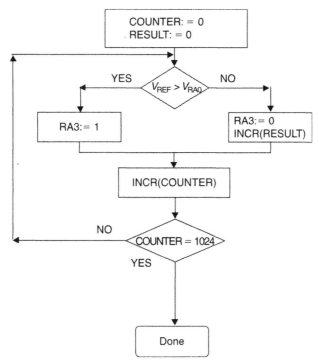

Figure 15-13: This is the $\Delta-\Sigma$ **A/D conversion flowchart, from the circuit shown in Figure 15-12. You should ensure that the cycle time through the flowchart is constant. This code runs until a conversion is complete.**

voltage, *RA3* will be set low, taking charge out of the capacitor and lowering the capacitor voltage, and the resulting register is incremented.

This continues as long as necessary to get the required resolution. For 10 bits of resolution, 2^{10} (1024) laps through the loop are required.

You would take each integration result at a regular time interval. If you assume that the time interval of a conversion is 20 msec, you can easily calculate the conversion time versus bits. **Figure 15-14** shows this relationship graphically.

For instance, a 10-bit conversion would require 2^{10}, or 1024, samples. If the microcontroller conversion loop is 20 μs, one complete conversion would take a little more than 20 ms. **Figure 15-15** shows the room temperature test data for the circuit that uses a PIC16C623 from Microchip.

Figure 15-15 shows the plot of the voltage input versus the output code on the left axis and the output error on the right axis. This data represents the results of 1024 laps through the flowchart in Figure 15-13. The expected resolution of this configuration is 10 bits. The

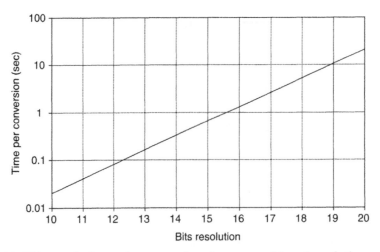

Figure 15-14: This graph shows the conversion time versus bits of resolution, assuming a 20 msec integration time while using the circuit shown in Figure 15-12.

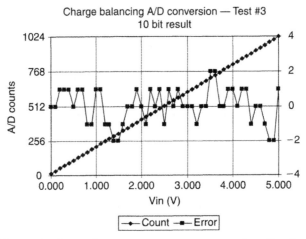

Figure 15-15: This graph shows the room temperature test data for the circuit shown in Figure 15-12 and a PIC16C623 microcontroller from Microchip. The input voltage range is 0.003 to 4.99 V. The maximum error found in the test was ±2 counts. In this 10-bit system, the ±2 LSBs is equivalent to ±9.8 mV. The data from the A/D counts (left y-axis) was only one sample per point on the graph. Results may vary from part to part. $V_{DD} = 5$ V.

maximum code error for this test was ±2 counts or 2-bits of uncertainty. Consequently, the effective number of bits of this A/D converter is 8 bits.

The A/D error was calculated assuming the codes for $Vin = 0.5$ V and $Vin = 4.5$ V are ideal. The test conditions are room temperature with one microcontroller. These results may vary from part to part.

Error Analysis of This $\Delta-\Sigma$ A/D Converter Implemented With a Controller

This low-cost $\Delta-\Sigma$ converter provides a good solution for ratiometric applications where having the absolute result is not critical. Additionally, a first-order, digital, FIR filter replaces the analog gain function. This FIR filter is a simple accumulator. In this example, V_{DD} is 5 V and the reference voltage is $\sim V_{DD}/2$. The resistors are 47 kΩ. This value of resistance minimizes the leakage errors across the resistors versus the RDS_{ON} error of the output pin, *RA3*. The capacitor value is equal to 100 nF.

RDS$_{ON}$ Error

This error comes from the drain-source resistance of the output FETs on the output pin, *RA3*. At room temperature, this resistance error is small and is typically less than 100 Ω. Compared to R_2, RDS$_{ON}$ introduces about 0.2% gain error. You can compensate for this error by increasing the resistor, R_1, by approximately 100 Ω. Additionally, the value of the RDS$_{ON}$ resistance will probably increase with rising temperature. Refer to these specifications in the product datasheet of your controller or processor.

RA0 Port Leakage Current

A typical specification for leakage current of the PIC16C623 is 1 nA at room temperature and 0.5 μA (max) over temperature. The leakage current from the port at *RA0* causes a voltage drop across the parallel combination of R_1 and R_2. With these two resistors equaling 47 kΩ, the error caused by this leakage current is $\sim$11 mV. This is also close to a 0.2% error. At room temperature, this error is negligible. Leakage current does increase with temperature. Refer to these specifications in the product datasheet of your controller or processor.

Nonsymmetrical Output Port (*RA3*)

When the output port is high, the FET resistance is dependent on the p-channel on-resistance. When the output port is low, the FET resistance is dependent on the n-channel on-resistance. The p-channel on-resistance is usually greater than the on-resistance of the n-channel FET. As a consequence, there is an additional offset contribution of 5.5 mV at room temperature. Refer to these specifications in the product datasheet of your controller or processor.

Voltage Reference

The internal voltage reference to the comparator is a simple internal voltage divider. If this is the case, the absolute value of this voltage is dependent on internal resistor matching and power supply voltage. Assuming the power supply is an accurate 5 V, the voltage error of this reference, part to part, is significant. However, once you remove the initial error of the internal voltage reference through calibration, it is ratiometric to the power supply. This is the biggest error in the circuit but is easily reduced with an external voltage reference. The design equations for this circuit are:

$$V_{IN(CM)} = V_{RA0}$$
$$V_{IN(P\ TO\ P)} = V_{RA3(P\ TO\ P)}(R_1/R_2)$$

[15-11]

Where

$V_{IN(CM)}$ is equal to $(V_{IN(MAX)} - V_{IN(MIN)})/2 + V_{IN(MIN)}$,
V_{RA0} is the voltage applied to the comparator's inverting input,
$V_{IN\ (P\ TO\ P)}$ is equal to $(V_{IN(MAX)} - V_{IN(MIN)})$, and
$V_{RA3(P\ TO\ P)}$ is equal to $V_{RA3(MAX)} - V_{RA3(MIN)}$.

Refer to these specifications in the product datasheet of your controller or processor.

Other Input Ranges

Figure 15-12 shows a configuration that uses a 0 V to 5 V input range. The resistor network (R_1 and R_2) and the reference voltage to the noninverting input of the comparator determine the input range for this circuit. If the ratio of R_1 and R_2 is changed, the input range can be increased or decreased in accordance with the relationship between R_1 and R_2. You can implement further adjustments by adding an additional resistor to this input structure that is biased to ground or the power supply.

Input Range of 2 V to 3 V

You can increase or decrease the input range of this converter by adjusting the ratio of R_1 and R_2. In **Figure 15-16** these resistors reduce the input range from ±2.5 V. In Figure 15-12 the range is ±500 mV. In both cases, the input range is centered around the comparator reference voltage, 2.5 V. This type of input range is best suited for sensors with smaller output voltage ranges, such as the buffered output of a pressure sensor or load cell.

The resistors are determined by comparing the desired input range to the voltage range of *RA3*. Assuming that the reference voltage in this problem is 2.5 V, the input range changes ±500 mV and the voltage at *RA3* changes by ±2.5 V. The ratio of these two voltage ranges is 5:1. Consequently, during one integration period, the difference between the current through R_2 and R_1 must always be less than zero. In this manner, the *RA3* gate

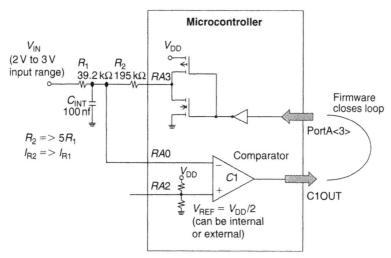

**Figure 15-16: This is the configuration for the microcontroller of a
Δ−Σ conversion with a ±500 mV range centered around 2.5 V.**

will be capable of driving the capacitor, C_{INT}, past the reference voltage applied to the noninverting input of the comparator. Figure 15-16 shows this circuit.

The design equations for this circuit are:

$$V_{IN(CM)} = V_{RA0}(1 + R_1/R_3)$$
$$V_{IN(P\ TO\ P)} = V_{RA3(P\ TO\ P)}(R_1/R_2)$$

[15-12]

Where

$V_{IN(CM)}$ is equal to $(V_{IN(MAX)} - V_{IN(MIN)})/2 + V_{IN(MIN)}$
V_{RA0} is the voltage applied to the comparator's inverting input
$V_{IN(P\ TO\ P)}$ is equal to $(V_{IN(MAX)} - V_{IN(MIN)})$
$V_{RA3(P\ TO\ P)}$ is equal to $V_{RA3(MAX)} - V_{RA3(MIN)}$

Input Range of 10 V to 15 V

You can apply an offset adjustment by adding an additional resistor to the input structure of the A/D converter. In **Figure 15-17**, R_1 and R_2 are equal and configured to allow for an input range of ±2.5 V as shown in Figure 15-12. The addition of R_3, which is referenced to ground, provides a level shift to the input range of 10 V.

With this circuit configuration, a 5 V (full-scale) current through R_1 is equal to V_{REF}/R_1. If R_3 draws the same current to ground, the integrating capacitor will not be charged. In

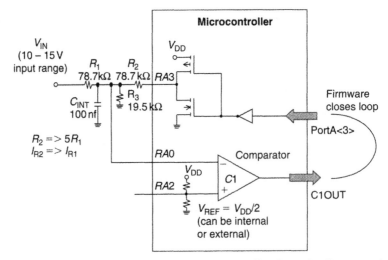

Figure 15-17: Configuration of the microcontroller for a Δ−Σ conversion with a 10 V to 15 V range centered around 12.5 V.

this manner, a 2.5 V offset is implemented with $R_3 = R_1$. To achieve a 10 V offset, R_3 must be equal to $4 \times R_1$, per Figure 15-17.

The design equations for this circuit are:

$$V_{IN(CM)} = V_{RA0}(1 + R_1/R_3)$$

$$V_{IN(P\ TO\ P)} = V_{RA3(P\ TO\ P)}(R_1/R_2)$$

[15-13]

Where

  $V_{IN(CM)}$ is equal to $(V_{IN(MAX)} - V_{IN(MIN)})\ /2 + V_{IN(MIN)}$
  V_{RA0} is the voltage applied to the comparator's inverting input
  $V_{IN(P\ TO\ P)}$ is equal to $(V_{IN(MAX)} - V_{IN(MIN)})$
  $V_{RA3(P\ TO\ P)}$ is equal to $V_{RA3(MAX)} - V_{RA3(MIN)}$

Input Range of ± 500 mV

The circuit in Figure 15-17 uses the scaling technique discussed in the circuit shown in Figure 15-16 and the offset shift technique discussed in the circuit shown in Figure 15-16. With this circuit, the input range is $\pm 500\,mV$. You achieve this by making $R_2 = 5R_1$. There is a level shift of $-2.5\,V$ to the signal-input range. You can implement this with a resistor, R_3, to the positive supply, per **Figure 15-18**. The magnitude of this level shift is achieved by making $R_3 = R_1$.

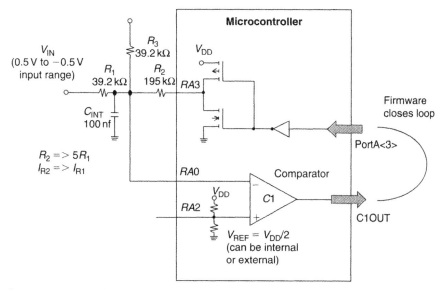

Figure 15-18: Configuration of the microcontroller for a $\Delta-\Sigma$ conversion with a ±500 mV range centered around ground.

The design equations for this circuit are:

$$V_{IN(CM)} = V_{RA0}(1 + R_1/R_3) \qquad\qquad [15\text{-}14]$$
$$V_{IN(P\ TO\ P)} = V_{RA3(P\ TO\ P)}(R_1/R_2)$$

Where

> $V_{IN(CM)}$ is equal to $(V_{IN(MAX)} - V_{IN(MIN)})/2 + V_{IN(MIN)}$
> V_{RA0} is the voltage applied to the comparator's inverting input
> $V_{IN(P\ TO\ P)}$ is equal to $(V_{IN(MAX)} - V_{IN(MIN)})$
> $V_{RA3(P\ TO\ P)}$ is equal to $V_{RA3(MAX)} - V_{RA3(MIN)}$

You can use this circuit to measure the current through a shunt resistor. The main error term at room temperature is comparator offset. In systems with a known "zero-current" state, the offset can be measured and removed through calculation or removed by adding or subtracting the offset of the result counter.

Final Thoughts

So, you might have thought that you were using digital devices when you chose to work with a microcontroller or microprocessor. You might also think that the bridge to analog will never happen with your devices. But in fact, the peripherals like counters, I/O ports, and comparators allow you to get much closer to analog than you might have imagined.

The combinations of the tools that we discussed in this chapter will not replace the high-precision or high-speed analog circuits. However, in your designs you need to consider how much is good enough. If you conclude that the controller can get you the analog functionality that you want, then have at it.

References

Cox, D., "Implementing ohmmeter/temperature sensor," AN512, Microchip Technology, Inc.

Curtis, K., "Analog design in a digital world using mixed-signal controllers," AN823, Microchip Technology, Inc.

Palacherla, A., "Using PWM to generate analog output," AN538, Microchip Technology, Inc.

Richey, R., "Resistance and capacitance meter using a PIC16C622," AN611, Microchip Technology, Inc.

Stein, D., "D/A conversion using PWM and R2R ladders to generate sine and DTMF waveforms," AN655, Microchip Technology, Inc.

CHAPTER 16

What's All This Error Budget Stuff, Anyhow?

Robert A. Pease

> *Well, I stated at the start of the story the reason it's important to do an error budget on even a simple circuit—and then I showed the size of trouble you can get into if you don't. I rest my case. /rap*

I was just on the phone explaining to a young engineer how to do an error budget analysis on some fairly simple circuits. Later, I mentioned this while I was visiting my friend Martin, and he said he had been quite surprised when he found that many engineers in Europe were quite unfamiliar with the concept of an error budget. How can you design a good circuit without being aware of which components will hurt your accuracy?

When I was a kid engineer back in 1962, my boss George Philbrick gave me a book on differential amplifiers by Dr. R. David Middlebrook, and he asked me to do a book review. I studied the book, and it was full of hundreds of partial differential equations. If you wanted the output of a circuit with fourteen components, you could see a complete analysis of how each component would affect the output offset and gain. Each equation filled up a whole page. It did this several times.

Yet the book didn't offer any insights into what's important. I mean, is $\beta \times d(R1)$ more important than $R1 \times d(\beta)$? In retrospect, I'm glad I didn't submit any critique of that book. I would've done more harm than good. Such a mess! Even now, it would be hard to write a critique on a book that was so true but so unhelpful.

Things are much simpler now that people are mostly (but not entirely) designing with op-amps. The best thing is that the output offset and DC gain and AC gain errors are largely orthogonal. An "operational" amplifier does perform, largely, an "operation" based on what task you ask it to perform when you "program it" with Rs and Cs. If the offset

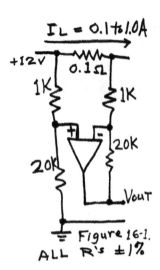

Figure 16-1: Conventional differential amplifier.

varies, the gain does not, and vice versa. We all agree that it's very helpful that you can compute what the performance will be with almost no interaction. No partial derivatives.

Now let's take a look at a couple of applications—real circuits—and their tolerances within an error budget. Here is an amplifier to magnify the $I \times R$ drop of current through a 0.1-Ω resistor and bring it back down to ground. **Figure 16-1** shows a conventional differential amplifier, with the common mode up at $+12$ V. The gain of -20 will bring the 1.0 A $\times$ 0.1 Ω signal down to a ground level. If the current is 0.1 A, the output will be 0.2 V, small scale. A full-scale current of 1 A will bring the output up to 2.0 V, which is suitable to send to a detector or analog-to-digital converter.

Let's select an op-amp like the LMC6482A, with low offset voltage less than 1.0 mV. (There are other versions of this amplifier with less than 0.35 mV, but let's select an intermediate model.) This 1 mV does cause 21 mV of output error. This op-amp has less than 20 pA of I_B at all temperatures, so at least that's negligible. (Bipolar op-amps might have small I_B errors, but you'd have to check it.)

Now let's see what the resistors add. Assuming that all Rs have a 1% tolerance, the gain of (2.0 V per A) has a tolerance of $\pm3\%$. This would cause ±60 mV at full scale, but only ±6 mV at small scale (0.1 A). This might be acceptable.

Then let's consider the common-mode errors. If $R4$ has a 1% tolerance and it has 11.4 V across it, the 1% tolerance could cause a 114-mV error. By symmetry, a 1% error of each of $R1$, $R2$, $R3$ can cause another 114 mV! Added together, the common mode could cause an output error of 456 mV! That's about $\pm1/4$ of full scale—even for small signals. That doesn't look so good to me!

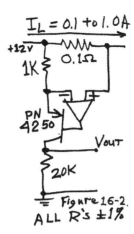

$I_L = 0.1$ to 1.0A

+12V

1K

0.1Ω

PN
4250

Vout

20K

Figure 16-2.
ALL R's ±1%

Figure 16-2: Alternative circuit.

It's true that if adjacent 1-kΩ resistors are inserted, they're likely to match within ±1/2%, so the probable error between the pair might cause ±60 mV and the ±1/2% matching between the 20 kΩ would cause another 60 mV. That added to the 21 mV from the V_{OS} would add to 141 mV.

Some textbooks teach you that you should add these errors arithmetically to 141 mV. Others point out that they could be added in an RMS way, so that 60 + 60 + 21 mV = 87 mV. Typically, this might be true. But the worst case of 141 or 456 mV might be more realistic. I mean, if you're going to build 1000 circuits and most of them are better than 141 mV, what are you going to do with the 400 circuits that are worse than 141 mV? And that's still 7% of full scale....

You could go shopping for 0.1% resistors, but they aren't cheap. You could put in a trimpot to trim the error (to no offset error) for small signals. But as you might have noticed, a trimpot has to be properly trimmed. And if that pot is accessible, it could someday be mistrimmed and it would have to be corrected in some awkward calibration cycle. Most people want to avoid that trimpot. Before we decide that this 141 mV is unacceptable, let's look at another circuit.

Figure 16-2 shows an alternative circuit with the same gain, 2.0 V per A, using a PN4250 or 2N4250, a high-beta pnp transistor. What does the error budget look like? The same op-amp causes just 20 mV of output error. The 1% resistor tolerances cause the same gain error, 60 mV at full scale, or 6 mV at small scale. The newly added transistor adds (?1/3%) max from its alpha, or less than 7 mV, at full scale.

What is the offset error due to common-mode rejection ratio (CMRR) or due to resistor mismatch? Nothing. Zero. The transistor doesn't care about the voltage across it. There are no resistors with 12 V across them.

So the offset error is ±20 mV, due primarily to the amplifier's VOS (which could be reduced), not ±400 mV. This little circuit has greatly reduced errors compared to Figure 16-1, even if Figure 16-1 had a couple bucks of 0.1% resistors. This might be acceptable. Even the offset errors could be reduced to 7 mV by selecting the LMV841 or LMC7701.

So we have seen that circuits with similar functions can have completely different error budgets. I love to recommend amplifiers with high CMRR. But depending on cheap 1% resistors can hurt your error budget a lot more than you'd suspect.

This article was first printed in *Electronic Design Magazine*, June 8, 2006, and is reprinted with the kind permission of Penton Publishing.

CHAPTER 17

What's All This V_{BE} Stuff, Anyhow?

Robert A. Pease

Part 1

This V_{BE} topic has come up many, many times since junction transistors were introduced in the 1950s. Usually scholars like to use lots of exponential equations, and they seem to pretend that I_S is constant. Then they show the old, trite curve where Ic bends like a hockey stick at $V_{BE} = 0.6$ volts on a linear-linear scale. They pretend the transistor has no collector current below $V_{BE} = 0.4$ volts. This does not help a user or engineer understand how things change versus Ic or temperature. They ignore the way that the transistor's current shrinks exponentially, all the way down to just a few millivolts of V_{BE}, and does not magically stop below a certain "threshold." This analysis helps me a lot; how about you? Bob Widlar used graphical techniques to design transistor circuits that ran some transistors on a small number of nanoamperes or of millivolts. /rap

The other day, I was walking past the applications engineering area when I heard a grouchy debate between a couple of guys over in the corner. As they saw me walk by, they called out, "Bob, come on over here, and maybe you can solve this problem for us." I looked at their problem.

"Bob, we were trying to use the standard diode equation to compute the tempco of a transistor's V_{BE}, and it doesn't seem to make any sense." I looked at their standard equation:

$$I_C = i_S \times e^{(q V_{BE}/kt)} \qquad [17\text{-}1]$$

Yes, there was a term for temperature, t, in there, but it wasn't a very prominent term. Obviously they had tried to see how this equation responded to temperature. They were puzzled because it does *not* respond properly to temperature. It doesn't give anything *like* $-2\,\text{mV}/°\text{C}$. I began assisting them by explaining, "When they give you this equation in school, they neglect to tell you that the i_S isn't a constant but rather a very wild function of temperature. This function is so wild that they won't tell it to you, because it's not very useful. You can't successfully differentiate it versus temperature. So you're better off *not* having such an unusable equation."

They responded, "Okay, what are we supposed to use?" I replied, "Ah, let's do a graphical approach. Let me make up a couple of sketches." First I scribbled out **Figure 17-1**, showing the log of collector current versus V_{BE}.

I went on to explain, "That schoolbook linear plot of V_{BE} versus I_c isn't very useful, because it just shows a severe knee. I never use that one. Look at the middle line of this plot. It shows that at room temperature, the slope of the log of I_c versus V_{BE} is quite linear over seven, eight, or nine decades of current. Only at high currents does the curve bend, due to emitter resistance. And only at very small currents do you get errors due to leakages. So, in the whole midrange, you get a wide range of conformity to the slope of 60 mV/decade." The two guys agreed with what I had said.

After this, I pointed to the upper line. "At a hot temperature such as +127°C, the curve is very similar. But at a shallower slope, the millivolts per decade is *worse*, very close to 80 mV/decade. Indeed, this number of millivolts per decade is predicted by the diode equation." They further agreed that my explanations seemed correct. Plus I showed the guys that the lower sloping line is sort of like the curve for −73°C, but it's at a slope of 40 mV/decade—a rather higher gain, with a higher gm. Fine.

Also, it's possible to see that all the curves tend to converge or extrapolate to a single high point at a *very* high base-emitter voltage, perhaps +1.24 V, at a *very* high current, maybe 10,000 Amperes. Based on this outrageously high theoretical current at an absurd voltage, one could (theoretically) compute what the V_{BE} is really doing—not very accurately, or usefully.

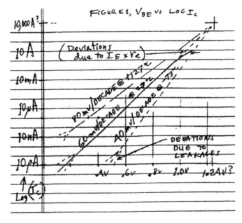

Figure 17-1

But I pointed out that this curve is just good for giving a ballpark overview of what goes on. Yes, in concept this could be used for computing the actual V_{BE} of the transistor, at various currents and temperatures. But it's too crude and too broad to be useful. What we want to use is closer to **Figure 17-2.**

I sketched away madly to get this figure, showing the plot of V_{BE} versus temperature. This illustrates the bias of transistors at various constant currents versus temperature. *"This,"* I said, "is *useful*—and let me show you where and why." I stated that it was based on the real data for a real standard transistor, and it's what I use to compute biases for real precision linear circuits, such as band-gap references or temperature sensors. This and a slide rule (or a little handheld scientific calculator) lets me compute the operating points I need.

I pointed out the middle, solid, sloping line. "This line is based on some measured data. This transistor, when used in a band-gap reference, has a Magic Voltage of about 1.240 V. That's where the band-gap runs flattest. So this line is drawn in order to go through 1.240 V DC at absolute zero temperature. That's where the V_{BE} extrapolates to—if the transistor were cooled off—and that is not real data.

"The other point of calibration is where it goes through 0.640 V of V_{BE} at 10 μA at room temperature, about +27°C. That's a simple, factual, measured data point." Then one guy asked, "But why +27°C? Why not +25?" I replied that +27°C is, with an accuracy better than 0.2°C, exactly 300°Kelvin. Therefore, it makes the math much easier to work with, at +200, +300, and +400°Kelvin. They agreed.

Furthermore, I pointed out that the voltage represented by this line is just the nominal V_{BE} of the transistor versus temperature at a constant emitter current. This has a nominal

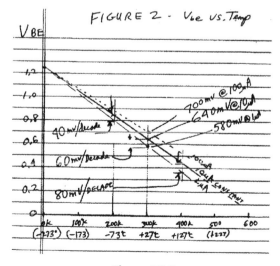

Figure 17-2

slope which is quite close to −2.00 mV/°C. *This* is a very useful thing to know—the bias at which the transistor runs at −2.00 mV/°C—because we will soon see that at many other operating currents, the tempco is *not* −2.00.

Next I stated to them that *if* the voltage between the solid, slanting line and the horizontal line at 1.240 V is studied, you can see that it's a voltage proportional to absolute temperature (VPTAT). Therefore, when we want to build a band-gap reference that's 1.240 V, all we have to do is *add* to the V_{BE} a voltage that's VPTAT. Then we can make a band-gap reference.

This is all you have to do: If you have a VPTAT that's 60 mV at room temperature, and you can amplify this with a gain of 10, you can add that onto a V_{BE} to make a band-gap reference—as Mr. Widlar proved about 30 years ago. They agreed that made sense as well.

I had to admit that the solid, sloping line appears to be nominally linear, and I drew it as more or less linear—but it's *not* truly linear. The V_{BE} curve actually is bowed downward at both hot and cold temperatures, perhaps as much as 2 to 4 mV. But for many uses, that's a negligible error, which is easy to make corrections for later.

I explained further: "Let's take a look at the upper, dotted line of Figure 17-2 (and **Figure 17-3**). This is for the transistor running at 100 μA. It, too, extrapolates back toward that point at absolute zero. This line does *not* have a slope of −2.000 mV/°C, but instead −1.800. This line isn't parallel to the other line. It's set above it by 60 mV/decade at room temp, by 80 mV at +127°C, and by 40 mV at −73°C. This difference is very accurately a VPTAT."

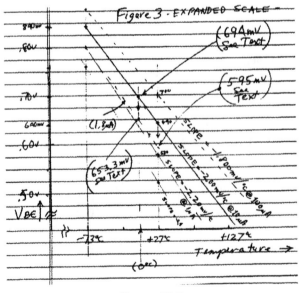

Figure 17-3

The *lower* dashed line is the line for a bias of 1 μA. It has a slope of −2.200 mV/°C. It's offset by 60 mV/decade at room temperature, more when hot and less when cold. The tiny segment of the line is at 0.1 μA and has a slope of −2.40 mV/°C (Figure 17-3).

So, all lines for V_{BE} at a constant current are all fanned out, radiating from that point at absolute zero. THIS is the curve from which it's easy to compute temperature coefficients and operating points. Now, an expanded plot, Figure 17-3, depicts only the central portion of Figure 17-2.

Let's say we want to estimate a V_{BE} at some other biases. I'll take you through some examples. The main point is, though, that you can fairly easily compute the bias for any normal situation.

Okay, we agree that we know the V_{BE} at those specified conditions: 640 mV at 10 μA and +27°C. Let's say I want to compute the V_{BE} of the transistor at the same current but at a different temperature, such as 0°C. In the example shown, the temperature coefficient of V_{BE} is −2.000 mV/°C. A shift of −27°C will cause the V_{BE} to increase by (−27) × (−2.0) = +54 mV, up to 694.0 mV. That's not very hard. For any change of temperature, at a constant bias current, simply multiply the change in temperature by the tempco of V_{BE}. But the tempco of −2.0 mV/°C only applies at 10 μA in this example. At any other current, the tempco will be different. (More on this later.)

What if we want to start from our initial conditions and move to a different current, such as 1.8 μA at +27°C? For this case, where things are at a constant temperature, you can use the diode equation:

$$IC1 = I_S \times e^{qV_{BE}/kt} 2 \qquad\qquad [17\text{-}2]$$

or its inverse:

$$V_{BE}1 - V_{BE}2 = kt/q \; \ln(IC1)/(IC2) \qquad\qquad [17\text{-}3]$$

The ratio of currents is 0.18, and the natural log of 0.18 is −1.7148. At +27°C, the factor kt/q = 26.06 mV per factor of e, which is the same factor as 60.0 mV/decade.

Therefore, the delta V_{BE} will be −1.7148 × 26.06 mV, or −44.7 mV. The V_{BE} will decrease from 640 mV to (640 − 44.7) = 595.3 mV. This isn't a surprise. Any time the collector current of a transistor changes at a constant temperature, the V_{BE} changes in a nice logarithmic way. But that 26.06 mV is only at that value at +27°C. At all other temperatures, it's different, as a linear function of absolute temperature.

Another useful way to look at it is that any time you change the current by a factor of 10 at room temp (about +27°C), the V_{BE} will shift by 60 mV, up or down, as appropriate.

For many cases where decades of current are the important factor, the multiples of 60 mV make calculations simple. No computers or calculators are required.

Now let's consider the case where you want to compute the V_{BE} when both the current and the temperature are changed. There are two ways to compute this. And both of these computations had better give the same answer.

Let's say we want to compute the V_{BE} at 1.8 µA at 0°C. You could first change the temperature of the 10-µA transistor to 0° at constant current and then change the current at a constant temperature.

Let's do that: We just agreed that the V_{BE} would be 694 mV at 10 µA at 0°C. How much will V_{BE} change if we then go to 1.8 µA? At 0°C, kt/q isn't 26.06 mV, but 273/300 × 26.06 mV, or 23.712 mV, as the temperature has decreased by that factor. Therefore, as we decrease the current by a factor of 0. 18, the VBE changes by −1.7148 × 23.712 mV, or −40.7 mV, so the V_{BE} decreases to 653.3 mV.

What if we arrive at this point by the other route of first decreasing the current, *then* decreasing the temperature? We just computed that the V_{BE} at +27°C and at 1.8 µA was 595.3 mV. What is the tempco of V_{BE} at *this* current? It isn't −2.000 mV/°C, as it is at 10 µA. And, and it isn't −2.200 mV/°C, as it is at 1 µA. It's at an intermediate value. These slopes are all proportional to absolute temperature, as they intercept absolute zero at 1240 mV. So the slope of (1240 mV − 595.3 mV)/300°C is 644.7 mV/300°C, or −2.149 mV/°C. If you multiply this tempco by a −27°C change, the shift will be 58.02 mV. When you add this to 595.3 mV, the answer is 653.32 mV. So, fortunately, we get the correct answer when we compute it either way.

If you need to know the tempco of V_{BE}, it normally changes −200 µV/°C every time the current is reduced by a factor of 10. Thus while the transistor of this example had −2.000 mV/°C at 10 µA, it has −2.200 mV/°C at 1 µA, −2.400 mV/°C at 0.1 µA, −2.6 mV/°C at 10 nA, and −3.0 mV/°C at 100 PA. Although most people don't bias transistors down there, that does *not* mean that the tempco isn't surprisingly well defined down there, and it's a *lot* bigger than just −2.0 mV/°C!

What other factors should we take into account when we want to compute V_{BE}? With monolithic npn transistors, it's fairly safe to assume that the transistors' V_{BE}s are fairly well matched and predictable. We need to only take into account a difference of about 5 or 10 mV if the transistors are designed with similar geometries. That's even if no special care is taken to match them perfectly. With discrete transistors from the same batch, the matching might be similar, or it might be *poor* if the transistors came from different batches. There could be a lot of deviations, but you can't count on that.

As mentioned earlier, the curvature of V_{BE} versus temperature will cause the V_{BE} to be 1 or 2 mV smaller, at 0°C and also at +70°C, compared to the linear predictions. It could easily be 3 or 4 mV lower at −55°C or +150°C—it really is quite close to a parabolic error.

Additionally, Early Effect will normally cause a low-beta transistor (beta = 50 or 100) to run 1 or 2 mV lower in V_{BE} if the VCE is as high as 20 V rather than 0.6 V. On high-beta transistors (beta = 200 or 400), the decrease in V_{BE} may easily run 3 or 4 mV. (At another time, we can discuss the complete ramifications of this Early Effect. Suffice it to say here, transistors with high beta might have smaller *current* errors, but they tend to have correspondingly poorer *voltage* errors.)

Of course, if you run the transistor at high currents where $V = IE \times RE$ is significant, that effect can be additive (approximately) and is usually fairly linear and predictable (not to mention self-heating). If the IC or IE are small, the leakages could cause significant deviations. Also, if IE becomes quite small, some transistors could have a rapid fall-off of beta, so you cannot be sure the base current is negligible any more! And if you ever let the transistor saturate, the V_{BE} can rise or fall considerably, depending on how the transistor was made. Still, these graphical techniques can do a pretty good job of helping you to estimate the V_{BE} of a bipolar npn transistor—and of a discrete pnp, too.

Now you could write a fancy equation to compute all this, but I prefer a graphical approach. That way, I get good insights into what's going on, and I don't get fooled by computational mistakes.

Part 2

For a given V_{BE}, *the collector current of a transistor is well behaved for many values of* V_{CE}, *extending nicely plus and minus a couple hundred millivolts above and below* V_{BE} = 0.7 *volts. But "The Technical Books" all talk about the "forward active region" that only has* V_{CE} *greater than* V_{BE}. *Not so! This essay just points out the obvious fact that this region extends quite a good way down from* V_{CE} = V_{BE}, *down well below 0.7 volts, down toward* V_{CE} = 0.25 *volts. It also corrects a couple other myths. /rap*

I've been debating with a guy who argues that a transistor won't work as a transistor unless its V_{CE} is bigger than its V_{BE} (see **Figure 17-4**). He keeps reading this in books. Also, he points out that if the base and collector are nominally tied together to make a diode, you might think that it's okay. But actually, he says, the $I \times R$ in the collector path makes the V_{CE} lower than the base voltage, so it won't work. Well, I've been looking in some of those books, and they sometimes do say that. But when they do, they're wrong.

When a transistor's V_{CE} is slightly less than its V_{BE}, it keeps right on working like a transistor. Can I prove this? Sure. Look in the NSC linear Databook at circuits such as

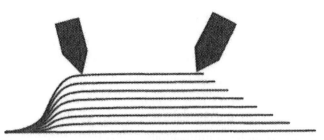

The upper curve represents a 2N3904 biased up with I_B = 7 µA, V_{BE} = 700 mV, and I_C = 1.05 mA. As V_{CE} decreases from 0.7 V (at the right-hand arrow) to V_{CE} = 0.25 V (at the left-hand arrow), this transistor obviously is acting like a transistor in its active region, even though V_{CE} is well below V_{BE}. Of course!

Figure 17-4

the LM10. The LM10 wouldn't work on a 1.1-V power supply if the transistors aren't working well with V_{CE} as low as 350 or 250 or even 150 mV, which is far below V_{BE}. Of course, you have to be a good engineer to make these circuits work well.

NSC guys (like Bob Widlar) have been doing this for 40 years. Look at the V_{CE} curves of any transistor. When V_{CE} falls below V_{BE}, it's not a disaster. Put a transistor on a curve tracer. Apply a bias like 1 µA per step to the base. When you change the V_{CE} from +1.0 V to 0.6 V to 0.5 or 0.4 V, I_C doesn't change much, does it?

Okay, maybe when you get V_{CE} down to 0.35 V the gain starts to degrade some. But above that, at room temperature, it's not a big deal. There is no demarcation between $V_{CE} > V_{BE}$ and $V_{CE} < V_{BE}$. No inflection. The beta doesn't even change more than perhaps 2% per volt, and it does so smoothly.

Now run the temperature up to 125°C. Can you design a circuit that works up there? It's not easy. But if you don't need a lot of swing, some specialized circuits work just fine. Look at the LM4041-1.2 or the LM185. Many of their V_{CE}s are about 0.3 V, yet they work hot and cold.

How about 160°C? How about 260°C? I can't, but Widlar could, and did, in the LM12. After all, in the old days, a pentode could run with a very low V_{PLATE}—much lower than V_{SCREEN}. It's hard to comprehend this, but after a while you get to understand and believe it. It's an analogous situation that the output voltage is so low, you can't believe it will work. But it's true. It does work.

Next Topic

When V_{BE} = approximately zero, changes in V_{BE} certainly have no effect on IC, right? Wrong. In Widlar's LM12, some of the transistors are so biased, when V_{BE} = 0, the

V_{BE} can still influence the collector current (and vice versa) whether the V_{BE} is a few millivolts positive or negative.

Admittedly, you can't see this easily in a silicon transistor at room temperature. But you can see this in a silicon transistor at 220°C or in a germanium transistor at room temperature, which is about the same idea. Go ahead and measure it. When I did, I was impressed by Bob Widlar's brilliance.

Also, the beta of a transistor can still be important, even when V_{BE} is about zero. That's because as V_{BE} moves up and down a few millivolts compared to zero, the base current needed might be small but finite—not negligible. The base current and its changes are necessary. And if you start at $I_B = 0$ and pull the base negative, the collector current can decrease.

I must remind you that high-beta transistors (300 and up) still have disadvantages in terms of voltage gain or mu. When the beta gets too high, and because mu is inversely proportional to beta, the voltage gain is hurt. I remember a test that asked how much voltage gain a particular amplifier design has. The answer was supposed to be 20,000. But the gain was really 9000, as the betas were too high and the Early Effect was too strong. I passed the test after I explained my solution.

As a rule of thumb, I use mu × beta = 2 million. On some devices, that product is only as good as 1 million, or even 4 million on LM194. If the beta gets better, the mu = $1/h_{RB}$ gets worse and the voltage gain suffers. Be careful not to allow in transistors with too high beta in circuits where poor mu could cause poor performance. Beta is often important. Too little of it can do harm; so can TOO MUCH.

Part 1: This article was printed in *Electronic Design's Analog Applications Supplement*, June 26, 2000, and is reprinted with the kind permission of Penton Publishing.

Part 2: This article was printed in *Electronic Design Magazine*, June 21, 2007, and is reprinted with the kind permission of Penton Publishing.

CHAPTER 18

The Zoo Circuit

Jim Williams

Here is a classic example of how to adapt an old circuit to new requirements. Jim took one of my old (power-wasteful) circuits and adapted it and reengineered it to run at very low power. Nice engineering! With no degradation of accuracy. /rap

This chapter is dedicated to the memory of Professor Jerrold R. Zacharias, who saved my ass.

History, Mistakes, and Some Monkeys Design a Circuit

A couple of years ago I was asked to design a circuit for a customer. The requirements were not trivial, and the customer was having difficulty. I worked on this problem for some time and was asked to present my solution in a formal design review at the customer's location.

When I say "formal," I mean it! I came expecting to talk circuits with a few guys over a pizza. Upon arrival, I was taken to a large and very grand room, reminiscent of a movie theater. About 150 engineers were in attendance. There was every audiovisual machine known to humanity at the ready, and I was almost embarrassed to report that I had no slides, overheads, charts, or whatever (although the piece of chalk I grasped was nice). A "senior technical management panel," positioned in a boxed-off section adjacent to the lectern, was to present a prepared list of questions. A video camera duly recorded the proceedings. The whole thing was chaired by somebody who introduced himself as "Dr. So-and-So, senior vice president of engineering." Everybody in the place talked in whispers and nodded his head a lot. I found myself alternating between intimidation and amusement.

I gave a fairly stiff presentation, clutching my dear little piece of chalk the whole time. Things seemed to go okay but not great, and then the panel began with their prepared list of questions. The first question went something like, "Can you explain, precisely, where the ideas for this and that piece of the circuit came from? Can you detail what design procedures, programs, and methodologies were helpful?"

I considered various acceptable answers but decided to simply tell the truth: "Most of the ideas came from history and making mistakes, and the best source of help was some monkeys at the San Francisco Zoo."

You could have heard a pin before it dropped. There was absolute silence for a bit, and then some guy stood up and asked me to elaborate "a little." Everybody cracked up, the mood shifted, and we finally began to really *talk* about the circuit.

This customer originally came to me with a need for a "CMOS voltage-to-frequency converter." The performance requirements were as follows:

Output frequency	0 to 10 kHz
Input voltage	0 to 5 V
Linearity	0.04%
Drift	100 ppm/°C
PSRR	100 ppm/V
Temperature range	0° to 55°C
Step response	<5 cycles of output frequency
Output pulse	5 V CMOS-compatible
Power supply	Single 9 V battery (6.5 to 10 V)
Power consumption	200 µA maximum
Cost	< $6.00/100,000 pieces

These people had been working on a design for several months. It functioned but was described as wholly unsatisfactory. I asked why they needed CMOS and was assured that "the low-power requirement is nonnegotiable." Without further comment, I asked them to send me their breadboard. It arrived the next morning, and it looked like **Figure 18-1**.

This is probably the most obvious way to design a V/F converter. The 9 V battery is regulated to 5 V by *IC1* and a −5 V rail is derived by *IC2*. The input voltage causes current flow into *A1*'s summing point. *A1* responds by integrating negative, as shown in **Figure 18-2**, trace A. When *A1*'s output goes low enough, A2 trips high (see trace *B* in Figure 18-2), turning on the CD4066 switch and resetting the integrator. Local positive feedback around A2 (A2's positive input is trace *C*) "hangs up" the reset, ensuring a complete integrator discharge. When the positive feedback decays, *A1* begins to ramp again. The ramp slope, and hence the repetition frequency, depends on the input voltage-dependent current into *A1*'s summing point.

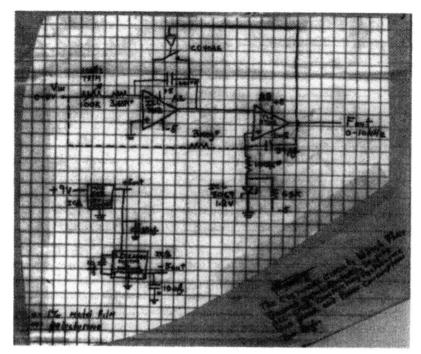

Figure 18-1: The customer's circuit, which they deemed unsatisfactory. Despite all-CMOS construction, performance was poor and power consumption too high.

As soon as I saw the schematic, I knew I couldn't salvage any portion of this design. A serious drawback to this approach is *A1*'s integrator reset time. This time, "lost" in the integration, results in significant linearity error as the operating frequency approaches it. The circuit's 6 μsec reset (see Figure 18-2, traces *A* and *B*) interval introduces a 0.6% error at 1 kHz, rising to 6% at 10 kHz. Also, variations in the reset time contribute additional errors. I added the 3 M resistor (shown in dashed lines) in a half-hearted attempt to improve these figures. This resistor causes *A2*'s trip point to vary slightly with input, partially compensating for the integrator's "lost" reset time. This Band-Aid did improve linearity by more than an order of magnitude, to about 0.4%, but it ain't the way to go.

There are other problems. Quiescent current consumption of this entirely CMOS circuit is 190 μA, rising to a monstrous 700 μA at 10 kHz. Additionally, the polystyrene capacitor's drift alone is –120 ppm/°C, eating up the entire budget. The 1.2 V reference and the input resistor-trimmer could easily double this figure. There are a host of other problems, but what is really needed is an approach with inherently better linearity and lower power consumption.

There are many ways to convert a voltage to a frequency. The "best" approach in an application varies with desired precision, speed, response time, dynamic range, and other considerations.

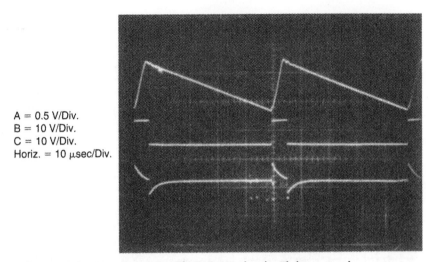

A = 0.5 V/Div.
B = 10 V/Div.
C = 10 V/Div.
Horiz. = 10 μsec/Div.

Figure 18-2: Waveforms for Fig. 18-1's circuit. Finite reset time prevents good linearity performance.

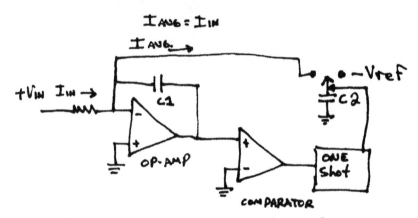

Figure 18-3: Conceptual charge-dispensing-type voltage-to-frequency converter.

Figure 18-3's concept potentially achieves high linearity by enclosing Figure 18-1's integrator in a charge-dispensing loop.

In this approach, $C2$ charges to $-V_{ref}$ during the integrator's ramping time. When the comparator trips, $C2$ is discharged into $A1$'s summing point, forcing its output high. After $C2$'s discharge, $A1$ begins to ramp and the cycle repeats. Because the loop acts to force the average summing currents to zero, the integrator time constant and reset time do not affect frequency. Gain-drift terms are V_{ref}, $C2$, and the input resistor. This approach yields high linearity (typically 0.01%) into the megahertz range.

Figure 18-4 is conceptually similar except that it uses feedback current instead of charge to maintain the op-amp's summing point. Each time the op-amp's output trips the

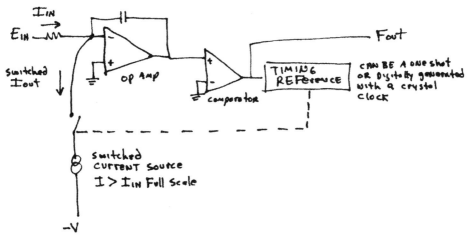

Figure 18-4: Current balance voltage-to-frequency converter.

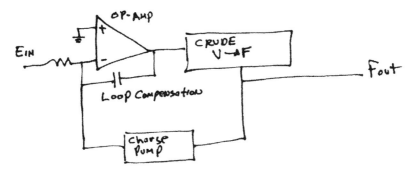

Figure 18-5: Loop-charge pump voltage-to-frequency converter.

comparator, the current sink pulls current from the summing point. Current is pulled from the summing point for the timing reference's duration, forcing the integrator positive. At the end of the current sink's period, the integrator's output again heads negative. The frequency of this action is input related.

Figure 18-5 uses DC loop correction. This arrangement offers all the advantages of charge and current balancing except that response time is slower. Additionally, it can achieve exceptionally high linearity (0.001%), output speeds exceeding 100 MHz, and very wide dynamic range (160 dB). The DC amplifier controls a relatively crude V/F converter. This V/F converter is designed for high speed and wide dynamic range at the expense of linearity and thermal stability. The circuit's output switches a charge pump whose output, integrated to DC, is compared to the input voltage.

The DC amplifier forces the V/F converter operating frequency to be a direct function of input voltage. The DC amplifier's frequency compensation capacitor, required because

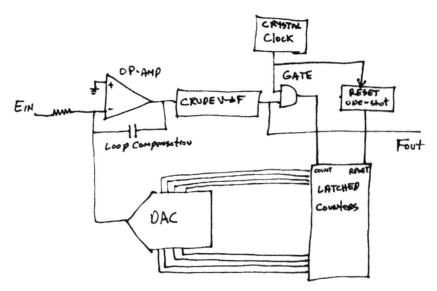

Figure 18-6: Loop-DAC voltage-to-frequency converter.

of loop delays, limits response time. **Figure 18-6** is similar except that the charge pump is replaced by digital counters, a quartz time base, and a DAC. Although it is not immediately obvious, this circuit's resolution is not restricted by the DAC's quantizing limitations. The loop forces the DAC's LSB to oscillate around the ideal value. These oscillations are integrated to DC in the loop compensation capacitor. Hence, the circuit will track input shifts much smaller than a DAC LSB. Typically, a 12-bit DAC (4096 steps) will yield one part on 50,000 resolution. Circuit linearity, however, is set by the DAC's specification.

If you examine these options, Figure 18-3 looks like the winner for the customer's application. The specifications call for step response inside five cycles of output frequency. This eliminates the circuits in Figures 18-4, 18-5, and 18-6 with their DC amplifiers' response time lag. Figure 18-4 requires a timing reference and a precision switched current source, implying some degree of complexity. In theory, Figure 18-3's approach can meet all the specifications without undue complexity.

This technique is not new. I first saw it back in 1964 in a copy of the *GE Transistor Manual*. T. P. Sylvan used a discrete op-amp and a unijunction transistor to form the loop. Hewlett-Packard built rack-mounted V/F converters in the early 1960s that also relied on this approach. In 1972, R. A. Pease developed a commercially produced modular version (Teledyne-Philbrick Model 4701) using a single op-amp that routinely achieved 0.01% linearity with commensurate drift performance. Pease's circuit is particularly relevant, and a version of it is shown in **Figure 18-7**.

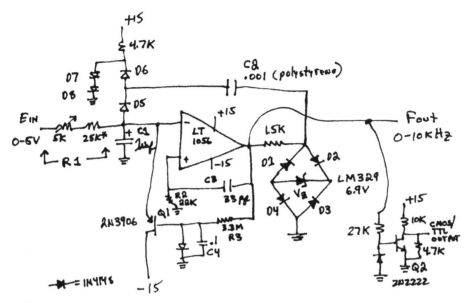

Figure 18-7: A version of Pease's elegant voltage-to-frequency converter circuit.

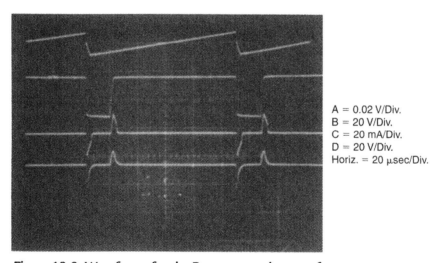

A = 0.02 V/Div.
B = 20 V/Div.
C = 20 mA/Div.
D = 20 V/Div.
Horiz. = 20 μsec/Div.

Figure 18-8: Waveforms for the Pease-type voltage-to-frequency converter.

Assume *C1* sits at a small negative potential. *A1*'s negative input is below its zero-biased positive input, and its output is high. The zener bridge clamps high (at $V_Z + V_{D4} + V_{D2}$) and *C2* charges via *D6*, *D7*, and *D8*. The input voltage forces current through *R1*, and *C1* begins to charge positively (trace *A*, **Figure 18-8**). When *C1* crosses zero volts, *A1*'s output (trace *B*) goes low and the zener bridge clamps negative, discharging *C2* (*C2*'s current is trace *C*) via the *D5–C1* path. The resultant charge removal from *C1* causes it to rapidly discharge (trace *A*). *R2–C3* provides positive feedback to *A1*'s positive input (trace *D*), reinforcing this action and hanging up *A1*'s output long enough for a complete *C2* discharge. When the *R2–C3* feedback decays, *A1*'s output returns high and the cycle repeats. The frequency of this sequence is directly proportional to the input voltage derived current through *R1*. Drift terms include *R1*, *C2*, and the zener, as well as residual diode mismatches. In theory, all the diode drops cancel and do not contribute toward drift. The *R2–C3* "one-shot" time constant is not critical, as long as it allows enough time for *C2* to completely discharge. Similarly, "integrator"*C1*'s value is unimportant as long as it averages *A1*'s negative input to zero.

Q1 and associated components form a startup loop. Circuit startup or input overdrive can cause the circuit's AC-coupled feedback to latch. If this occurs, *A1* goes negative and wants to stay there. *R3* and *C4* slowly charge negative, biasing *Q1*. *Q1* turns on, pulling *C1* toward the –15 V rail, initiating normal circuit action. Once the circuit starts, *C4* assumes a small positive potential and *Q1* goes off. *Q2*, a simple level shifter, furnishes a logic-compatible output.

Pease's 1972 circuit is a very elegant, practical incarnation of Figure 18-3. With care, it will meet all the customer's requirements except two. It requires a split ±15 V supply and pulls well over 10 mA.The job now boils down to dealing with these issues.

Figure 18-9 shows my first attempt at adapting Pease's circuit to my customer's needs. Operation is similar to Pease's circuit. When the input current-derived ramp (trace *A*, **Figure 18-10**) at *C1A*'s negative input crosses zero, *C1A*'s output (trace *B*) drops low, pulling charge through *C1*. This forces the negative input below zero. *C2* provides positive feedback (trace *D* is the positive input), allowing a complete discharge for *C1* (*C1* current is trace *C*). When *C2* decays, *C1A*'s output goes high, clamping at the level set by *D1*, *D2*, and V_{ref}. *C1* receives charge, and recycling occurs when *C1A*'s negative input again arrives at zero. The frequency of this action is related to the input voltage. Diodes *D3* and *D4* provide steering and are temperature compensated by *D1* and *D2*. *C1A*'s sink saturation voltage is uncompensated but small. (These temperature coefficient assumptions are first order and will require more care later.) Although the LT1017 and LT1034 have low operating currents, this circuit pulls almost 400 μA. The AC current paths include *C1*'s charge-discharge cycle and *C2*'s branch. The DC path through *D2* and V_{ref} is particularly costly. *C1*'s charging must occur quickly enough for 10 kHz operation, meaning the clamp seen by *C1A*'s output must have low impedance at this frequency. *C3* helps, but significant current still must come from somewhere to keep impedance low. *C1A*'s current-limited output ($\approx$30 μA source) cannot do the job unaided, and the resistor from the supply is required. Even if *C1A* could supply the necessary current, V_{ref}'s settling time would be an issue. Dropping *C1*'s value will reduce impedance requirements

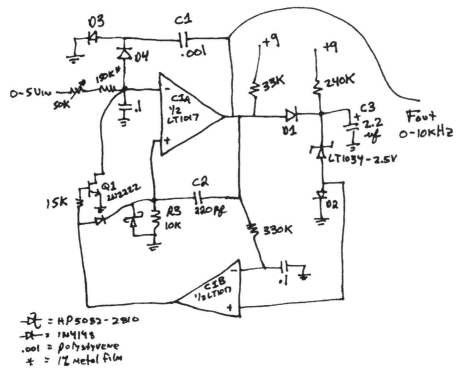

Figure 18-9: My first cut at adapting Pease's circuit.

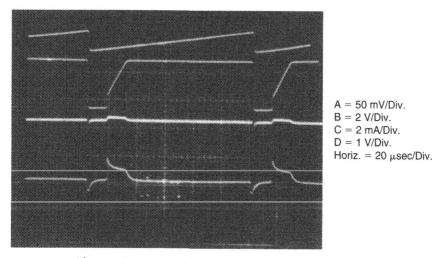

A = 50 mV/Div.
B = 2 V/Div.
C = 2 mA/Div.
D = 1 V/Div.
Horiz. = 20 μsec/Div.

Figure 18-10: Waveforms for the circuit in Fig. 18-9.

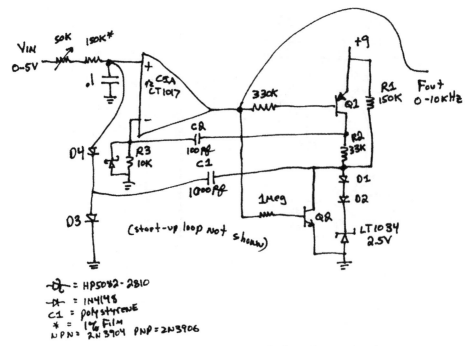

Figure 18-11: The second try. Q1 and Q2 switch the reference, saving some power.

proportionally and would seem to solve the problem. Unfortunately, such reduction magnifies the effects of stray capacitance at the *D3* to *D4* junction. It also mandates increasing R_{in}'s value to keep scale factor constant. This lowers operating currents at *C1A*'s negative input, making bias current and offset more significant error sources.

C1B, *Q1*, and associated components form a startup loop that operates in similar fashion to the one in Pease's circuit (Figure 18-7).

Figure 18-11 shows an initial attempt at dealing with these issues. This scheme is similar to Figure 18-9, except that *Q1* and *Q2* appear. V_{ref} (the top of D1) receives switched bias via *Q1* instead of being on all the time. *Q2* provides the sink path for *C1*. These transistors invert *C1A*'s output, so its input pin assignments are exchanged. *R1* provides a light current from the supply, improving reference settling time. This arrangement decreases supply current to about 300 μA, a significant improvement. Several problems do exist, however. *Q1*'s switched operation is really effective only at higher frequencies. In the lower ranges, *C1A*'s output is low most of the time, biasing *Q1* on and wasting power. Additionally, when *C1A*'s output switches, *Q1* and *Q2* simultaneously conduct during the transition, effectively shunting *R2* across the supply. Finally, the base currents of both transistors flow to ground and are lost. **Figure 18-12** shows the waveform traces for this circuit. The basic temperature

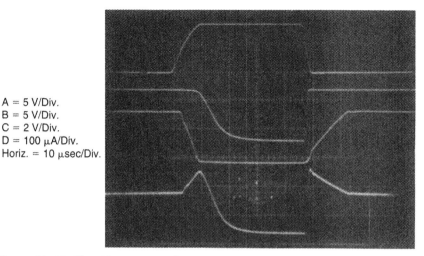

A = 5 V/Div.
B = 5 V/Div.
C = 2 V/Div.
D = 100 μA/Div.
Horiz. = 10 μsec/Div.

Figure 18-12: Fig. 18-11's waveforms. Traces *A, B, C,* and *D* are *C1A* output, *Q1* collector, *Q2* collector, and *R2* current, respectively. *Q1* to *Q2* simultaneous conduction problem is evident in trace *D*.

compensation is as before except that *Q2*'s saturation term replaces the comparator's. This temperature compensation scheme looks okay, but we're still hand waving.

Figure 18-13 is better. *Q1* is gone, *Q2* remains, but *Q3, Q4,* and *Q5* have been added. V_{ref} and its associated diodes are biased from *R1*. *Q3*, an emitter-follower, is used to source current to *C1*. *Q4* temperature compensates *Q3*'s V_{be}, and *Q5* switches *Q3*.

This method has some distinct advantages. The V_{ref} string can operate at greatly reduced current because of *Q3*'s current gain. Also, Figure 18-11's simultaneous conduction problem is largely alleviated because *Q5* and *Q2* are switched at the same voltage threshold out of *C1A*. *Q3*'s base and emitter currents are delivered to *C1*. *Q5*'s currents are wasted, although they are much smaller than *Q3*'s. *Q2*'s small base current is also lost. The values for *C2* and *R3* have been changed. The time constant is the same, but some current reduction occurs due to *R3*'s increase.

Operating wave forms are shown in **Figure 18-14**, and include *C1*'s output (trace *A*), *Q5*'s collector (trace *B*), *Q2*'s collector (trace *C*), *Q2*'s collector current (trace *D*), *C1*'s current (trace *E*), and *Q3*'s emitter current (trace *F*). Note that the current steering is clean, with no simultaneous conduction problems.

This circuit's 200 μA power consumption was low enough to make other specifications worth checking. Linearity came in at 0.05% and dropped to 0.02% when I added a 1 M resistor (dashed lines) across *C1*. The *D4–Q2* path cannot *fully* switch *C1* because of junction drop limitations. The resistor squeezes the last little bit of charge out of *C1*, completing the discharge and improving linearity.

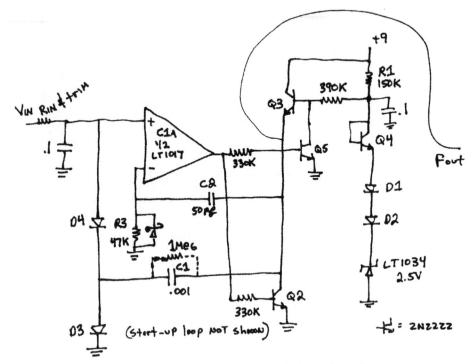

Figure 18-13: A better scheme for switching the reference.

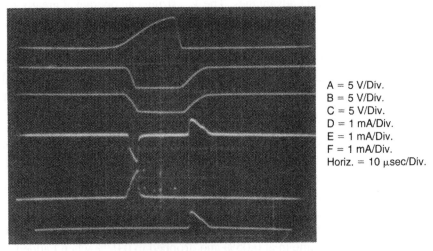

A = 5 V/Div.
B = 5 V/Div.
C = 5 V/Div.
D = 1 mA/Div.
E = 1 mA/Div.
F = 1 mA/Div.
Horiz. = 10 μsec/Div.

Figure 18-14: Fig. 18-13's operation. Traces *D*, *E*, and *F* reveal
no simultaneous conduction problems.

Power supply rejection ratio (PSRR) was not good enough. Supply shifts show up as current changes through *R1*. The LT1034 is relatively insensitive to this, but the *Q4*, *D1*, *D2* trio shift value. As such, I measured 0.1%/V PSRR. *R1* really needs to be a current source, or some compensation mechanism must be used.

Temperature compensation was next. Now it was time to stop hand waving and take a hard look. *Q4* supposedly compensates *Q3*, with *D1* and *D2* opposing *D3* and *D4*. Unfortunately, these devices operate under different dynamic and DC conditions, making precise cancellation difficult. In practice, *R1*'s value should be established to source the current through *Q4–D1–D2*, which provides optimum circuit temperature coefficient. Assuming perfect cancellation and no LT1034 or input resistor drift, we still must deal with *Q2*'s V_{ce} saturation term. At 100 mV saturation, *Q2* will drift about $+0.3\%/°C$ (see the Motorola 2N2222 datasheet), causing about a $-300\,\mu V/°C$ shift in the voltage *C1* discharges toward. This works out to about $-100\,ppm/°C$ (*C1* charges to 3 V) temperature coefficient, which will force a similar *positive* shift in output frequency. *C1*, a polystyrene type, drifts about $-120\,ppm/°C$, contributing further overall positive temperature coefficient (as *C1*, or the voltage it charges to, gets smaller, the circuit must oscillate faster to keep the summing point at zero). So the best case is about $220\,ppm/°C$, and reality dictates that all the other junctions won't match precisely. Temperature testing confirmed all this. Initially, the breadboard showed about $275\,ppm/°C$ and, by varying *R1*, bottomed out at about $200\,ppm/°C$. This certainly wasn't production-worthy engineering but pointed the way toward a solution.

How could I reduce the temperature coefficient and fix the PSRR? Additionally, power consumption was still marginal, although linearity was close. Replacing *R1* with a current source offered hope for PSRR, but reliable temperature compensation and lower power needed another approach. I pined for inspiration but got nothing. I was stuck.

Something that *had* inspired me for a couple of months was a physician I'd been dating. We really had a good time together—a couple of playful kids. There was much dimension to this woman, and I really enjoyed just how relaxed I felt being with her. Things were going quite nicely, and I sometimes allowed myself the luxury of wondering what would become of us.

One weekday afternoon, we played hooky and went to the San Francisco Zoo. The weather was gorgeous, no crowds, and the Alfa ran great. (On our second date it had thrown a fan belt.) We saw bears, elephants, tigers, birds, and ate lots of junk food. The lions got fed; they were *loud* and *hungry*. Strolling around, eating cheeseburgers, and doing just fine, we came to the monkeys.

These guys are actors; they love an audience. There was the usual array of grinning, simian catcalls, cheeping, squawking, lots of jungle bar performances, wondrous feats of balance, and other such theatrics. One character particularly caught my eye. He did a little routine between two parallel rails. First, he hung by his hands as shown in **Figure 18-15**.

Then, very quickly, he flipped over, simultaneously rotating, so he ended up inverted (see **Figure 18-16**).

He did this over and over at great speed; it was his act. Standing there, watching the little fellow do his inverting routine between the rails, I saw my circuit problems simply melt. I felt very lucky. I had a good lady, and a good circuit too.

If you look inside a CMOS logic inverter, the output stage looks like **Figure 18-17**.

The MOS output transistors connect the output terminal to the supply or ground rail. The input circuitry is arranged so only one transistor is on at a time; simultaneous conduction cannot occur. Typically, channel-on resistance is 100 to 200 Ω. There are no junction effects; the transistor channels are purely ohmic. The device's input pin appears almost purely capacitive, drawing only picoamperes of bias current.

Figure 18-18 shows what happens when the CMOS inverter is dropped into the gizzard of Figure 18-13's circuit. *C1* is charged and discharged via the CMOS inverter's ohmic output transistors. *Q3* now drives the inverter's supply pin, and *Q2* goes away. Along with

Figure 18-15: The zoo monkey on parallel rails.

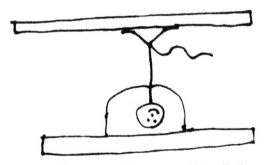

Figure 18-16: The zoo monkey on parallel rails, inverted.

Q2's departure goes its 100 ppm/°C temperature coefficient error. Also, *Q2*'s base current is eliminated, along with *Q5*'s base and collector current losses.

This scheme promises both lower temperature drift and lower power. Assuming ideal junction compensation, the remaining uncompensated drift terms are *C1*'s −120 ppm temperature coefficient and the input resistor. Unfortunately, this configuration does nothing to fix the PSRR problem. The only realistic fix for that is to replace *R1* with a current source. The current source doesn't have to be very stable but must run with only 2 V of headroom because the circuit has to work down to 6.5 V. The simplest alternative is the monolithic LM134. This three-terminal, resistor-programmable device will function with only 800 mV across it, although it does have a 0.33%/°C temperature coefficient.

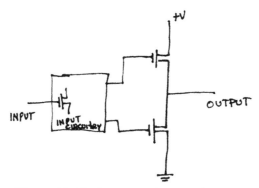

Figure 18-17: Conceptual CMOS inverter.

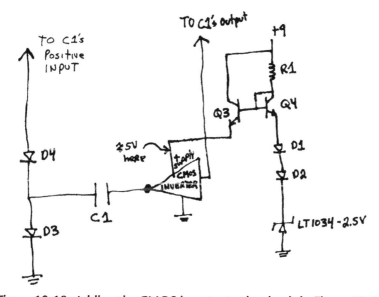

Figure 18-18: Adding the CMOS inverter to the circuit in Figure 18-13.

This temperature coefficient seemed small enough to avoid causing any trouble. The LT1034 shouldn't care, but what about *D1*, *D2*, and *Q4*?

When I calculated the effect of current-source shift with temperature on these devices, I realized I had just inherited the world. It came out *positive* 180 ppm/°C! This tends to cancel the capacitor's −120 ppm/°C term. Additionally, increasing the LT1034s reference voltage by about 50% would pull the compensation down to +120 ppm/°C, further reducing drift. This also aids overall temperature coefficient by making the residual junction mismatches a smaller percentage of the total reference voltage. The current source's low headroom capability allows this, while maintaining operation down to $V_{supply} = 6.2$ V. The sole uncompensated term is the input resistor, which can be specified for low temperature drift.

Figure 18-19 is the final circuit. It meets or exceeds every customer specification.

A 0 to 5 V input produces a 0 to 10 kHz output with a linearity of 0.02%. Gain drift is 40 ppm/°C, and PSRR is inside 40 ppm/V. Maximum current consumption is 145 μA,

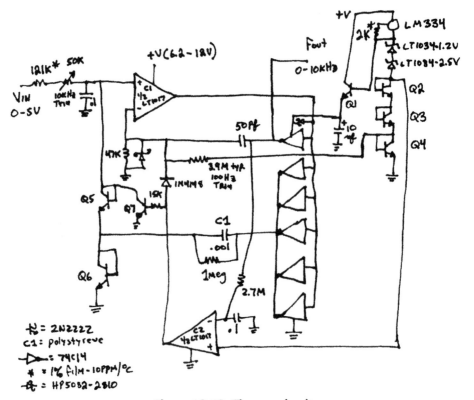

Figure 18-19: The zoo circuit.

descending to 80 μA for $V_{in} = 0$. Other specifications appear in Table 18-1. Much of this circuit should be familiar by now. Some changes have occurred, but nothing too drastic. The diodes have been replaced with transistors for lower leakage and more consistent matching. Also, paralleling the CMOS inverters provides lower-resistance switching. The startup loop has also been modified.

To maintain perspective, it's useful to review circuit operation. Assume *C1*'s positive input is slightly below its negative input (*C2*'s output is low). The input voltage causes a positive-going ramp at *C1*'s positive input (trace A, **Figure 18-20**). *C1*'s output is low, biasing the CMOS inverter outputs high. This allows current to flow from *Q1*'s emitter through the inverter supply pin to the 0.001 μF capacitor. The 10 μF capacitor provides high-frequency bypass, maintaining a low impedance at *Q1*'s emitter. Diode connected *Q6* provides a path to ground. The voltage that the 0.001 μF unit charges to is a function of *Q1*'s emitter potential and *Q6*'s drop. When the ramp at *C1*'s positive input goes high enough, *C1*'s output goes high (trace B) and the inverters switch low (trace C). The Schottky clamp prevents CMOS inverter input overdrive. This action pulls current from *C1*'s positive input capacitor via the *Q5*–0.001 μF route (trace D). This current removal resets *C1*'s positive input ramp to a potential slightly below ground, forcing *C1*'s output to go low. The 50 pF capacitor connected to the circuit output furnishes AC positive feedback, ensuring that *C1*'s output remains positive long enough for a complete discharge of the 0.001 μF capacitor. As in Figure 18-13, the 1 MΩ resistor completes *C1*'s discharge.

The Schottky diode prevents *C1*'s input from being driven outside its negative common-mode limit. When the 50 pF unit's feedback decays, *C1* again switches low and the entire cycle repeats. The oscillation frequency depends directly on the input voltage–derived current.

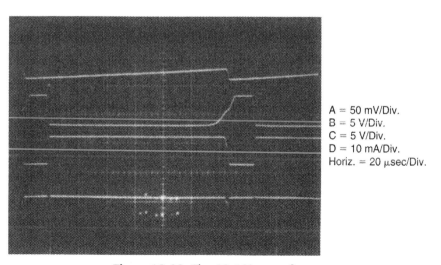

A = 50 mV/Div.
B = 5 V/Div.
C = 5 V/Div.
D = 10 mA/Div.
Horiz. = 20 μsec/Div.

Figure 18-20: Fig. 18-19's waveforms.

Q1's emitter voltage must be carefully controlled to get low drift. *Q3* and *Q4* temperature compensate *Q5* and *Q6* while *Q2* compensates *Q1*'s V_{be}. The two LT1034 s are the actual voltage reference, and the LM334 current source provides excellent supply immunity (better than 40 ppm/V PSRR) and aids circuit temperature coefficient. It does this by utilizing the LM334's 0.3%/°C temperature coefficient to slightly temperature modulate the voltage drop in the *Q2* to *Q4* trio. This correction's sign and magnitude directly oppose that of the −120 ppm/°C 0.001 µF polystyrene capacitor, aiding overall circuit stability.

The *Q1* emitter-follower delivers charge to the 0.001 µF capacitor efficiently. Both base and collector current end up in the capacitor. The paralleled CMOS inverters provide low-loss SPDT reference switching without significant drive losses. Additionally, the inverter specified is a Schmitt input type, minimizing power loss due to *C1*'s relatively slow rising edges. The 0.001 µF capacitor, as small as accuracy permits, draws only small transient currents during its charge and discharge cycles. The 50 pF−47 K positive feedback combination draws insignificantly small switching currents. **Figure 18-21**, a plot of supply current versus operating frequency, reflects the low-power design. At zero frequency, the LT1017's quiescent current and the 35 µA reference stack bias accounts for all current drain. There are no other paths for loss. As frequency scales up, the charge–discharge cycle of the 0.001 µF capacitor introduces the 7 µA/kHz increase shown. A smaller value capacitor would cut power, but the effects of stray capacitance, charge imbalance in the 74C14, and LT1017 bias currents would introduce accuracy errors. For example, if *C1* is reduced to 100 pf (along with other appropriate changes), the circuit consumes only 90 µA at 10 kHz, but linearity degrades to .05%.

Circuit startup or overdrive can cause the circuit's AC-coupled feedback to latch. If this occurs, *C1*'s output goes high. *C2*, detecting this via the inverters and the 2.7 M−0.1 µF lag, also goes high. This lifts *C1*'s negative input and grounds the positive input with *Q7*, initiating normal circuit action.

Because the charge pump is directly coupled to *C1*'s output, response is fast. **Figure 18-22** shows the output (trace *B*) settling within one cycle for a fast input step (trace *A*).

To calibrate this circuit, apply 50 mV and select the value at *C1*'s input for a 100 Hz output. Then apply 5 V and trim the input potentiometer for a 10 kHz output.

Table 18-1 shows what the customer ended up getting.

The zoo circuit made my customer happy, even if it is almost entirely bipolar. The inverter is the only piece of CMOS in the thing. I'm fairly certain the customer wouldn't mind if I had used 12AX7s[1] as long as it met specifications. It runs well in production, and they make lots of them, which makes my boss and the stockholders happy.

---

[1] For those tender of years, 12AX7s are thermionically activated FETs, descended from the work of Lee De Forest.

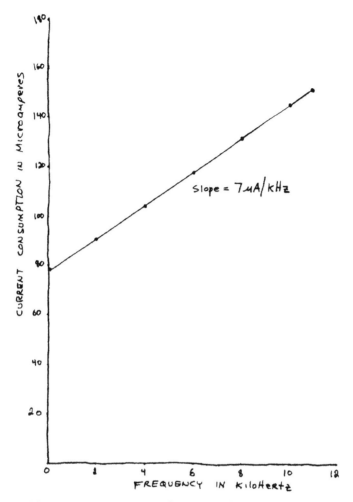

Figure 18-21: Current consumption versus frequency for Fig. 18-19.

This circuit has received some amount of attention in the technical community. I am aware of some spectacularly complex mathematical descriptions of it, along with some arcane explanations of its behavior. Similarly, it has been shown that the circuit could have only been arrived at with the aid of a computer. Given this undue credit, the least I could do is come clean about the circuit's humble origins.

I hope it was as much fun to read about the circuit as it was to build it.

Updated versions of this circuit, which draw as little as 8 μA, can be found in J. Williams' "Circuitry for Signal Conditioning and Power Conversion" (LTC Application Note 75, March 1999, pp. 1–4).

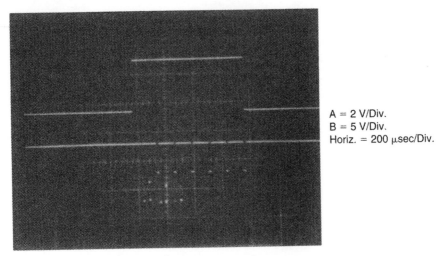

A = 2 V/Div.
B = 5 V/Div.
Horiz. = 200 μsec/Div.

Figure 18-22: Figure 18-18's step response.

Table 18-1: Summary: voltage-to-frequency converter

Output frequency	0 to 10 kHz
Input voltage	0 to 5 V
Linearity	0.02%
Drift	40 ppm/°C
PSRR	40 ppm/V
Temperature range	0 to 70°C
Step response	1 cycle of output frequency
Output pulse	5 V CMOS-compatible
Power supply	Single 9 V battery (6.2 to 12 V)
Power consumption	145 μA maximum, 80 μA quiescent
Cost	< $6.00/100,000 pieces

References

General Electric Co, "Voltage-to-frequency converter," *General Electric Transistor Manual*, GEC, Syracuse, New York, 1964, p. 346.

Pease, R.A., "A new ultra-linear voltage-to-frequency converter," *NEREM Record*, vol. I, 1973, p. 167.

Pease, R.A., assignee to Teledyne, "Amplitude to frequency converter," U.S. patent 3,746,968, filed September 1972.

Williams, J., "Circuitry for signal conditioning and power conversion," LTC Application Note 75, March 1999, pp. 1–4.

———, "Designs for high-performance voltage-to-frequency converters," Linear Technology Corp., Application Note 14, 1986.

———, "Micropower circuits for signal conditioning," *10 kHz Voltage-to-Frequency Converter,* Linear Technology Corp., Application Note 23, 1987, pp. 10–13.

Analog-to-Digital Converter Specification Definitions and Formulas

Bonnie Baker

Acquisition Time—The acquisition time for the SAR converter is the time required for the sampling mechanism to capture one input voltage for one digital output (**Figure A-1**). The acquisition time for the delta-sigma converter lasts during the time that multiple samples are acquired for one digital output.

Average Noise Floor—In a Fast Fourier Transform (FFT) representation of converter data, the average noise floor is a calculated root-sum-square (rss) combination of all the bins within the FFT plot, excluding the input signal and signal harmonics.

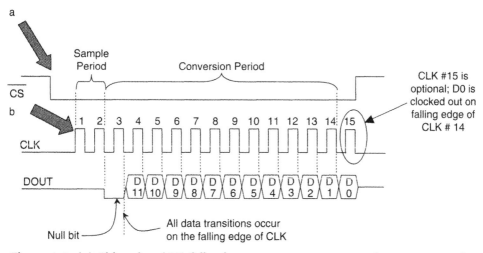

Figure A-1: (a) Chip select (CS) falls, the converter powers up and starts to sample. (b) The falling edge of the clock closes the sample switch and the converter starts to convert the signal.

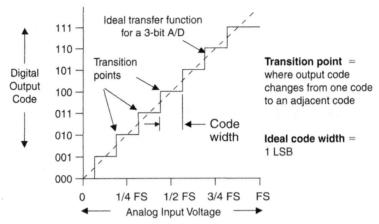

**Figure A-2: The ideal transfer function has zero offset error,
zero gain error, zero DNL error, and zero INL error.**

Bipolar Input Mode (Single-Ended or Differential Inputs)—An input range that uses two input pins and allows negative and positive analog inputs. A negative input is the difference between the two input pins. In this configuration, neither pin goes below or above the power supply rails.

Code Width—The voltage difference between two transition points. The ideal code width is equal to 1 LSB. See **Figure A-2**.

Common-Mode Rejection (CMR)—The degree of rejection of a common-mode signals (ac or DC) across the differential input stage.

Conversion Time—After sampling the signal, the conversion time is the time required for a SAR A/D converter to complete a single conversion. The conversion time does not include the acquisition time or the multiplexer setup time. The conversion time for a given device is less than the throughput time.

Cycle Latency (as it relates to Delta-Sigma A/D Converters)—For delta-sigma ADCs, cycle-latency is equal to the number of *complete* data cycles between the initiation of the input signal conversion and the initiation of the next signal conversion. The unit of measure for this definition of latency is (n)-cycle latency, where n is a whole number.

Data Rate or Data Output Rate—The rate at which conversion results are available from a converter. For a SAR converter, the data rate is equal to the sampling frequency, f_S. With a delta-sigma converter, the data rate is equal to the modulator frequency (f_{MOD}) divided by the decimation ratio.

Delta-Sigma Converter ($\Delta-\Sigma$)—The delta-sigma converter is a one-bit sampling system. In this system, multiple bits are sent through a digital filter, where there is a fair degree of mathematical manipulation performed. With most of the industry's converters, the digital filter is usually a Finite Impulse Response (FIR) filter.

Differential Input—The A/D converter has two inputs per channel. These inputs subtract the input signals, which provide a single digital output code.

Differential Nonlinearity (DNL)—The maximum deviation in code width from the ideal 1 LSB ($FS/2^n$) code width. You calculate the difference between all of the pairs of transitions in the transfer function. **Figure A-3** illustrates the ideal transfer function as a solid line and the DNL error as a dashed line.

Digital Code Out—The digital output code of a converter is equal to $2^n \times V_{IN}/V_{FS}$, where n is the number of bits, V_{IN} is the input analog voltage, and V_{FS} is the full-scale input voltage range.

Digital Interface—SPI™ is a three- or four-wire interface. With this interface, the chip is a slave device. I^2C is a two-wire, Philips standard interface.

Effective Number of Bits (ENOB)—For SAR converters with the unit of measure for SINAD is dB and the unit of measure for ENOBs is bits. You can change SINAD into ENOB with the following calculation:

$$ENOB = (SINAD - 1.76)/6.02$$

For delta-sigma converters:

$$ENOB = M - \log^2(\sigma)$$

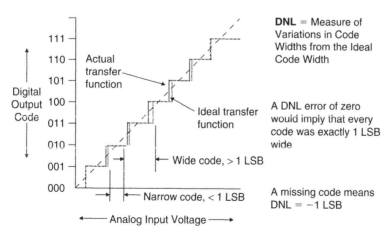

Figure A-3: The differential nonlinearity is the difference between an ideal code width and the measured code width.

where M is equal to the converter resolution and σ is equal to one standard deviation (or rms) of multiple DC input conversions.

Full-Scale Input (FS)—With A/D converters the input signal is analog. The full-scale input voltage is determined by the voltage reference value that is applied the converter reference pin. In many cases, the full-scale input range is equal to the voltage reference referred to ground. In other cases, the full-scale input range is equal to twice the voltage reference value, referred to ground. Refer to the specific ADC datasheet for details.

Gain Error (Full-Scale Error)—The difference between the ideal slope between zero and full scale and the actual slope between the measured zero point and full scale. You zero out the offset errors with this error calculation. See **Figure A-4**.

Ideal A/D Converter Transfer Function—An analog voltage is mapped into an n-bit digital value with no offset, gain, or linearity errors. See Figure A-2.

Idle Tones—These tones are caused by the interaction between the delta-sigma A/D converter modulator and digital filter. Idle tones come from two sources. One is inherent in the voltage being measured, such as when the modulator output repeats a pattern that cannot be filtered by the digital filter. This type of pattern occurs at 0 V, one-half the FSR, three-fourths of the FSR, etc. The second source of idle tones is the chopping frequency being sampled into the measurement. This sampled frequency produces a digital pattern of codes that oscillate at a slow frequency within the pass band. As the name implies, idle tones can appear as a frequency in the output conversion data with multiple DC input conversions at a constant data rate. Patented techniques are available to reduce idle tone concerns.

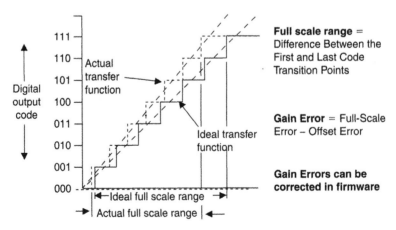

Figure A-4: Gain error is the difference between the ideal gain curve and the actual gain curve with offset removed.

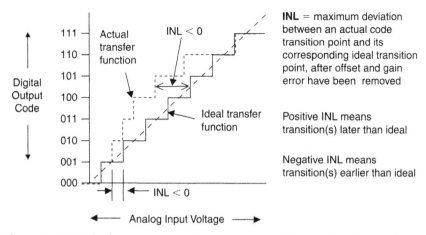

Figure A-5: INL is the aggregate of DNL errors and is equal to the maximum deviation for an ideal A/D converter transfer function.

Integral Nonlinearity (INL)—The maximum deviation of a transition point from the corresponding point of the ideal transfer curve, with offset and gain errors zeroed. See **Figure A-5**.

Internal Buffer—If the A/D converter has an input buffer at its input, this provides a high-impedance input that "isolates" the input signal from the converter.

Latency—The two types of latency with ADCs are cycle-latency and latency time. In both cases, the input signal is a step response over the full-scale range and the output digital code is fully-settled. Cycle-latency is equal to the number of **complete** data cycles between the initiation of the input-signal conversion and the initiation of the next signal conversion. The cycle-latency of a SAR converter is equal to zero. The latency time is between the time where the signal acquisition begins to the time next conversion is started.

Least Significant Bit (LSB)—The least significant bit is the bit representation of the smallest analog input signal that is converted. It is also referred to as the *furthest right bit* in a binary digital word.

Monotonic—Implies that an increase (or decrease) in the analog voltage input will always produce no change or an increase (or decrease) in digital code. Monotonicity does not imply there are no missing codes. See **Figure A-6**.

Most Significant Bit (MSB)—The most significant bit is often thought as the furthest left bit in a binary digital word.

No Missing Code—Implies that an increase (or decrease) in the analog voltage input will always increase (or decrease) in digital output converter code. A converter with no missing code is also monotonic.

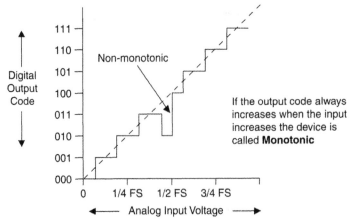

Figure A-6: This curve is nonmonotonic because an increase in the analog voltage can produce a smaller digital output code.

Normal-Mode Rejection—Normal-mode Rejection (NMR) is the degree of rejection of a common-mode signal (DC or ac) across the different input stage. This specification is the ratio of the changing input common-mode signal to the resulting digital output. NMR is the same as Common-mode Rejection (CMR).

Number of Converter Bits (n)—The number of output codes of an A/D converter produces 2^n possible codes, where n is the number of converter bits. The number of converter bits is also known as the *converter resolution*.

Offset Error—The difference between the first measured transition point and the first ideal transition point. See **Figure A-7**.

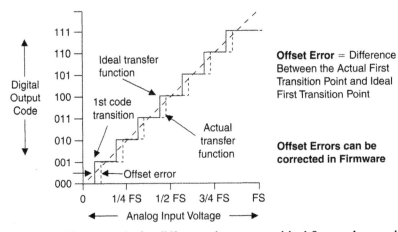

Figure A-7: Offset error is the difference between an ideal first code transition and a measured first code transition.

Quantization Noise—The noise that an A/D converter generates as a consequence of dividing the input signal into discrete buckets. The width of these buckets is equal to the LSB size of the converter. The quantization noise of a converter determines the maximum signal-go-noise ratio equal to $SNR_{IDEAL} = 6.02n + 1.76\,dB$. If it is assumed the response of this error is triangular across the analog input signal, the rms value of the triangular signal is equal to the magnitude of the signal divided by $\sqrt{3}$.

Resolution—When describing the general performance of a converter, resolution is the number of possible output bits an A/D converter can produce in one conversion. Resolution also is the smallest analog increment corresponding to a 1 LSB converter change. This specification determines the smallest analog input signal that can be resolved.

Sample and Hold—The analog switched input to a circuit with a sample and hold function closes (samples) for a short duration to capture (hold) the analog input voltage.

Sampling Time—The time required to accurately sample an analog input signal.

Sample Rate—The speed that a converter can continuously convert several conversions. Typically specified as samples per second (sps) or Hertz (Hz).

Settling Time (as it relates to Delta-Sigma A/D Converters)—The settling time of a delta-sigma converter is typically viewed as the time required for an ideal step input to converge, within an error margin, to a final digital output value. This error band is expressed as a predefined percentage of the total output voltage step. The latency time of a conversion is between the time where the signal acquisition begins to the time that data is available to download from the converter. In contrast to the cycle latency specification, the settling or latency time is never equal to zero.

Signal-to-Noise Ratio (SNR)—A calculated value that represents the ratio of signal power to noise power. The ideal SNR of an A/D converter is $6.02n + 1.76\,dB$.

Signal-to-Noise Ratio plus Distortion (SINAD or SNR+D)—The calculated combination of SNR and total harmonic distortion (THD). SINAD is the ratio of the RMS amplitude of the fundamental input frequency of the input signal to the RMS sum of all other spectral components below one half of the sampling frequency (excluding DC). The theoretical maximum for SINAD is equal to the SNR or $6.02\,n + 1.76\,dB$.

Single-Ended Inputs—An A/D converter that is configured for one input voltage that is referenced to ground.

Spurious Free Dynamic Range (SFDR)—The distance in dB on an FFT plot from the fundamental input signal to the first spur.

Straight Binary Code—With the lowest input voltage, the digital count begins with all zeros and counts up sequentially all ones with a full-scale input. Straight binary is a digital coding scheme for unipolar voltages only (**Figure A-8**).

Successive Approximation Register Converter (SAR)—The modern SAR converter uses a capacitive array at the analog input. You can manufacture this capacitive array and the remainder of the device in a CMOS process, making it easy to integrate it with microcontrollers or microprocessors.

Throughput Time—The time required for the converter to sample, acquire, digitize, and prepare for the next conversion.

Total Harmonic Distortion (THD)—The rms sum of the powers of the harmonic components (spurs) ratioed to the input signal power.

Total Unadjusted Error—The root-sum-squares (rss) of offset, gain, and integral nonlinearity errors.

Median analog voltage (V)	Digital code
0.9375 FS (15/16 FS)	1111
0.875 FS (14/16 FS)	1110
0.8125 FS (13/16 FS)	1101
0.75 FS (12/16 FS)	1100
0.6875 FS (11/16 FS)	1011
0.625 FS (10/16 FS)	1010
0.5625 FS (9/16 FS)	1001
0.5 FS (8/16 FS)	1000
0.4375 FS (7/16 FS)	0111
0.375 FS (6/16 FS)	0110
0.3125 FS (5/16 FS)	0101
0.25 FS (4/16 FS)	0100
0.1875 FS (3/16 FS)	0011
0.125 FS (2/16 FS)	0010
0.0625 FS (1/16 FS)	0001
0	0000

Figure A-8: The straight binary code (also known as unipolar straight binary code) representation of zero volts is equal to a digital (0000). The analog full-scale minus one LSB digital representation is equal to (1111). With this code, there is no digital representation for analog full scale.

Transition Point—The analog input voltage at which the digital output switches from one code to the next.

Two's Complement—See **Figure A-9**.

Unipolar Input Mode (Single-Ended Input)—An input range that only allows positive analog input signals.

Voltage Reference (also know as Analog Voltage Reference)—The input range (V_{IN}) and LSB size is determined by the voltage reference (V_{REF}) to the converter. Depending on the converter, $V_{IN} = V_{REF}$ or $V_{IN} = 2V_{REF}$. LSB $= V_{REF}/2^n$ or LSB $= 2V_{REF}/2^n$ (where n is the number of bits).

Median voltage (V)	Code
0.875 FS (7/8 FS)	0111
0.75 FS (6/8 FS)	0110
0.625 FS (5/8 FS)	0101
0.5 FS (4/8 FS)	0100
0.375 FS (3/8 FS)	0011
0.25 FS (2/8 FS)	0010
0.125 FS (1/8 FS)	0001
0	0000
−0.125 FS (−1/8 FS)	1111
−0.25 FS (−2/8 FS)	1110
−0.375 FS (−3/8 FS)	1101
−0.5 FS (−4/8 FS)	1100
−0.625 FS (−5/8 FS)	1011
−0.75 FS (−6/8 FS)	1010
−0.875 FS (−7/8 FS)	1001
−1 FS	1000

Figure A-9: The two's-complement (also known as binary two's-complement) representation of zero volts is also equal to a digital (0000). The analog positive full-scale minus one LSB digital representation is equal to (0111) and the analog negative full-scale representation is (1000).

References

Albanus, J., *Coding Schemes Used With Data Converters*, SBAA042, Texas Instruments.

IEEE Standard for Terminology and Test Methods for Analog-to-Digital Converters, IEEE-STD-1241-2000.

Oljaca, M., and Hendrick, T., *Data Converters for Industrial Power Measurements*, SBAA117B, Texas Instruments.

Texas Instruments, *Understanding Data Converters*, SLAA013.

Capacitor Coefficients for Lowpass Sallen-Key Filters

Robert A. Pease

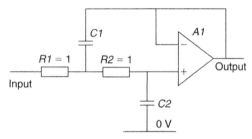

Figure B-1: Schematic diagram of Sallen-Key filter for second order. Note that C2 is connected from the amplifier's + input to ground.

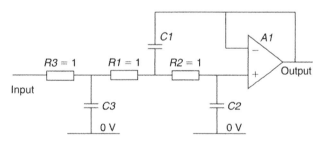

Figure B-2: Schematic diagram of Sallen-Key filter for third order. Note that C2 is still connected from the amplifier's + input to ground. The capacitor C3 is connected from the nexus of R1 and R3 to ground. This circuit is used when odd orders are desired.

Table B-1: Capacitor values for 0.01-dB Chebyshev lowpass Sallen-Key filters.

Order (n)	C_1	C_2
2	1.4826	0.7042
4	1.4874	1.1228
	3.5920	0.2985

(*Continued*)

Table B-1: (Continued)

Order (n)	C_1	C_2
6	1.8900	1.5249
	2.5820	0.5953
	7.0522	0.1486
8	2.3652	1.9493
	2.7894	0.8196
	4.1754	0.3197
	11.8920	0.08672

(Refer to the circuits of Figures 6-10 and B-1, normalized to 1 ohm at $f_{3dB} = 0.159\,Hz$; capacitors in Farads. Reprinted from *Electronics*, McGraw-Hill, Inc., Aug. 18, 1969.)

Table B-2: Capacitor Values for 0.1-dB Chebyshev lowpass Sallen-Key filters.

Order (n)	C_1	C_2	C_3
2	1.638	0.6955	
3	6.653	0.1345	1.825
4	1.900	1.241	
	4.592	0.2410	
5	4.446	0.3804	2.520
	6.810	0.1580	
6	2.553	1.776	
	3.487	0.4917	
	9.531	0.1110	
7	5.175	0.5693	3.322
	4.546	0.3331	
	12.73	0.08194	
8	3.270	2.323	
	3.857	0.6890	
	5.773	0.2398	
	16.44	0.06292	
9	6.194	0.7483	4.161
	4.678	0.4655	
	7.170	0.1812	
	20.64	0.04980	
10	4.011	2.877	
	4.447	0.8756	
	5.603	0.3353	
	8.727	0.1419	
	25.32	0.04037	

(Refer to the circuits of Figures 6-10, 6-11, B-1, and B-2, normalized to 1 ohm at $f_{3dB} = 0.159\,Hz$; capacitors in Farads. Reprinted from *Electronics*, McGraw-Hill, Inc. Aug. 18, 1969.)

Table B-3: Capacitor values for 1-dB Chebyshev lowpass Sallen-Key filters.

Order (n)	C_1	C_2	C_3
2	2.218	0.6061	
3	16.18	0.06428	2.567
4	3.125 7.546	1.269 0.1489	
5	8.884 11.55	0.2540 0.09355	3.935
6	4.410 6.024 16.46	1.904 0.3117 0.06425	
7	10.29 7.941 22.25	0.4012 0.1993 0.04684	5.382
8	5.756 6.792 10.15 28.94	2.538 0.4435 0.1395 0.03568	
9	12.33 8.281 12.68 36.51	0.5382 0.2813 0.1038 0.02808	6.853
10	7.125 7.897 9.952 15.50 44.98	3.170 0.5630 0.1962 0.08054 0.02269	

(Refer to the circuits of Figures 6-10 and 6-11, B-1, and B-2, normalized to 1 ohm at $f_{3dB} = 0.159\,Hz$; capacitors in Farads. Reprinted from *Electronics*, McGraw-Hill, Inc., Aug. 18, 1969.)

Table B-4: Capacitor values for Bessel lowpass Sallen-Key filters.

Order (n)	C_1	C_2	C_3
2	0.9066	0.6800	
3	1.423	0.2538	0.9880
4	0.7351 1.012	0.6746 0.3900	
5	1.010 1.041	0.3095 0.3100	0.8712
6	0.6352 0.7225 1.073	0.6100 0.4835 0.2561	

(*Continued*)

Table B-4: (Continued)

Order (n)	C_1	C_2	C_3
7	0.8532	0.3027	0.7792
	0.7250	0.4151	
	1.100	0.2164	
8	0.5673	0.5540	
	0.6090	0.4861	
	0.7257	0.3590	
	1.116	0.1857	
9	0.7564	0.2851	0.7070
	0.6048	0.4352	
	0.7307	0.3157	
	1.137	0.1628	
10	0.5172	0.5092	
	0.5412	0.4682	
	0.6000	0.3896	
	0.7326	0.2792	
	1.151	0.1437	

(Refer to the circuits of Figures 6-10, 6-11, B-1, and B-2, normalized to 1 ohm at $f_{3dB} = 0.159\,Hz$; capacitors in Farads. Reprinted from *Electronics*, McGraw-Hill, Inc., Aug. 18, 1969.)

Index

Printed and bound by CPI Group (UK) Ltd, Croydon, CR0 4YY

03/10/2024

01040331-0005